LINEAR ALGEBRA

This book is in the
ADDISON-WESLEY SERIES IN MATHEMATICS

LINEAR ALGEBRA

by
G. HADLEY
University of Chicago
and
University of the Andes

ADDISON-WESLEY PUBLISHING COMPANY, INC.
READING, MASSACHUSETTS · PALO ALTO · LONDON

Copyright © 1961
ADDISON-WESLEY PUBLISHING COMPANY, INC.

Printed in the United States of America

ALL RIGHTS RESERVED. THIS BOOK, OR PARTS THERE-
OF, MAY NOT BE REPRODUCED IN ANY FORM WITH-
OUT WRITTEN PERMISSION OF THE PUBLISHERS.

Library of Congress Catalog Card No. 61–5305

Second Printing, March 1964

PREFACE

In recent years, linear mathematical models have assumed an important role in almost all the physical and social sciences, and as was to be expected, this development has stimulated a remarkable growth of interest in linear algebra. It is therefore surprising that the number and variety of volumes written on linear algebra seem not to have kept pace with the diversified needs of those in such fields as mathematics, engineering, economics, operations research, and business. This text, however, represents an effort to meet the needs not only of those studying mathematics, but also of those working in the physical and social sciences. It is intended to be a reasonably rigorous, but not abstract, text. It was written with the intention that it could be read and used by those with a limited mathematical background.

An attempt has been made to introduce new ideas slowly and carefully, and to give the reader a good intuitive "feeling" for the subject. The abstract axiomatic development, while it has many things to recommend it, did not seem appropriate here and was not used. Many numerical examples are given and, insofar as possible, each important idea is illustrated by an example, so that the reader who does not follow the theoretical development may assimilate the material by studying the example.

For simplicity, all scalars in the text are assumed to be real numbers, although it is pointed out in appropriate places that the results hold for complex numbers as well. However, the author believes that students, especially engineers and physicists, should have the opportunity to solve problems involving complex numbers, and therefore such problems are included at the end of the appropriate chapters. It is interesting to note that in many mathematics texts which are more general in their presentation and allow the scalars to be elements of a field, it is not possible to find a single problem requiring the use of complex numbers.

Those who expect to make use of linear algebra must have some awareness of the problems involved in making numerical computations. Consequently, numerical techniques are discussed in somewhat greater detail than is usual.

A novel feature of the text is the inclusion of a chapter covering certain topics in convex sets and n-dimensional geometry. It also contains an elementary discussion of some of the properties of sets of linear inequalities.

The author considers the problems at the end of each chapter to be very important, and the reader should examine all and work a fair number of them. They contain additional theoretical developments as well as routine exercises.

This text might be used in a variety of ways: for a one-semester or one-quarter course in linear algebra; as a text or reference for part of a course in engineering mathematics, mathematical physics, or mathematical economics; as a supplementary text for courses in linear programming, quantum mechanics, classical mechanics; etc.

The author is especially indebted to Professor J. H. Van Vleck, who first impressed upon him the importance of gaining a firm intuitive grasp of any technical subject. With respect to the text itself, the suggestions of Professors H. Houthakker, H. Wagner (especially his insistence on numerous examples), and of several other (unknown) reviewers were helpful. Jackson E. Morris provided a number of the quotations which appear at the beginning of the chapters. The School of Industrial Management, Massachusetts Institute of Technology, very generously provided secretarial assistance for typing the manuscript.

G. H.

CONTENTS

Chapter 1. Introduction 1

1-1 Linear models 1
1-2 Linear algebra 2
1-3 Leontief's interindustry model of an economy 3
1-4 Linear programming 6
1-5 Graphical solution of a linear programming problem in two variables 8
1-6 Regression analysis 10
1-7 Linear circuit theory 12
1-8 Other linear models 14
1-9 The road ahead 14

Chapter 2. Vectors 17

2-1 Physical motivation for the vector concept 17
2-2 Operations with vectors 19
2-3 The scalar product 22
2-4 Generalization to higher dimensions 25
2-5 Generalized vector operations 28
2-6 Euclidean space and the scalar product 30
2-7 Linear dependence 34
2-8 The concept of a basis 39
2-9 Changing a single vector in a basis 41
2-10 Number of vectors in a basis for E^n 43
2-11 Orthogonal bases 45
2-12 Generalized coordinate systems 48
2-13 Vector spaces and subspaces 50

Chapter 3. Matrices and Determinants 60

3-1 Matrices 60
3-2 Matrix operations 61
3-3 Matrix multiplication—introduction 64
3-4 Matrix multiplication—further development 68
3-5 Vectors and matrices 71
3-6 Identity, scalar, diagonal, and null matrices 73
3-7 The transpose 76
3-8 Symmetric and skew-symmetric matrices 78
3-9 Partitioning of matrices 79
3-10 Basic notion of a determinant 85
3-11 General definition of a determinant 85
3-12 Some properties of determinants 88

3–13	Expansion by cofactors	90
3–14	Additional properties of determinants	93
3–15	Laplace expansion	95
3–16	Multiplication of determinants	99
3–17	Determinant of the product of rectangular matrices	100
3–18	The matrix inverse	103
3–19	Properties of the inverse	105
3–20	Computation of the inverse by partitioning	107
3–21	Product form of the inverse	111
3–22	Matrix series and the Leontief inverse	116

Chapter 4. Linear Transformations, Rank, and Elementary Transformations 132

4–1	Definition of linear transformations	132
4–2	Properties of linear transformations	135
4–3	Rank	138
4–4	Rank and determinants	140
4–5	Elementary transformations	144
4–6	Echelon matrices and rank	148

Chapter 5. Simultaneous Linear Equations 162

5–1	Introduction	162
5–2	Gaussian elimination	162
5–3	Cramer's rule	166
5–4	Rules of rank	167
5–5	Further properties	170
5–6	Homogeneous linear equations	173
5–7	Geometric interpretation	177
5–8	Basic solutions	178

Chapter 6. Convex Sets and n-Dimensional Geometry . . . 188

6–1	Sets	188
6–2	Point sets	189
6–3	Lines and hyperplanes	194
6–4	Convex sets	202
6–5	The convex hull	207
6–6	Theorems on separating hyperplanes	210
6–7	A basic result in linear programming	215
6–8	Convex hull of extreme points	217
6–9	Introduction to convex cones	219
6–10	Convex polyhedral cones	224
6–11	Linear transformations of regions	227

CHAPTER 7. CHARACTERISTIC VALUE PROBLEMS AND QUADRATIC
        FORMS 236

7–1   Characteristic value problems 236
7–2   Similarity 238
7–3   Characteristic value problems for symmetric matrices 239
7–4   Additional properties of the eigenvectors of a symmetric matrix . 242
7–5   Diagonalization of symmetric matrices 247
7–6   Characteristic value problems for nonsymmetric matrices . . . 249
7–7   Quadratic forms 251
7–8   Change of variables 253
7–9   Definite quadratic forms 254
7–10 Diagonalization of quadratic forms 255
7–11 Diagonalization by completion of the square 257
7–12 Another set of necessary and sufficient conditions for positive and
      negative definite forms 259
7–13 Simultaneous diagonalization of two quadratic forms 263
7–14 Geometric interpretation; coordinates and bases 264
7–15 Equivalence and similarity 267
7–16 Rotation of coordinates; orthogonal transformations 269

CHAPTER 1

INTRODUCTION

"... to pursue mathematical analysis while at the same time turning one's back on its applications and on intuition is to condemn it to hopeless atrophy."

R. Courant.

1-1 Linear models. A large part of the history of physical science is a record of the continuous human striving for a formulation of concepts which will permit description of the real world in mathematical terms. The more recent history of the social sciences (notably economics) also reveals a determined attempt to arrive at more quantitatively substantiated theories through the use of mathematics. To define mathematically some part of the real world, it is necessary to develop an appropriate mathematical model relating the one or more relevant variables. The purpose of the model might be, for example, to determine the distance of the earth from the sun as a function of time, or to relate the boiling point of water to the external pressure, or to determine the best way of blending raw refinery stocks to yield aviation and motor fuels. A model will consist of one or more equations or inequalities. These may involve only the variables, or they may involve variables and their derivatives (that is, differential equations), or values of variables at different discrete times (difference equations), or variables related in other ways (through integral or integro-differential equations, for example). It is not necessarily true that the variables can be determined exactly by the model. They may be random variables and, in this case, only their probability distributions can be found.

No model is ever an exact image of the real world. Approximations are always necessary. In some cases, models of a high degree of accuracy can be developed so that the values obtained will be correct to ten or more decimal places. In other situations, the best models available may yield values which differ by more than 100% from the results of actual physical measurements. In fact, at times we expect a model to serve only one purpose, i.e., to predict, in a qualitative manner, the behavior of the variables. The accuracy required from a model depends upon the ultimate use for which it was devised.

The real world can frequently be represented with sufficient accuracy by so-called linear models. Linearity is a very general concept: There are linear equations in the variables, linear ordinary and partial differential equations, linear difference equations, linear integral equations, etc. All linear models have the properties of additivity and homogeneity. *Additivity* means: If a variable x_1 produces an effect α_1 when used alone, and a variable x_2 produces an effect α_2 when used alone, then x_1, x_2 used together produce the effect $\alpha_1 + \alpha_2$. *Homogeneity* implies that if a variable x_1 produces an effect α_1, then for any real number λ, λx_1 produces an effect $\lambda \alpha_1$. These remarks must be rather vague at present. The precise mathematical definition of linearity will be given later.

From a mathematical point of view, linear models are of great advantage. The mathematics of nonlinear systems almost always presents considerable difficulty to analytic treatment and usually requires the use of digital computers to obtain numerical solutions. In many cases, even large-scale digital computers are of no assistance. However, it is often relatively easy to work with linear models and to find analytic or numerical solutions for the quantities of interest. Both factors, ease of manipulation and sufficiently accurate approximation of the real world, have made linear models the most popular and useful tools in the physical and social sciences.

1–2 Linear algebra. Almost all linear models lead to a set of simultaneous linear equations or inequalities, although the original model may consist of a set of linear differential or difference equations. The variables in the set of simultaneous linear equations will not necessarily be the physical variables in the original model; however, they will be in some way related to them. We know from elementary algebra that a set of m simultaneous linear equations has the form

$$
\begin{aligned}
a_{11}x_1 + \cdots + a_{1n}x_n &= r_1, \\
a_{21}x_1 + \cdots + a_{2n}x_n &= r_2, \\
&\vdots \\
a_{m1}x_1 + \cdots + a_{mn}x_n &= r_m.
\end{aligned}
\qquad (1\text{--}1)
$$

The coefficients a_{ij} are usually known constants. In some cases the r_i are given constants, and the x_j, $j = 1, \ldots, n$, are the variables which must satisfy the equations (1–1). In other instances, both the r_i and x_j are variables, and the r_i are related to the x_j by (1–1).

The general concept of linearity, introduced in the preceding section, can be made more concrete for a set of equations such as (1–1). The contribution of variable x_j to r_i is $\alpha_{ij} = a_{ij}x_j$, and the sum of the contributions of all the variables yields r_i. If x_j is changed to $\hat{x}_j = \lambda x_j$, the contribution to r_i of $\hat{x}_j$ is $a_{ij}\hat{x} = \lambda a_{ij}x_j = \lambda \alpha_{ij}$. This equality expresses the homogeneity property. If the contribution to r_i of x_j were $\alpha_{ij} = a_{ij}x_j^2$, the expression would not be homogeneous since the contribution of $\hat{x}_j = \lambda x_j$ would be $\lambda^2 \alpha_{ij}$. Linear models in a given set of variables cannot involve any powers of the variables other than the first; neither can they involve such functions as $\log x_j$ or $\exp x_j$. Referring again to Eq. (1-1), we note that the contribution to r_i of x_j and x_k when used together is $a_{ij}x_j + a_{ik}x_k = \alpha_{ij} + \alpha_{ik}$, that is, the sum of the individual contributions of x_j and x_k. This expression indicates the additivity property which is characteristic of linearity. This property rules out the possibility that products of the variables, such as $x_j x_k$, can appear in linear models since the contribution of x_k depends on the value of x_j, and the combined contribution of x_j and x_k is not the sum of the contributions of x_j, x_k when used separately.

Linear algebra developed from studies of systems of linear equations, such as Eq. (1-1), as it became necessary to derive theoretical results for such systems and to invent simplified notations for their manipulation. Further generalizations have led to a branch of mathematics with wide applicability in the physical and social sciences. The techniques of linear algebra can be extended to all linear models; in addition, they also simplify operations with a wide variety of nonlinear models. The purpose of this text is to provide a succinct and comprehensive introduction to linear algebra which will enable the reader to use the subject in his own particular field. No attempt will be made to illustrate in detail the usefulness of the subject for any particular field; however, we shall discuss occasionally applications of the techniques to specific models, such as, for example, linear programming.

Before turning directly to the subject matter, we shall mention briefly several linear models and note that the theory of linear algebra can be used to great advantage in their analysis. We do not expect that the reader will understand all (or even any) of the models. They are presented only for the purpose of illustrating the range of problems to which linear algebra is applicable.

1-3 Leontief's interindustry model of an economy. In the early 1930's, Professor Wassily Leontief of Harvard University developed an interesting linear model of the national economy. This model assumes that the economy consists of a number of interacting industries, each of which is imagined to produce only a single good and to use only one process of production. For example, steel manufacture and agriculture might be treated as industries. To produce its given good, each industry will have to purchase goods from other industries, for example: The automobile industry purchases steel from the steel industry and tires from the rubber industry. In addition to selling its good to other industries, a given industry will, in general, be called upon to meet exogenous demands from consumers, the government, or foreign trade.

Suppose that there are n different industries. Imagine also that, in a given year, each industry produces just enough to meet the exogenous demand and the demand of other industries for its good. Let x_i be the quantity of good i produced by industry i in a given year. Assume that in this year industry j will require y_{ij} units of good i, and that the exogenous demand for i will be b_i. Thus, if industry i produces exactly enough to meet the demand, we obtain

$$x_i = y_{i1} + y_{i2} + \cdots + y_{in} + b_i. \tag{1-2}$$

Equation (1-2) allows for the possibility that industry i may use some of its own good. We obtain a balance equation (1-2) for each industry i.

If industry j is going to produce x_j units of good j, we have to know how many units of good i will be required. Clearly, the answer depends on the technology of the industry. Here, Leontief makes the important assumption that the amount of good i required to produce good j is directly proportional to the amount of good j produced, that is,

$$y_{ij} = a_{ij}x_j, \tag{1-3}$$

where a_{ij}, the constant of proportionality, depends on the technology of industry j.

Substitution of (1-3) into (1-2) yields

$$x_i - a_{i1}x_1 - a_{i2}x_2 - \cdots - a_{in}x_n = b_i$$

for each i, and we obtain the following set of simultaneous linear equations:

$$\begin{aligned}(1 - a_{11})x_1 - a_{12}x_2 - \cdots - a_{1n}x_n &= b_1, \\ -a_{21}x_1 + (1 - a_{22})x_2 - \cdots - a_{2n}x_n &= b_2, \\ &\vdots \\ -a_{n1}x_1 - a_{n2}x_2 - \cdots + (1 - a_{nn})x_n &= b_n.\end{aligned} \tag{1-4}$$

This is a set of n equations in n unknowns; intuitively, we suspect that we should be able to solve for a unique set of x_j which will satisfy these equations. Thus, by specifying the exogenous demands b_i, we expect to determine how much each industry in the economy should produce in order to meet the exogenous demands plus the demands of other industries. Naturally, the relation between the x_j and the b_i depends on the technologies of the industries, that is, on the a_{ij}.

Leontief also shows that the same type of analysis can be used to determine the prices which should prevail in the hypothetical economy. The technological coefficient a_{ij} can be viewed as the number of units of product

i required to produce one unit of product j. Let p_j be the price of one unit of j. Then the cost of materials required to turn out one unit of j is

$$a_{1j}p_1 + \cdots + a_{nj}p_n.$$

The difference between the price of one unit of j and the cost of materials required to produce this one unit is called the *value added* by industry j and will be denoted by r_j. Thus*

$$p_j - \sum_{i=1}^{n} a_{ij}p_i = r_j, \qquad j = 1, \ldots, n. \tag{1-5}$$

The value added may include labor, profit, etc. Equation (1–5) represents a set of n simultaneous linear equations in the n prices p_j, that is,

$$\begin{aligned}
(1 - a_{11})p_1 - a_{21}p_2 - \cdots - a_{n1}p_n &= r_1, \\
-a_{12}p_1 + (1 - a_{22})p_2 - \cdots - a_{n2}p_n &= r_2, \\
&\vdots \\
-a_{1n}p_1 - a_{2n}p_2 - \cdots + (1 - a_{nn})p_n &= r_n.
\end{aligned} \tag{1-6}$$

This set of equations looks quite similar to (1–4). The same coefficients a_{ij} appear. However, now they appear in transposed order. Intuitively, it would seem that once the value added is specified for each product in the economy, the prices have been determined.

We have presented the bare outlines of a static model of an economy (static because, for the period under consideration, nothing is allowed to change with time). The model is linear. It permits us to determine the quantities to be produced in terms of the exogenous demands, and the prices in terms of the values added. This model has been found useful in studies to determine whether the United States economy could meet certain wartime demands with a given labor supply, and in investigating the influence of a price change in one industry on prices in other industries. The basic Leontief model has been generalized in many different ways which we need not discuss here.

---

* The upper-case Greek sigma ($\sum$) is a summation sign and, by definition,

$$\sum_{i=1}^{n} x_i = x_1 + x_2 + \cdots + x_n, \qquad \sum_{i=1}^{n} x_i^2 = x_1^2 + \cdots + x_n^2,$$

$$\sum_{i=1}^{n} a_{ij}p_i = a_{1j}p_1 + \cdots + a_{nj}p_n.$$

This notation will be used frequently to simplify the writing of summations.

1-4 Linear programming. Linear programming, a linear model developed within the last twelve years, has attracted wide attention. It has been applied to a wide variety of problems, for example: programming of petroleum refinery operations, determination of optimal feed mixes for cattle, assignment of jobs to machines in manufacturing, etc. We shall show how linear programming problems can arise in practice by studying a typical, albeit oversimplified, example.

Let us consider a shop with three types of machines, A, B, and C, which can turn out four products, 1, 2, 3, 4. Any one of the products has to undergo some operation on each of the three machines (a lathe, drill, and milling machine, for example). We shall assume that the production is continuous, and that each product must first go on machine A, then B, and finally C. Furthermore, we shall assume that the time required for adjusting the setup of each machine to a different operation, when production shifts from one product to another, is negligible. Table 1–1 shows: (1) the machine hours for each machine per unit of each product; (2) the machine hours available per year; (3) the profit realized on the sale of one unit of any one of the products. It is assumed that the profit is directly proportional to the number of units sold; we wish to determine the yearly optimal output for each product in order to maximize profits.

TABLE 1–1

DATA FOR EXAMPLE

Machine type	Products				Total time available per year
	1	2	3	4	
A	1.5	1	2.4	1	2000
B	1	5	1	3.5	8000
C	1.5	3	3.5	1	5000
Unit profit	5.24	7.30	8.34	4.18	

Examination of Table 1–1 shows that the item with the highest unit profit requires a considerable amount of time on machines A and C; the product with the second-best unit profit requires relatively little time on machine A and slightly less time on machine C than the item with the highest unit profit. The product with the lowest unit profit requires a considerable amount of time on machine B and relatively little time on C. This cursory examination indicates that the maximum profit will not be achieved by restricting production to a single product. It would seem that at least two of them should be made. It is not too obvious, however, what the best product mix is.

Suppose x_j is the number of units of product j produced per year. It is of interest to find the values of x_1, x_2, x_3, x_4 which maximize the total profit. Since the available machine time is limited we cannot arbitrarily increase the output of any one product. Production must be allocated among products 1, 2, 3, 4 so that profits will be maximized without exceeding the maximum number of machine hours available on any one machine.

Let us first consider the restrictions imposed by the number of machine hours. Machine A is in use a total of

$$1.5x_1 + x_2 + 2.4x_3 + x_4 \text{ hours per year,}$$

since 1.5 hours are required for each unit of product 1, and x_1 units of product 1 are produced; and so on, for the remaining products. Hence the total time used is the sum of the times required to produce each product. The total amount of time used cannot be greater than 2000 hours. Mathematically, this means that

$$1.5x_1 + x_2 + 2.4x_3 + x_4 \leq 2000. \tag{1-7}$$

It would not be correct to set the total hours used equal to 2000 (for machine A) since there may not be any combination of production rates that would use each of the three machines to capacity. We do not wish to predict which machines will be used to capacity. Instead, we introduce a "less than or equal to" sign; the solution of the problem will indicate which machines will be used at full capacity.

For machines B and C we can write

$$x_1 + 5x_2 + x_3 + 3.5x_4 \leq 8000 \quad \text{(machine } B\text{)}, \tag{1-8}$$

$$1.5x_1 + 3x_2 + 3.5x_3 + x_4 \leq 5000 \quad \text{(machine } C\text{)}. \tag{1-9}$$

Since no more than the available machine time can be used, the variables x_j must satisfy the above three inequalities. Furthermore, we cannot produce negative quantities; that is, we have either a positive amount of any product or none at all. Thus the additional restrictions

$$x_1 \geq 0, \quad x_2 \geq 0, \quad x_3 \geq 0, \quad x_4 \geq 0 \tag{1-10}$$

require that the variables be non-negative.

We have now determined all the restrictions on the variables. If x_j units of product j are produced, the yearly profit z is

$$z = 5.24x_1 + 7.30x_2 + 8.34x_3 + 4.18x_4. \tag{1-11}$$

We wish to find values of the variables which satisfy restrictions (1–7) through (1–10) and which maximize the profit (1–11). This is a linear programming problem.

A general linear programming problem seeks to determine the values of r non-negative variables x_j which will yield the largest value of z,

$$z = c_1 x_1 + \cdots + c_r x_r, \tag{1–12}$$

for those sets of non-negative x_j which satisfy a set of m linear inequalities or equations of the form

$$a_{i1} x_1 + \cdots + a_{ir} x_r \{\leq = \geq\} b_i, \qquad i = 1, \ldots, m. \tag{1–13}$$

One and only one of the signs $\leq$, $=$, $\geq$ holds for each constraint, but the sign can vary from one constraint to another. The value of m can be greater than, less than, or equal to, r. The linear function (1–12) is called the *objective function*; the linear inequalities (1–13) are the *constraints*. A set of non-negative variables $(x_1, \ldots, x_r)$ which satisfies (1–13) is a *feasible solution*; an *optimal* feasible solution maximizes z in (1–12). A linear programming problem is solved when an optimal feasible solution to the problem has been found.

1–5 Graphical solution of a linear programming problem in two variables. Linear programming problems involving only two variables can be solved graphically. Consider, for example:

$$\begin{aligned} 3x_1 + 5x_2 &\leq 15, \\ 5x_1 + 2x_2 &\leq 10, \\ x_1, x_2 &\geq 0, \\ \max z = 5x_1 &+ 3x_2. \end{aligned} \tag{1–14}$$

First, we shall find the sets of numbers (x_1, x_2) which are feasible solutions to the problem. We introduce an $x_1 x_2$-coordinate system and note that any set of numbers (x_1, x_2) represents a point in the $x_1 x_2$-plane. All points (x_1, x_2) lying on or to the right of the x_2-axis have $x_1 \geq 0$. Similarly, all points lying on or above the x_1-axis have $x_2 \geq 0$. Hence any point lying in the first quadrant has $x_1, x_2 \geq 0$ and thus satisfies the non-negativity restrictions. Any point which is a feasible solution must lie in the first quadrant.

To find the set of points satisfying the constraints, we must interpret geometrically such inequalities as $3x_1 + 5x_2 \leq 15$. If the equal sign holds, then $3x_1 + 5x_2 = 15$ is the equation for a straight line, and any point on this line satisfies the equation. Now consider the point $(0, 0)$, that is, the origin. We observe that $3(0) + 5(0) = 0 < 15$; hence the origin also satisfies the inequality. In fact, any point lying on or below the line $3x_1 + 5x_2 = 15$ satisfies $3x_1 + 5x_2 \leq 15$. However, no point lying above the line satisfies the inequality. Therefore, the set of points satisfying the inequality $3x_1 + 5x_2 \leq 15$ consists of all the points in the x_1x_2-plane lying on or below the line $3x_1 + 5x_2 = 15$. However, not all these points satisfy the non-negativity restriction and this inequality; only the points in the first quadrant lying on or below the line $3x_1 + 5x_2 = 15$ fulfill both conditions. By analogy, all points in the first quadrant lying on or below the line $5x_1 + 2x_2 = 10$ satisfy the inequality $5x_1 + 2x_2 \leq 10$ and the restriction of non-negativity.

The set of points satisfying both inequalities ($3x_1 + 5x_2 \leq 15$, $5x_1 + 2x_2 \leq 10$) and the non-negativity restriction is represented by the darkly shaded region in Fig. 1–1. Any point in this region is a feasible solution, and only the points in this region are feasible solutions.

Nothing has been said so far about the objective function: To solve our problem, we must find the point or points in the region of feasible solutions which will yield the maximum value of the objective function. For any fixed value of z, $z = 5x_1 + 3x_2$ is a straight line. Any point on this line will give the same value of z. For each different value of z, we obtain a different line. It is important to note that all the lines representing the different values of z are parallel since the slope of any line $z = c_1x_1 + c_2x_2$ is $-c_1/c_2$, and hence is independent of z. In our problem, c_1, c_2 are fixed, and the lines are parallel.

We wish to find the line of maximum z which has at least one point in common with the region of feasible solutions. The lines in Fig. 1–1 repre-

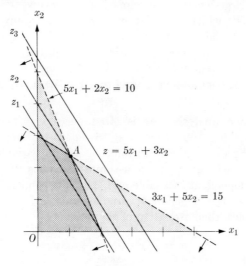

FIGURE 1–1

sent the objective function for three different values of z. Clearly, z_1 is not the maximum value of z: The line can be moved up (this increases z) and still have some points in the region of feasible solutions. Although $z_3 > z_2$ and z_1, the line representing z_3 has no point in common with the region of feasible solutions and thus does not satisfy our premise. Hence z_2 is the maximum value of z; the feasible solution which yields this value is the corner A of the region of feasible solutions.

In Fig. 1-1, the approximate values of the variables for the optimal solution are $x_1 = 1$, $x_2 = 2.4$. To find the exact values, we note that the point representing the optimal solution is the intersection of the lines $3x_1 + 5x_2 = 15$, $5x_1 + 2x_2 = 10$. Solving these two equations simultaneously, we obtain $x_1 = 1.053$, $x_2 = 2.368$. Substitution of these values into the objective function yields the maximum value of $z = 12.37$.

In our example of a linear programming problem, several features have emerged which merit further discussion. First of all, there is an infinite number of feasible solutions which form a region in the $x_1 x_2$-plane. This region has straight boundaries and some corners; geometrically speaking, it is a convex polygon. For any fixed value of z, the objective function is a straight line. The lines corresponding to different values of z are parallel. The maximum value of z is represented by the line with the largest value of z which has at least one point in common with the polygon of feasible solutions. The optimal solution occurred at a corner of the polygon. Interestingly enough, the same characteristics are found in general linear programming problems with a feasible solution $(x_1, \ldots, x_r)$ representing a point in an r-dimensional space.

In 1947, George Dantzig developed an iterative algebraic technique (simplex method) which provides exact solutions to any linear programming problem (it is not an approximation method). Since a very considerable number of arithmetical operations is required to find an optimal feasible solution to a problem involving many variables and constraints, large-scale digital computers are used to solve these problems.

1-6 Regression analysis. It is often necessary (particularly in the fields of engineering and economics) to determine empirically the best formula relating one variable to another set of variables. Economists and engineers often start with the assumption that a linear relation exists between the "dependent" variable and the "independent" variables. For example, if y is the dependent variable and $x_1, \ldots, x_n$ are the independent variables, then a relation of the form

$$y = \alpha_1 x_1 + \cdots + \alpha_n x_n \tag{1-15}$$

might be used to explain y in terms of the x_j. The α_j are assumed to be constants; the best α_j are to be determined from experimental or historical data. For example, we may wish to relate the demand for automobiles, y, in a given year to the national income for the previous year, x_1, the change in national income over the last two years, x_2, the demand for automobiles in the past year, x_3, etc. If, after determining the α_j, comparison shows that the computed demand y (1–15) is a sufficiently close approximation of the actual demand, then formula (1–15) may be used to predict from current data sales of automobiles for the coming year.

We have not yet defined the meaning of the "best α_j" in (1–15) or shown how these values are computed. Let us suppose that, from a series of experimental or historical data, we obtain $k(k > n)$ sets of data which yield y for a given set of values of $x_1, \ldots, x_n$. The ith set of data is denoted by $y(i), x_1(i), \ldots, x_n(i)$. For any given set of α_j in (1–15), let $\hat{y}(i)$ be the value of y computed by using $x_1(i), \ldots, x_n(i)$ in formula (1–15). The criterion most frequently used to determine the best set of α_j requires that the sum of the squares of the differences between the measured values of y and the value computed for the same set of x_j by means of Eq. (1–15) be minimal; that is, we wish to find the α_j which minimize z when

$$z = \sum_{i=1}^{k} [y(i) - \hat{y}(i)]^2. \tag{1–16}$$

But

$$\hat{y}(i) = \sum_{j=1}^{n} \alpha_j x_j(i), \tag{1–17}$$

and

$$z = \sum_{i=1}^{k} \left[y(i) - \sum_{j=1}^{n} \alpha_j x_j(i) \right]^2$$

$$= \sum_{i=1}^{k} \left\{ y^2(i) - 2y(i) \sum_{j=1}^{n} \alpha_j x_j(i) + \left[\sum_{j=1}^{n} \alpha_j x_j(i) \right]^2 \right\}. \tag{1–18}$$

Thus, z has been expressed in terms of the α_j and the given data, and hence the α_j are the only variables.

The set of α_j which minimizes z in (1–18) must satisfy* the following set of n simultaneous linear equations:

$$\sum_{j=1}^{n} u_{qj} \alpha_j = v_q, \qquad q = 1, \ldots, n, \tag{1–19}$$

---

* The reader familiar with calculus will note that the necessary conditions to be satisfied by a set of α_j which minimizes z are $\partial z/\partial \alpha_j = 0, j = 1, \ldots, n$. These n partial derivatives yield the n linear equations (1–19).

where

$$v_q = \sum_{i=1}^{k} y(i)x_q(i), \qquad u_{qj} = \sum_{i=1}^{k} x_q(i)x_j(i). \qquad (1\text{--}20)$$

Thus, to determine the α_j, a set of linear equations must be solved.

The material discussed in this section is part of statistical regression analysis and is used extensively in that branch of economics called econometrics. In engineering, the technique for determining the best α_j is often referred to as finding the best "least-squares" fit. The regression or least-squares techniques lead to a set of simultaneous linear equations to be solved for the best α_j. Our starting point was a linear model, since we assumed that y could be related to the x_j by (1–15).

1–7 Linear circuit theory. Linear circuit theory provides an excellent example of a linear model used in engineering. For d-c (direct-current) circuits, the fundamental linearity assumption is known as Ohm's law which states that the voltage drop E across a conductor is proportional to the current I flowing through the conductor. The constant of proportionality is called the resistance R. Resistance is a property of the material through which the current is passing. Thus $E = IR$. The equation indicates the homogeneity property of linearity. The additivity property is also present: If the same current passes through a resistance network of two resistors connected in series, the voltage drop across the network is the sum of the individual voltage drops across each resistor.

Let us consider the d-c circuit in Fig. 1–2. Suppose that we wish to find the values of all the labeled currents. The symbol $\;^-E\!\!\mid\!\!\mid\!^+\;$ repre-

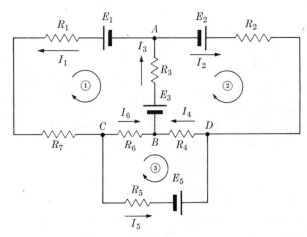

Figure 1–2

sents a battery; if the current moves from the plus terminal of the battery to the minus terminal, a voltage drop of E occurs (any internal resistance of the battery is here neglected). The currents in each part of the circuit can be determined by two rules known as Kirchhoff's laws. These laws simply state that (1) the algebraic sum of the currents at any node (point A in Fig. 1–2) is zero, that is, there is conservation of current at a node; (2) the algebraic sum of the voltages (drops and rises) around a loop is zero.

We wish to determine the six currents shown in Fig. 1–2. We shall assume arbitrarily that the currents flow in the direction of the arrows. If our choice of directions should prove incorrect, the solution will yield negative values for the current. Intuitively, we feel that six independent equations will be needed to determine the six currents. Let us obtain three equations using the three loops shown in Fig. 1–2. From loop 1

$$E_1 - I_3R_3 + E_3 - I_6R_6 - I_1R_7 - I_1R_1 = 0,$$

or (1–21)

$$I_3R_3 + I_6R_6 + I_1(R_1 + R_7) = E_1 + E_3.$$

Moving in a direction opposite to the assumed current flow, we obtain a voltage rise which is the negative of the voltage drop occurring in the direction of the current flow. For loop 2

$$-E_2 + I_2R_2 + I_4R_4 - E_3 + I_3R_3 = 0,$$

or (1–22)

$$I_2R_2 + I_4R_4 + I_3R_3 = E_2 + E_3;$$

for loop 3

$$E_5 - I_5R_5 + I_6R_6 - I_4R_4 = 0,$$

or (1–23)

$$I_5R_5 - I_6R_6 + I_4R_4 = E_5.$$

Many other loops can be considered, such as a combination of 1 and 2, or 2 and 3, etc. However, these do not yield independent equations. All other loop equations can be obtained from (1–21) through (1–23) (prove!).

The remaining three equations are found by means of the first Kirchhoff law concerning the conservation of flow at the nodes. Using nodes A, B, C, we obtain

$$I_1 + I_2 - I_3 = 0 \quad \text{(node } A\text{)}, \tag{1–24}$$

$$I_3 - I_4 - I_6 = 0 \quad \text{(node } B\text{)}, \tag{1–25}$$

$$I_5 + I_6 - I_1 = 0 \quad \text{(node } C\text{)}. \tag{1–26}$$

Equations (1–21) through (1–26) provide a set of six simultaneous linear equations to be solved for the currents. Here again a linear model leads to sets of linear equations.

Linear circuit theory is also very important for the analysis of a-c (alternating current) circuits where the currents and voltages vary sinusoidally with time. Again, we obtain sets of simultaneous linear equations which, however, involve complex rather than real numbers. The theory of linear a-c circuits provides an excellent example of the usefulness of complex numbers in engineering. Although they are not introduced specifically in this book (except in some of the problems), the material developed applies to both complex and real numbers.

1–8 Other linear models. Linear models and hence linear algebra are valuable tools in the solving of a wide range of problems. Linear algebra is used, for example, in:

(1) *Linear differential and difference equations* to deal with sets of simultaneous linear differential or difference equations;

(2) *Linear vibration theory* to determine the characteristic frequencies of vibration (of atoms in molecules, for example).

(3) *Statistics, probability theory*, and in *related areas* dealing with noise and stochastic processes (the theory of *Markov processes* in probability theory provides an excellent example);

(4) *Transformation theory in quantum mechanics* to carry out transformations from one representation of an operator to another (here, complex numbers appear and have to be dealt with);

(5) *Establishing sufficient conditions in classical theory of maxima and minima* (this theory finds important applications in economics, especially in the classical theory of production and consumer behavior);

(6) *Rigid body mechanics* to simplify the theoretical treatment of the motion of a rigid body, such as a gyroscope.

This list and the examples given in the preceding sections do not exhaust by any means the applicability of linear models; however, they furnish sufficient proof for the importance and usefulness of linear models and of linear algebra.

1–9 The road ahead. The n variables in a set of simultaneous linear equations (1–1) can be thought of as a point in an n-dimensional space. In Chapter 2, a particularly useful n-dimensional space, the euclidean space, is introduced. The algebraic properties of points in this space are studied. Chapter 3 discusses matrices and determinants. Matrices enable us to deal efficiently with linear models from both a theoretical and practical standpoint, since they greatly simplify the mathematical manipulation of linear relations.

Chapter 4 continues the discussion of matrices and presents the notion of linear transformations. This concept supplies the key to the general meaning of linear models. The notion of rank is also introduced. Chapter 5 deals with the theory of simultaneous linear equations. It develops the conditions which determine whether a set of equations has a solution, and whether this solution is unique.

Chapter 6 discusses the geometry of n dimensions and introduces the notions of lines, planes, spheres, regions, etc. It outlines the theory of convex sets and its applicability to the study of many linear models of which linear programming is an outstanding example. The chapter concludes with a brief discussion of a particular class of convex sets known as convex cones. These are useful in examining generalized Leontief models in economics and in linear programming.

Finally, Chapter 7 discusses the subject of characteristic values. This material facilitates greatly the study of linear vibrations, linear differential and difference equations, and Markov processes—to mention only a few applications. Nonlinear expressions, called quadratic forms, are also introduced which are important in developing sufficient conditions for maxima and minima of a function of a number of variables. The techniques of linear algebra provide a very powerful tool for analyzing quadratic forms.

REFERENCES

Leontief models:

(1) W. W. LEONTIEF, *The Structure of the American Economy, 1919–1939.* 2d ed., New York: Oxford, 1951.

(2) R. DORFMAN, P. SAMUELSON, and R. SOLOW, *Linear Programming and Economic Analysis.* New York: McGraw-Hill, 1958.

(3) O. MORGENSTERN, Ed., *Economic Activity Analysis.* New York: Wiley, 1954.

Linear programming:

(1) S. GASS, *Linear Programming: Methods and Applications.* New York: McGraw-Hill, 1958.

(2) G. HADLEY, *Linear Programming.* Reading, Mass.: Addison-Wesley, 1961.

Regression analysis:

(1) D. FRASER, *Statistics: An Introduction.* New York: Wiley, 1958. This entry is only an example. Almost any statistics texts discusses this topic.

(2) L. KLEIN, *A Textbook of Econometrics.* Evanston: Row, Peterson, 1953.

Linear network analysis:

(1) E. GUILLEMIN, *Introductory Circuit Theory.* New York: Wiley, 1953.

(2) P. LECORBEILLER, *Matrix Analysis of Electrical Networks.* New York: Wiley, 1950.

Linear differential equations:

(1) W. KAPLAN, *Ordinary Differential Equations.* Reading, Mass.: Addison-Wesley, 1958.

Linear difference equations:

(1) F. HILDEBRAND, *Methods of Applied Mathematics.* Englewood Cliffs: Prentice-Hall, 1952.

Linear vibrations:

(1) L. BRILLOUIN, *Wave Propagation in Periodic Structures.* New York: Dover, 1953.

(2) H. CORBEN and P. STEHLE, *Classical Mechanics.* New York: Wiley, 1950.

Markov processes:

(1) J. KEMENY and J. SNELL, *Finite Markov Chains.* Princeton: Van Nostrand, 1960.

Transformation theory in quantum mechanics:

(1) J. FRENKEL, *Wave Mechanics, Advanced General Theory.* New York: Dover, 1950.

Maxima and minima with applications to economics:

(1) P. SAMUELSON, *Foundations of Economic Analysis.* Cambridge: Harvard University Press, 1955.

Rigid body mechanics:

(1) H. GOLDSTEIN, *Classical Mechanics.* Reading, Mass.: Addison-Wesley, 1951.

PROBLEMS

Discuss the physical or economic meaning of linearity for each of the four models presented in this chapter.

CHAPTER 2

VECTORS

"... Arrows of outrageous fortune."

Shakespeare—Hamlet.

2–1 Physical motivation for the vector concept. Vectors are frequently used in many branches of pure and applied mathematics and in the physical and engineering sciences. The vector analysis applied by the physicist or engineer to problems in their fields differs in many ways from the n-dimensional vector spaces used in pure mathematics. Both types, however, have a common intuitive foundation. The need for a vector concept arose very naturally in mechanics. The force on a body, for example, cannot in general be completely described by a single number. Force has two properties, magnitude and direction, and therefore requires more than a single number for its description. Force is an example of a vector quantity. At the most elementary level in physics, a vector is defined as a quantity which has both magnitude and direction. A large number of physical quantities are vectors and, interestingly enough, the same laws of operation apply to *all* vectors.

Vectors are often represented geometrically by a line with an arrowhead on the end of it. The length of the line indicates the magnitude of the vector, and the arrow denotes its direction. When representing vectors geometrically as directed line segments, it is desirable to introduce a coordinate system as a reference for directions and as a scale for lengths. Familiar rectangular coordinates will be used. Usually, the coordinate axes will be named x_1, x_2, x_3. Some vectors lying in a plane are shown in Fig. 2–1.

We shall often represent a vector by a directed line segment. It should be stressed that a vector *is not a number*. If we are considering vectors lying in a plane, then two numbers are needed to describe any vector: one for its magnitude and another giving its direction (the angle it makes with one of the coordinate axes). If vectors in three-dimensional space are being studied, three numbers are needed to describe any vector: one number for its magnitude, and two numbers to denote its orientation with respect to some coordinate system. In physics, the symbol associated with a vector usually has a mnemonic connotation, such as **f** for force, **a** for acceleration, etc. Since we shall not be referring to any particular physical quantities, we shall simply use arbitrary lower-case *boldface* symbols for vectors, **a**, **b**, **x**, **y**. Boldface type implies that the symbol does not stand for a number.

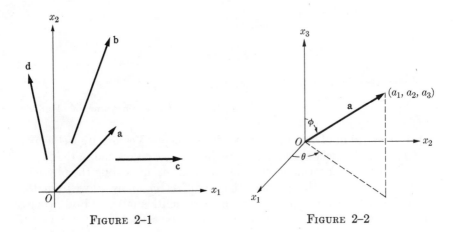

FIGURE 2-1   FIGURE 2-2

In general, a vector may originate at any point in space and terminate at any point. Physical quantities such as force are, of course, independent of where one places the vector in a coordinate system. They depend only on the magnitude and direction of the vector. For this reason, it is convenient to have a vector start always at the origin of the coordinate system (**a** in Fig. 2-1). In this book we shall adopt the convention that all vectors begin at the origin of the coordinate system rather than at some other point in space. This convention provides some simplifications in dealing with vectors.

Consider the vector in Fig. 2-2. It will be observed that by specifying point (a_1, a_2, a_3), that is, the point where the head of the vector terminates, we have completely characterized the vector. Its magnitude (in some physical units) is

$$[a_1^2 + a_2^2 + a_3^2]^{1/2},$$

and its direction is characterized by the two angles θ and ϕ, where

$$\tan \theta = \frac{a_2}{a_1}, \qquad \cos \phi = \frac{a_3}{[a_1^2 + a_2^2 + a_3^2]^{1/2}}.$$

Thus, there is a one-to-one correspondence between all points in space and all vectors which emanate from the origin. For any given point (a_1, a_2, a_3), a corresponding unique vector **a** can be drawn from the origin to this point. Conversely, for any given vector **a**, there is a unique point (a_1, a_2, a_3) which is the point where the vector terminates. Because of this correspondence between vectors and points we can write

$$\mathbf{a} = (a_1, a_2, a_3). \tag{2-1}$$

Equation (2–1) means that **a** is the vector drawn from the origin to the point (a_1, a_2, a_3) with coordinates a_1, a_2, a_3.

The correspondence between vectors and points in space is fundamental and important because of its practical value in vector computations; it also offers the key to more abstract generalizations of the vector concept.

EXAMPLES: (1) **a** $= (1, 2, 3)$ is the vector drawn from the origin to the point $x_1 = 1$, $x_2 = 2$, $x_3 = 3$. (2) **a** $= (3, 4)$ is a vector lying in a plane drawn from the origin to the point $x_1 = 3$, $x_2 = 4$. (3) **a** $= (1, 0, 0)$ is the vector drawn along the x_1-axis to $x_1 = 1$, $x_2 = 0$, $x_3 = 0$.

2–2 Operations with vectors. We now shall consider the operations that can be performed with vectors. First, from an intuitive point of view, two vectors are equal if they have the same magnitude and direction. According to our point representation, this means that both vectors are drawn from the origin to the same point in space, that is, they are coincident. Thus, if
$$\mathbf{a} = (a_1, a_2, a_3), \quad \mathbf{b} = (b_1, b_2, b_3),$$
then
$$\mathbf{a} = \mathbf{b},$$
if and only if
$$a_1 = b_1, \quad a_2 = b_2, \quad a_3 = b_3. \tag{2-2}$$
Equality of two vectors in space therefore implies that three equations in terms of real numbers must hold.

The next operation with vectors, which arises naturally in physics, is that of changing the magnitude of a vector without changing its direction, for example: The force on a body is doubled without altering the direction of application. Figure 2–3 illustrates this point. It shows, in terms of the point notation, that if the magnitude of vector **a** $= (a_1, a_2)$ is changed by a multiple λ, the new vector, $\lambda\mathbf{a}$, is
$$\lambda\mathbf{a} = (\lambda a_1, \lambda a_2).$$
To multiply the magnitude of **a** by λ without changing its direction, each

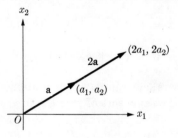

FIGURE 2–3

coordinate has to be multiplied by λ. The same procedure applies to three dimensions:
$$\lambda \mathbf{a} = (\lambda a_1, \lambda a_2, \lambda a_3). \tag{2-3}$$

In the preceding section we have been assuming that $\lambda > 0$. We can easily generalize the notation and allow λ to be negative. If $\lambda < 0$, the magnitude of the vector is changed by a factor $-\lambda$, and its direction along its line of application is completely reversed. Frequently we shall refer to a real number as a scalar (in contradistinction to a vector or matrix). The operation of multiplying any given vector by a real number λ will be referred to as multiplying a vector by a scalar. The rule for multiplying a vector by a scalar, when using point notation, is given by (2–3).

In geometrical language, the magnitude of a vector $\mathbf{a}$, which we shall denote by $|\mathbf{a}|$, is often called its "length," since the length of the line from the origin to the terminal point of the vector represents its magnitude. By our definition, the magnitude of $\lambda \mathbf{a}$ is $|\lambda \mathbf{a}| = |\lambda| \, |\mathbf{a}|$, where $|\lambda|$ is the absolute value of λ.

EXAMPLES: (1) If $\mathbf{a} = (2, 3)$, $6\mathbf{a} = (12, 18)$. (2) If $\mathbf{a} = (1, -1)$, $-4\mathbf{a} = (-4, 4)$. Illustrate these graphically.

Another very important operation is the addition of vectors. Again we turn to mechanics for our intuitive foundations. It is well known that if two forces act on a particle (a proton, for example) to produce some resultant motion, the same motion can be produced by applying a single force. This single force can, in a real sense, be considered to be the sum of the original two forces. The rule which we use in obtaining magnitude and direction of a single force which replaces the original two forces is rather interesting: If $\mathbf{a}$, $\mathbf{b}$ are the original forces, then the single force $\mathbf{c}$, which we shall call the sum of $\mathbf{a}$, $\mathbf{b}$, is the diagonal of the parallelogram with sides $\mathbf{a}$, $\mathbf{b}$. This is illustrated in Fig. 2–4. The addition of vectors follows what, in elementary physics, is known as the *parallelogram law*.

The rule for addition of vectors is very simple when point representation is used. We merely add corresponding coordinates as shown in Fig. 2–4. If
$$\mathbf{a} = (a_1, a_2), \quad \mathbf{b} = (b_1, b_2),$$
then
$$\mathbf{c} = \mathbf{a} + \mathbf{b} = (a_1 + b_1, a_2 + b_2) = (c_1, c_2).$$

Examination of the parallelogram law, when applied to three dimensions, shows that precisely the same sort of result holds. If
$$\mathbf{a} = (a_1, a_2, a_3), \quad \mathbf{b} = (b_1, b_2, b_3),$$
$$\mathbf{c} = (c_1, c_2, c_3) = \mathbf{a} + \mathbf{b} = (a_1 + b_1, a_2 + b_2, a_3 + b_3). \tag{2-4}$$

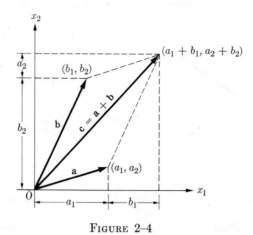

FIGURE 2-4

To add three vectors, the first two are added to obtain the resultant, and the third is then added to the resultant of the first two. If

$$\mathbf{a} = (a_1, a_2, a_3), \qquad \mathbf{b} = (b_1, b_2, b_3), \qquad \mathbf{c} = (c_1, c_2, c_3);$$
$$\mathbf{d} = \mathbf{a} + \mathbf{b} + \mathbf{c} = (a_1 + b_1 + c_1,\ a_2 + b_2 + c_2,\ a_3 + b_3 + c_3) = (d_1, d_2, d_3). \tag{2-5}$$

In the same way any number of vectors can be added. Note that $\mathbf{a} + \mathbf{b} = \mathbf{b} + \mathbf{a}$.

To subtract $\mathbf{b}$ from $\mathbf{a}$, we add $(-1)\mathbf{b}$ to $\mathbf{a}$, that is,

$$\mathbf{a} - \mathbf{b} = \mathbf{a} + (-1)\mathbf{b} = (a_1 - b_1,\ a_2 - b_2,\ a_3 - b_3). \tag{2-6}$$

The concept of addition of vectors brings up a new idea, that of the resolution of a vector into components. Consider the vectors

$$\mathbf{a} = (a_1, a_2), \qquad \mathbf{a}_1 = (a_1, 0), \qquad \mathbf{a}_2 = (0, a_2).$$

In this example, $\mathbf{a}_1$ is a vector of length (magnitude) $|a_1|$ lying along the x_1-axis, and $\mathbf{a}_2$ a vector of length $|a_2|$ lying along the x_2-axis. It will be observed that

$$\mathbf{a} = \mathbf{a}_1 + \mathbf{a}_2 = (a_1, 0) + (0, a_2) = (a_1, a_2). \tag{2-7}$$

The vectors $\mathbf{a}_1$, $\mathbf{a}_2$ are called the *vector components* of $\mathbf{a}$ along the coordinate axes (see Fig. 2-5).

The concept of resolving a vector into its vector components along the coordinate axes is a very useful one. However, it can be developed even

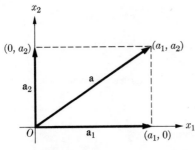
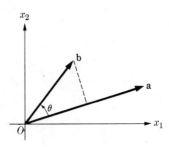

FIGURE 2–5 FIGURE 2–6

further. Applying the rule for multiplication by a scalar, we can write

$$\mathbf{a}_1 = (a_1, 0) = a_1(1, 0), \qquad \mathbf{a}_2 = (0, a_2) = a_2(0, 1).$$

Hence
$$\mathbf{a} = (a_1, a_2) = a_1(1, 0) + a_2(0, 1). \tag{2-8}$$

The vector $(1, 0)$ lies along the x_1-axis and has length one. The vector $(0, 1)$ lies along the x_2-axis and has length one. The vectors $(1, 0)$ and $(0, 1)$ are called unit vectors and will be denoted by $\mathbf{e}_1$, $\mathbf{e}_2$, respectively. The numbers a_1, a_2 are called the components of $\mathbf{a}$ along the x_1, x_2-axes. They are not the vector components $\mathbf{a}_1$, $\mathbf{a}_2$; the latter are obtained by multiplying the components by the corresponding unit vectors. Note that a_i, the component of $\mathbf{a}$ along the ith coordinate axis, is the ith coordinate of the point (a_1, a_2), $i = 1, 2$.

Similarly, working with three dimensions, we can write

$$\begin{aligned}\mathbf{a} &= (a_1, a_2, a_3) = a_1(1, 0, 0) + a_2(0, 1, 0) + a_3(0, 0, 1)\\ &= a_1\mathbf{e}_1 + a_2\mathbf{e}_2 + a_3\mathbf{e}_3.\end{aligned} \tag{2-9}$$

Equation (2–8) shows that any vector lying in a plane can be written as the sum of scalar multiples of the two unit vectors. In three dimensions, any vector can be written as the sum of scalar multiples of the three unit vectors:

$$\mathbf{e}_1 = (1, 0, 0), \qquad \mathbf{e}_2 = (0, 1, 0), \qquad \mathbf{e}_3 = (0, 0, 1).$$

2–3 The scalar product. We shall now discuss vector multiplication. Let us assume that we have two vectors $\mathbf{a}$, $\mathbf{b}$, and that θ is the angle between these vectors, as shown in Fig. 2–6. Consider the expression

$$|\mathbf{a}|\,|\mathbf{b}|\cos\theta. \tag{2-10}$$

The number $|\mathbf{b}|\cos\theta$ is, aside from sign, the magnitude of the perpendicular projection of $\mathbf{b}$ on the vector $\mathbf{a}$. It is called the *component* of

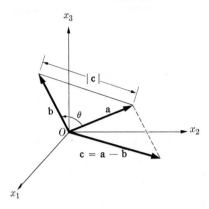

FIGURE 2–7

b along **a**. Thus, Eq. (2–10) represents the magnitude of **a** times the component of **b** along **a**. The number computed from (2–10) is called the scalar product of **a** and **b**, and we write

$$ab' = |a| \, |b| \cos \theta, \tag{2-11}$$

where the symbol **ab**′ denotes the scalar product of the two vectors. The scalar product of two vectors is a scalar, not a vector. This may seem to be a rather strange definition. However, it can be easily explained by an example taken from mechanics:

The work done in moving an object along a straight path a distance **r** (distance is a vector since it has both magnitude and direction), using a constant force **f**, is given by the product of the magnitude of the distance and the component of the force along **r**, that is, by $|r| \, |f| \cos \theta$, if θ is the angle between **f** and **r**. Therefore, the work w (a scalar) is the scalar product of **f**, **r** or $w = \mathbf{fr}'$.

The definition of a scalar product implies that $\mathbf{ab}' = \mathbf{ba}'$. Since the scalar product of two vectors is a scalar, it is not possible to form the scalar product of three or more vectors.

Next, we shall transform the scalar product into an expression which is much more suitable for extension to higher dimensions. We can write

$$\mathbf{a} = a_1 \mathbf{e}_1 + a_2 \mathbf{e}_2 + a_3 \mathbf{e}_3, \qquad \mathbf{b} = b_1 \mathbf{e}_1 + b_2 \mathbf{e}_2 + b_3 \mathbf{e}_3.$$

Consider also the vector $\mathbf{c} = \mathbf{a} - \mathbf{b}$:

$$\mathbf{c} = (a_1 - b_1)\mathbf{e}_1 + (a_2 - b_2)\mathbf{e}_2 + (a_3 - b_3)\mathbf{e}_3.$$

Then from Fig. 2–7 and the cosine law of trigonometry

$$|c|^2 = |a|^2 + |b|^2 - 2|a| \, |b| \cos \theta.$$

Therefore
$$\mathbf{ab}' = \tfrac{1}{2}[|\mathbf{a}|^2 + |\mathbf{b}|^2 - |\mathbf{c}|^2]. \tag{2-12}$$
But
$$|\mathbf{a}|^2 = \sum_{i=1}^{3} a_i^2, \quad |\mathbf{b}|^2 = \sum_{i=1}^{3} b_i^2, \quad |\mathbf{c}|^2 = \sum_{i=1}^{3} (a_i - b_i)^2.$$
Hence
$$\mathbf{ab}' = a_1 b_1 + a_2 b_2 + a_3 b_3. \tag{2-13}$$

Equation (2-13) is important since it shows that the scalar product of two vectors is computed by simply multiplying the corresponding components and adding the results.

The unit vectors $\mathbf{e}_1, \mathbf{e}_2, \mathbf{e}_3$ lie along the coordinate axes, which are assumed to be at right angles to each other. Hence

$$\mathbf{e}_1 \mathbf{e}_2' = |\mathbf{e}_1|\,|\mathbf{e}_2|\cos\frac{\pi}{2} = 0, \quad \mathbf{e}_1 \mathbf{e}_1' = |\mathbf{e}_1|\,|\mathbf{e}_1|\cos 0 = 1,$$

or, in general,

$$\mathbf{e}_i \mathbf{e}_j' = 0 \quad (i \neq j), \quad \mathbf{e}_i \mathbf{e}_i' = 1 \quad (i = 1, 2, 3). \tag{2-14}$$

The scalar product of a vector with itself is the square of the length of the vector:
$$\mathbf{aa}' = a_1^2 + a_2^2 + a_3^2 = |\mathbf{a}|^2. \tag{2-15}$$

EXAMPLES:

(1) If $\mathbf{a} = (2, 3, 1), \quad \mathbf{b} = (1, 7, 5)$:
$\mathbf{ab}' = 2(1) + 3(7) + 1(5) = 28.$

(2) If $\mathbf{a} = (a_1, a_2, a_3), \quad \mathbf{b} = (0, 1, 0) = \mathbf{e}_2$:
$\mathbf{ab}' = \mathbf{ae}_2' = a_1(0) + a_2(1) + a_3(0) = a_2.$

In the preceding sections we have attempted to give an intuitive development of vector operations. Depending on one's background, this may or may not have seemed to be a natural development. In any event, it is desirable to have some familiarity with the material so that generalizations will not seem completely unmotivated. Many operations with vectors which are very important in physics and engineering have not been considered here. We have only discussed those parts which form a foundation for n-dimensional generalizations. Before going on to these generalizations, let us note that vector analysis of the type used in physics and engineering goes back to Willard Gibbs (the founder of modern thermodynamics) and to the more clumsy quaternions used by Hamilton.

2–4 Generalization to higher dimensions.

We have seen that at the most elementary level in physics, a vector can be described as a physical quantity possessing both magnitude and direction. The behavior of the real world determines the laws governing operations with these vectors. Next we observed that, instead of characterizing a vector by magnitude and direction, an equally satisfactory description could be achieved by the terminal point of a vector of proper magnitude and direction emanating from the origin of the coordinate system. We then wrote $\mathbf{a} = (a_1, a_2, a_3)$; the a_i were called the components of the vector along the coordinate axes. The numbers a_i were also the coordinates of the point where the head of the vector terminated. From the laws for operating with vectors we derived rules applicable to operations with their components.

Let us suppose that we would like to divorce the concept of a vector from any physical connotations. In addition, we shall assume that we would like to develop a sound mathematical theory of vectors. Naturally, we hope to arrive at some useful generalizations of the theory. First we must decide upon a suitable definition of a vector. From our new point of view it is not very satisfactory to define a vector as a quantity having magnitude and direction, since this definition does not provide a very concrete concept and does not allow for an immediately obvious algebraic expression. The concept is especially fuzzy if we contemplate extending it to spaces of dimension greater than three. However, such an extension of the theory is one of the first generalizations which will come to mind.

The key to finding a definition which permits the development of a rigorous theory lies in focusing attention on the point representation of a vector. Our study of the intuitive foundations has shown us that a vector can be represented by an ordered array of numbers (a_1, a_2) or (a_1, a_2, a_3); hence, we may apply this concept when generalizing and simply define a vector as an ordered array of numbers (a_1, a_2) or (a_1, a_2, a_3). This is precisely what we shall do. From our new point of view, a vector will be nothing more or less than an *ordered array of numbers;* it will not have any physical meaning attached to it whatever. Once we take this step, it immediately becomes apparent that we are not limited to vectors containing at most three components. We may now generalize our notion of a vector to one containing any finite number of components, i.e., an ordered array of n numbers $(a_1, a_2, \ldots, a_n)$, which we shall call an ordered *n-tuple*. Indeed, this generalization will be incorporated into the theory.

Since we no longer plan to attach any physical meaning to vectors, we are free to define operations with these vectors in any way we choose. However, if the intuitive foundations of the subject are to be of any value, they should point the way to proper definitions. This is the procedure we shall follow. The definitions of operations with vectors in the generalized sense will be direct extensions of the results obtained from physical considerations.

When generalizing the theory of vectors, it is of considerable advantage to retain some of the geometrical terms and concepts which were so helpful in the intuitive approach. In this way, we may give to many of the general results obtained a clear geometrical interpretation in two and three dimensions. It is very easy to relate our new definition of a vector to geometrical concepts. Just as (a_1, a_2), (a_1, a_2, a_3) can be considered to be points in two- and three-dimensional spaces respectively, $(a_1, \ldots, a_n)$ can be considered to be a point in an n-dimensional space, where the a_i are the "coordinates" of the point. As a matter of fact, from here on we shall assume that the concepts of point and vector mean precisely the same thing. Our development will then proceed simultaneously along two paths, namely: (1) "rigorous" algebraic development; (2) geometric interpretation. At first the geometric interpretation will be somewhat intuitive. However, before we are finished, the idea of an n-dimensional space will have been removed from the intuitive realm, and we shall have defined precisely what we mean by one important n-dimensional space, called *euclidean space*. A number of useful properties of this space will be developed.

After this rather long introduction to the subject of generalizing the theory of vectors, we shall now proceed with the development of the theory. First, we shall repeat in a more formal way our new definition of a vector:

VECTOR: *An n-component vector* **a** is an ordered n-tuple of numbers written as a row $(a_1, a_2, \ldots, a_n)$ or as a column*

$$\begin{bmatrix} a_1 \\ a_2 \\ \vdots \\ a_n \end{bmatrix}.$$

The a_i, $i = 1, \ldots, n$, are assumed to be real numbers and are called the components of the vector.

Whether a vector is written as a row or a column is immaterial. The two representations are equivalent. Notational convenience determines which is most suitable. In the future we shall frequently use both *row* and *column*

---

* Frequently an n-component vector is referred to as an n-dimensional vector. This seems to be a very appropriate terminology. However, some mathematicians object to attaching the term "dimension" to a single vector or point in space, since dimension is a property of space, not of a point.

vectors. However, the column representation will be used more frequently than the row representation, and for this reason we shall in the remainder of this chapter usually think of vectors as columns of numbers. Since it is rather clumsy to print these columns, a row with square brackets enclosing the n numbers will be used to represent a column vector:

$$\begin{bmatrix} a_1 \\ a_2 \\ \vdots \\ a_n \end{bmatrix} \equiv [a_1, \ldots, a_n]. \tag{2-16}$$

Parentheses will enclose the n components of a row vector, that is, $(a_1, \ldots, a_n)$ will always represent a row vector. The notational difference between row and column vectors should be carefully noted.

The ordering of the numbers in the n-tuple which forms a vector is crucial. Different ordering represents different vectors. For example, in three dimensions, $(1, 2, 3)$ and $(3, 1, 2)$ are clearly not the same vector.

As has been already indicated, there is a complete equivalence between n-component vectors and the points in an n-dimensional space. An n-tuple will be called a point or a vector depending on whether, at any particular moment, we view it from an algebraic or geometric standpoint.* Although, on several occasions, we have mentioned a point in an n-dimensional space, it is expected that the reader has as yet only a vague intuitive idea of such a space. More precise concepts of an n-dimensional space, coordinates, etc., will be introduced shortly.

Several useful vectors are often referred to by name, and hence we shall define them at the outset. We shall begin with an important set of vectors, the unit vectors. For vectors having n components there are n unit vectors. They are:

$$\mathbf{e}_1 = [1, 0, \ldots, 0], \quad \mathbf{e}_2 = [0, 1, 0, \ldots, 0], \quad \ldots, \quad \mathbf{e}_n = [0, 0, \ldots, 1]. \tag{2-17}$$

It will be recalled that these vectors were introduced to advantage in our intuitive study of two and three dimensions.

UNIT VECTOR: *A unit vector, denoted by $\mathbf{e}_i$, is a vector with unity as the value of its ith component and with all other components zero.*

---

* When we speak of **a** as a vector, then the a_i, $i = 1, \ldots, n$ of (2–16) are called the components of **a**. If we are thinking of **a** as a point in an n-dimensional space, then the a_i are frequently called the coordinates of **a**. Consequently, component and coordinate are the respective algebraic and geometric names for the same element in the n-tuple (2–16).

The symbol $\mathbf{e}_i$ will always be used for a unit vector. The number of components in the unit vector will be clear from the context.

NULL VECTOR: *The null vector, or zero vector, written* $\mathbf{0}$, *is a vector all of whose components are zero.*

$$\mathbf{0} = [0, 0, \ldots, 0]. \tag{2-18}$$

SUM VECTOR: *A sum vector is a vector having unity as a value for each component; it will be written* $\mathbf{1}$.

$$\mathbf{1} = (1, 1, \ldots, 1). \tag{2-19}$$

In general, the sum vector will be used as a row vector rather than a column vector. The reason for calling it a sum vector will become clear later.

2–5 Generalized vector operations. Following the procedure suggested in the previous section, we shall define operations with vectors as straightforward generalizations of the results obtained in three dimensions. The method for generalizing is: (1) the operations are written using the point representation, that is, in component form; (2) the generalization follows immediately.

EQUALITY: *Two n-component vectors* $\mathbf{a}$, $\mathbf{b}$ *are said to be equal, written* $\mathbf{a} = \mathbf{b}$, *if and only if all the corresponding components are equal.*

$$a_i = b_i, \quad i = 1, \ldots, n. \tag{2-20}$$

Equality of two n-component vectors means that, in terms of real numbers, there are n equations which must hold. Two vectors cannot be equal unless they have the same number of components. Note that if $\mathbf{a} = \mathbf{b}$, then $\mathbf{b} = \mathbf{a}$.

EXAMPLES:
(1) If $\mathbf{a} = [a_1, a_2, a_3, a_4]$ and $\mathbf{b} = [b_1, b_2, b_3, b_4]$, then if $\mathbf{a} = \mathbf{b}$, $a_1 = b_1$, $a_2 = b_2$, $a_3 = b_3$, and $a_4 = b_4$.
(2) The vectors $\mathbf{a} = [0, 2, 1]$ and $\mathbf{b} = [0, 2, 2]$ are not equal since the third component of $\mathbf{a}$ is different from the third component of $\mathbf{b}$.
(3) If $\mathbf{a} = [2, 1]$ and $\mathbf{b} = [2, 1, 3]$, $\mathbf{a} \neq \mathbf{b}$ since $\mathbf{a}$, $\mathbf{b}$ do not have the same number of components.

Occasionally we shall find use for vector inequalities.

INEQUALITIES: *Given two n-component vectors* $\mathbf{a}$, $\mathbf{b}$, *then* $\mathbf{a} \geq \mathbf{b}$ *means* $a_i \geq b_i$, $i = 1, \ldots, n$, *and* $\mathbf{a} \leq \mathbf{b}$ *means* $a_i \leq b_i$, $i = 1, \ldots, n$.

Similarly $\mathbf{a} > \mathbf{b}$ means $a_i > b_i$ for all i, and $\mathbf{a} < \mathbf{b}$ means $a_i < b_i$ for all i.

EXAMPLES:

(1) $\mathbf{a} = [0, 1, 2]$, $\quad \mathbf{b} = [-1, 0, 1]$; $\quad \mathbf{a} > \mathbf{b}$ since $0 > -1, 1 > 0, 2 > 1$.

(2) $\mathbf{a} = [0, 1, 0]$, $\quad \mathbf{b} = [-1, 0, 1]$; $\quad \mathbf{a}$ is not $\geq \mathbf{b}$ (written $\mathbf{a} \not\geq \mathbf{b}$) since for the third components $0 < 1$.

MULTIPLICATION BY A SCALAR: *The product of a scalar λ and a vector $\mathbf{a} = [a_1, \ldots, a_n]$, written $\lambda\mathbf{a}$, is defined as the vector*

$$\lambda\mathbf{a} = [\lambda a_1, \lambda a_2, \ldots, \lambda a_n]. \qquad (2\text{-}21)$$

If $\mathbf{a} \geq \mathbf{b}$ and $\lambda > 0$, then $\lambda\mathbf{a} \geq \lambda\mathbf{b}$. This follows immediately, since $a_i \geq b_i$, and if $\lambda > 0$, then $\lambda a_i \geq \lambda b_i$. However, if $\mathbf{a} \geq \mathbf{b}$ and $\lambda < 0$, then $\lambda\mathbf{a} \leq \lambda\mathbf{b}$. This becomes equally clear if one considers the relations in terms of the vector components. Multiplying an inequality of real numbers by a negative number changes the direction of the inequality $(5 > 1, -5 < -1)$.

EXAMPLES:

(1) If $\mathbf{a} = [3, 4, 5]$; $\quad 3\mathbf{a} = [9, 12, 15]$.

(2) If $\mathbf{a} = [3, 1, 2]$, $\mathbf{b} = [4, 2, 4]$; $\quad \mathbf{a} < \mathbf{b}$. Also $-2\mathbf{a} > -2\mathbf{b}$ since $-2\mathbf{a} = [-6, -2, -4] > -2\mathbf{b} = [-8, -4, -8]$.

ADDITION: *The sum of two vectors $\mathbf{a} = [a_1, \ldots, a_n]$ and $\mathbf{b} = [b_1, \ldots, b_n]$, written $\mathbf{a} + \mathbf{b}$, is defined to be the vector*

$$\mathbf{a} + \mathbf{b} = [a_1 + b_1, a_2 + b_2, \ldots, a_n + b_n]. \qquad (2\text{-}22)$$

This definition applies only to vectors having the same number of components. The sum of two n-component vectors is another n-component vector.

Since addition is done by components, we immediately see that the addition of vectors possesses the commutative and associative properties of real numbers:

$$\mathbf{a} + \mathbf{b} = \mathbf{b} + \mathbf{a} \quad \text{(commutative property)}, \qquad (2\text{-}23)$$

$$\mathbf{a} + (\mathbf{b} + \mathbf{c}) = (\mathbf{a} + \mathbf{b}) + \mathbf{c} = \mathbf{a} + \mathbf{b} + \mathbf{c} \quad \text{(associative property)}. \qquad (2\text{-}24)$$

In the same way, we see that for scalars λ:

$$\lambda(\mathbf{a} + \mathbf{b}) = \lambda\mathbf{a} + \lambda\mathbf{b}, \qquad (2\text{-}25)$$

$$(\lambda_1 + \lambda_2)(\mathbf{a} + \mathbf{b}) = \lambda_1\mathbf{a} + \lambda_1\mathbf{b} + \lambda_2\mathbf{a} + \lambda_2\mathbf{b}. \qquad (2\text{-}26)$$

SUBTRACTION: *Subtraction of two vectors is defined in terms of operations already considered:*

$$\mathbf{a} - \mathbf{b} = \mathbf{a} + (-1)\mathbf{b} = [a_1 - b_1, \ldots, a_n - b_n] \qquad (2\text{-}27)$$

The concepts of addition and multiplication by a scalar can be combined to yield a *linear combination* of vectors.

LINEAR COMBINATION: *Given m n-component vectors $\mathbf{a}_1, \ldots, \mathbf{a}_m$, the n-component vector*

$$\mathbf{a} = \sum_{i=1}^{m} \lambda_i \mathbf{a}_i = \lambda_1 \mathbf{a}_1 + \cdots + \lambda_m \mathbf{a}_m \qquad (2\text{-}28)$$

is called a linear combination of $\mathbf{a}_1, \ldots, \mathbf{a}_m$ for any scalars λ_i, $i = 1, \ldots, m$.

EXAMPLES:
(1) $\mathbf{a} = [1, 3, 5]$, $\mathbf{b} = [2, 4, 6]$;
    $\mathbf{a} + \mathbf{b} = [(1 + 2), (3 + 4), (5 + 6)] = [3, 7, 11]$

(2) $\mathbf{a}_1 = [2, 3, 4, 7]$, $\mathbf{a}_2 = [0, 0, 0, 1]$, $\mathbf{a}_3 = [1, 0, 1, 0]$;
    $\mathbf{a}_1 + 2\mathbf{a}_2 + 3\mathbf{a}_3 = [2, 3, 4, 7] + 2[0, 0, 0, 1] + 3[1, 0, 1, 0] = [5, 3, 7, 9]$

(3) $\mathbf{a} - \mathbf{a} = \mathbf{0}$, $\mathbf{a} + \mathbf{0} = \mathbf{a}$; $\qquad\qquad\qquad\qquad\qquad$ (2-29)
    $\mathbf{a} + \mathbf{a}$ implies $a_i + a_i = 2a_i$;   thus $\mathbf{a} + \mathbf{a} = 2\mathbf{a}$.

2-6 Euclidean space and the scalar product. Using the operations defined in the foregoing section, we see that any n-component vector can be written as a linear combination of the n unit vectors (2-17):

$$\mathbf{a} = [a_1, \ldots, a_n] = a_1 \mathbf{e}_1 + \cdots + a_n \mathbf{e}_n. \qquad (2\text{-}30)$$

This is a straightforward generalization from two and three to higher dimensions. It will be recalled that, in operations with two and three dimensions, the unit vectors lay along the coordinate axes. We have, as yet, no coordinate system in our n-dimensional space. However, the preceding discussion suggests that *we use unit vectors to define a coordinate system*. If we imagine that a coordinate axis is "drawn" along each unit vector, we shall obtain a coordinate system for n-dimensional space. Furthermore, the ith component of $\mathbf{a}$ is the component of $\mathbf{a}$ along the ith coordinate axis (the same was true in three dimensions). The null vector $\mathbf{0}$ is the origin of our coordinate system. Whenever we write a vector $\mathbf{a}$ as $[a_1, \ldots, a_n]$, we automatically have a special coordinate system in mind, that is, the coordinate system defined by n unit vectors.

To complete the definition of our n-dimensional space by analogy to the properties of three-dimensional space, we must introduce the notion of distance and the related notions of lengths and angles. Distance is a very important concept since it forms the basis for notions of continuity and hence for analysis. A convenient way of introducing distance, lengths, and angles is to define the scalar product of two vectors.

SCALAR PRODUCT: *The scalar product of two n-component vectors* **a**, **b** *is defined to be the scalar*

$$a_1 b_1 + a_2 b_2 + \cdots + a_n b_n = \sum_{i=1}^{n} a_i b_i. \qquad (2\text{--}31)$$

It will be helpful to use different notations for the scalar product, depending on whether **a**, **b** are row or column vectors.* If **a**, **b** are both column vectors, we denote the scalar product by **a'b**. If **a**, **b** are both row vectors, the scalar product will be denoted by **ab'**. When **a** is a row vector and **b** is a column vector, the scalar product will be written **ab**. For the present, we assume that both vectors are column vectors; hence the scalar product can be written

$$\mathbf{a'b} = \sum_{i=1}^{n} a_i b_i. \qquad (2\text{--}32)$$

We see immediately that

$$\mathbf{a'b} = \mathbf{b'a}, \qquad (2\text{--}33)$$

$$\mathbf{a'(b + c)} = \mathbf{a'b} + \mathbf{a'c}, \quad \mathbf{(a + b)'c} = \mathbf{a'c} + \mathbf{b'c}, \qquad (2\text{--}34)$$

$$\mathbf{a'(\lambda b)} = \lambda(\mathbf{a'b}), \quad (\lambda \mathbf{a'})\mathbf{b} = \lambda(\mathbf{a'b}); \quad \lambda \text{ any scalar.} \qquad (2\text{--}35)$$

DISTANCE: *The distance from the vector (point)* **a** *to the vector (point)* **b**, *written* $|\mathbf{a} - \mathbf{b}|$, *is defined as*

$$|\mathbf{a} - \mathbf{b}| = [(\mathbf{a} - \mathbf{b})'(\mathbf{a} - \mathbf{b})]^{1/2} = \left[\sum_{i=1}^{n} (a_i - b_i)^2 \right]^{1/2}. \qquad (2\text{--}36)$$

Distance, as defined by (2–36), has the following properties:

(1) $|\mathbf{a} - \mathbf{b}| > 0$ unless $\mathbf{a} - \mathbf{b} = \mathbf{0}$; $\qquad (2\text{--}37)$

(2) $|\mathbf{a} - \mathbf{b}| = |\mathbf{b} - \mathbf{a}|$; $\qquad (2\text{--}38)$

(3) $|\mathbf{a} - \mathbf{b}| + |\mathbf{b} - \mathbf{c}| \geq |\mathbf{a} - \mathbf{c}|$ (triangle inequality). $\qquad (2\text{--}39)$

---

* Our notation for scalar products differs from others that are often used, such as **a** · **b** and (**a**, **b**). We have chosen our system in order to achieve consistency with the matrix notation introduced later.

These algebraic expressions mean: (1) the distance between two different points is always positive; (2) the distance from **a** to **b** is the same as the distance from **b** to **a**; (3) in two and three dimensional spaces, the sum of the lengths of two sides of a triangle is not less than the length of the third side. The proofs for (2-37) through (2-39), based on definition (2-36), are to be supplied in Problems 2-24 through 2-26.

It is important to note that we did not have to define distance by (2-36). There are many spaces and geometries in which distance is defined differently. Indeed, there are even geometries in which distance is not defined at all. For our purposes, and for many others, definition (2-36) is most suitable.

LENGTH: *The length or magnitude of a vector* **a**, *written* $|\mathbf{a}|$, *is defined as*

$$|\mathbf{a}| = [\mathbf{a'a}]^{1/2} = \left[\sum_{i=1}^{n} a_i^2\right]^{1/2}. \tag{2-40}$$

Note that length is a special case of distance since $|\mathbf{a}| = |\mathbf{a} - \mathbf{0}|$. The length of **a** is the distance from the origin to **a**.

The preceding definitions of scalar product, distance, and length are direct generalizations of the corresponding expressions for three dimensions. The appropriate definitions are fairly obvious. It is not quite so clear, however, how the angle between two vectors should be generalized. Here, it is important to note that, in the intuitive introduction, the angle between the two vectors was part of the definition of a scalar product.

In this section we have defined the scalar product without reference to angles. We shall now use the original definition of the scalar product to define the angle between two vectors:

ANGLE: *The angle* θ *between two vectors* $\mathbf{a} = [a_1, \ldots, a_n]$ *and* $\mathbf{b} = [b_1, \ldots, b_n]$, *where* $\mathbf{a}, \mathbf{b} \neq \mathbf{0}$, *is computed from*

$$\cos\theta = \frac{\mathbf{a'b}}{|\mathbf{a}|\,|\mathbf{b}|} = \frac{\sum_{i=1}^{n} a_i b_i}{[\sum_{i=1}^{n} a_i^2]^{1/2}[\sum_{i=1}^{n} b_i^2]^{1/2}}. \tag{2-41}$$

Note: The cosine of the angle between two vectors appears in statistics. If we have n sets of historic or experimental data $(y_1, x_1), \ldots, (y_n, x_n)$, and if we write

$$\mathbf{y} = [y_1 - \bar{y}, \ldots, y_n - \bar{y}], \quad \mathbf{x} = [x_1 - \bar{x}, \ldots, x_n - \bar{x}],$$

where

$$\bar{y} = \frac{1}{n}\sum_{i=1}^{n} y_i, \quad \bar{x} = \frac{1}{n}\sum_{i=1}^{n} x_i,$$

then the cosine of the angle between **y** and **x** is the correlation coefficient for the sets of data.

It can be easily shown that $\cos \theta$, as defined by (2–41), satisfies $-1 \leq \cos \theta \leq 1$; therefore, we can always compute an angle θ lying between 0 and π. To establish that $-1 \leq \cos \theta \leq 1$, is equivalent to showing that for any n-component vectors **a**, **b**

$$|\mathbf{a}'\mathbf{b}| \leq |\mathbf{a}|\,|\mathbf{b}|. \tag{2–42}$$

This property is called the *Schwarz inequality*. Problem 2–25 requires proof of (2–42).

In two and three dimensions, two non-null vectors are called orthogonal if the angle between them is $\pi/2$. In this case, the scalar product of the two vectors vanishes [see (2–11) and note that $\cos \theta = 0$]. We can generalize the concept of orthogonality to an n-dimensional space:

ORTHOGONALITY: *Two vectors* **a**, **b** (**a**, **b** $\neq$ **0**) *are said to be orthogonal if their scalar product vanishes, that is,* $\mathbf{a}'\mathbf{b} = 0$.

From (2–41) we see that if two vectors are orthogonal,* $\cos \theta = 0$, and the angle between the vectors is $\pi/2$. We notice immediately that the unit vectors are orthogonal since $\mathbf{e}'_i \mathbf{e}_j = 0$, $i \neq j$. This means that *the coordinate system in n dimensions defined by the unit vectors is an orthogonal coordinate system* analogous to the orthogonal coordinate systems for two and three dimensions.

Finally, we can define an n-dimensional space, often referred to as euclidean space:

EUCLIDEAN SPACE: *An n-dimensional euclidean space (or euclidean vector space) is defined as the collection of all vectors (points)* $\mathbf{a} = [a_1, \ldots, a_n]$. *For these vectors, addition and multiplication by a scalar are defined by (2–22) and (2–21), respectively. Furthermore, associated with any two vectors in the collection is a non-negative number called the distance between the two vectors; the distance is given by (2–36).*

The n-dimensional spaces which we shall discuss here will always be euclidean, represented by the symbol E^n. Ordinary two- and three-dimensional spaces of the types considered in the intuitive introduction are euclidean spaces. When $n = 3$, our n-dimensional space reduces to the familiar concept of a three-dimensional space.

The definition of a euclidean space encompasses definitions for operating with points in the space and for distance. Equation (2–41) provides the

---

* We can also say that **0** is orthogonal to every other vector, and that the angle between **0** and any other vector is $\pi/2$. This will remove the restriction **a**, **b** $\neq$ **0** in the definition of orthogonality.

definition of the angle between two points or vectors. Furthermore, we have seen that the unit vectors can be used to define an orthogonal coordinate system in this space. At this juncture, we could proceed to introduce the notions of lines and planes in E^n, etc. However, we shall defer the study of certain aspects of n-dimensional geometry to Chapter 6 and devote the remainder of the present chapter to a discussion of some properties of points or vectors in E^n.

Before passing on to the subject of linear dependence, let us pause and illustrate a use for the sum vector defined by (2–19). If we form the scalar product of the sum vector and any other vector $\mathbf{a}$, we obtain

$$\mathbf{1a} = \sum_{i=1}^{n} a_i. \qquad (2\text{–}43)$$

This scalar product is the sum of the components of $\mathbf{a}$. Any summation can be written as the scalar product of the sum vector and the vector formed from the elements in the summation. The reason for calling $\mathbf{1}$ the *sum vector* is now clear.

2–7 Linear dependence. If one vector in a set of vectors from E^n can be written as a linear combination of some of the other vectors in the set, then we say that the given vector is linearly dependent on the others, and *the set of vectors is also linearly dependent*. If no vector in a collection of vectors can be written as a linear combination of the others, then *the set of vectors is linearly independent*.

Linear dependence or independence are properties of the set of vectors and not of the individual vectors in the set. The differences between linearly dependent and linearly independent sets of vectors are very fundamental. The concepts of linear dependence and independence will appear repeatedly in our later developments.

In a very crude intuitive sense, linearly dependent sets of vectors contain an element of redundancy. Since at least one vector can be represented as a linear combination of the others, we could drop any such vector from the set without losing too much. In contrast, linearly independent vectors are essentially different from each other. No vector can be dropped from a linearly independent set of vectors without losing something. Although this intuitive discussion is very vague, the precise meaning of dependence and independence will become clear as the theory is developed further.

We shall now give a more symmetric definition of linear dependence, which makes it unnecessary to single out any one vector and attempt to write it as a linear combination of the others. This new definition, which is the standard mathematical definition of linear dependence, will be shown to be equivalent to the intuitive concept outlined in the first paragraph of this section.

LINEAR DEPENDENCE: *A set of vectors* $a_1, \ldots, a_m$ *from* E^n *is said to be linearly dependent if there exist scalars* λ_i *not all zero such that*

$$\lambda_1 a_1 + \lambda_2 a_2 + \cdots + \lambda_m a_m = 0. \tag{2-44}$$

If the only set of λ_i *for which (2-44) holds is* $\lambda_1 = \lambda_2 = \cdots = \lambda_m = 0$, *then the vectors are said to be linearly independent.*

The above definition implies that a set of vectors is linearly dependent if we can find, by one means or another, scalars not all zero such that multiplication of a vector by the appropriate scalar and addition of the resulting vectors will provide the null vector. If, however, no set of λ_i other than all $\lambda_i = 0$ exists, then the vectors are linearly independent. A set of vectors which is not linearly dependent must be linearly independent.

Let us show that this definition is equivalent to the intuitive concept discussed at the beginning of this section. We want to prove that *the vectors* $a_1, \ldots, a_m$ *from* E^n *are linearly dependent if and only if some one of the vectors is a linear combination of the others.* If one of the vectors is a linear combination of the others, the vectors can be labeled so that this one vector is a_m. Then

$$a_m = \lambda_1 a_1 + \cdots + \lambda_{m-1} a_{m-1},$$

or

$$\lambda_1 a_1 + \cdots + \lambda_{m-1} a_{m-1} + (-1) a_m = 0,$$

with at least one coefficient (-1) not zero. Thus, by (2-44), the vectors are linearly dependent. Now suppose that the vectors are linearly dependent. Then (2-44) holds, and at least one $\lambda_i \neq 0$. Label the λ_i so that $\lambda_m \neq 0$. Then

$$-\lambda_m a_m = \lambda_1 a_1 + \cdots + \lambda_{m-1} a_{m-1},$$

or

$$a_m = -\frac{\lambda_1}{\lambda_m} a_1 - \cdots - \frac{\lambda_{m-1}}{\lambda_m} a_{m-1},$$

and one vector has been written as a linear combination of the others. In this proof we have assumed, of course, that there are at least two vectors in the set.

Although when we speak of a set of vectors as being linearly dependent or independent, it is usually assumed that two or more vectors are in the set, we must, in order to make later discussions consistent, define what is meant by linear dependence and independence for a set containing a single vector. The general definition expressed by (2-44) includes this case. It says that a set containing a single vector **a** is linearly dependent if there exists a $\lambda \neq 0$ such that $\lambda \mathbf{a} = \mathbf{0}$. This will be true if and only if $\mathbf{a} = \mathbf{0}$. Thus a set containing a single vector **a** is linearly independent if $\mathbf{a} \neq \mathbf{0}$, and linearly dependent if $\mathbf{a} = \mathbf{0}$.

A vector **a** is said to be linearly dependent on a set of vectors $\mathbf{a}_1, \ldots, \mathbf{a}_m$ if **a** can be written as a linear combination of $\mathbf{a}_1, \ldots, \mathbf{a}_m$; otherwise **a** is said to be linearly independent of $\mathbf{a}_1, \ldots, \mathbf{a}_m$. It can be noted immediately that the null vector is not linearly independent of any other vector or set of vectors since

$$\mathbf{0} = 0\mathbf{a}_1 + 0\mathbf{a}_2 + \cdots + 0\mathbf{a}_m. \tag{2-45}$$

Thus the null vector is linearly dependent on every other vector, and no set of linearly independent vectors can contain the null vector.

If a set of vectors is linearly independent, then any subset of these vectors is also linearly independent. For suppose $\mathbf{a}_1, \ldots, \mathbf{a}_m$ are linearly independent, while, for example, $\mathbf{a}_1, \mathbf{a}_2, \mathbf{a}_3$ are linearly dependent $(3 < m)$. In this case, there exist $\lambda_1, \lambda_2, \lambda_3$ not all zero such that $\lambda_1 \mathbf{a}_1 + \lambda_2 \mathbf{a}_2 + \lambda_3 \mathbf{a}_3 = \mathbf{0}$. If we take $\lambda_4 = \cdots = \lambda_m = 0$, then $\sum_{i=1}^{m} \lambda_i \mathbf{a}_i = \mathbf{0}$, and one or more λ_i in the set $\lambda_1, \lambda_2, \lambda_3$ are not zero. This contradicts the fact that $\mathbf{a}_1, \ldots, \mathbf{a}_m$ are linearly independent. *Similarly, if any set of vectors is linearly dependent, any larger set of vectors containing this set of vectors is also linearly dependent.*

Given a set of vectors $\mathbf{a}_1, \ldots, \mathbf{a}_m$ from E^n. We say that the maximum number of linearly independent vectors in this set is k if it contains at least one subset of k vectors which is linearly independent, and there is no linearly independent subset containing $k + 1$ vectors. If the set $\mathbf{a}_1, \ldots, \mathbf{a}_m$ is linearly independent, then the maximum number of linearly independent vectors in the set is m. Unless a set of vectors contains only the null vector, the maximum number of linearly independent vectors in the set will be at least one.

Suppose that $k < m$ is the maximum number of linearly independent vectors in a set of m vectors $\mathbf{a}_1, \ldots, \mathbf{a}_m$ from E^n. Then, given any linearly independent subset of k vectors in this set, every other vector in the set can be written as a linear combination of these k vectors. To see this, label the vectors so that $\mathbf{a}_1, \ldots, \mathbf{a}_k$ are linearly independent. The set $\mathbf{a}_1, \ldots, \mathbf{a}_k$, $\mathbf{a}_r$ must be linearly dependent for any $r = k + 1, \ldots, m$. This implies that (2-44) holds and at least one $\lambda_i \neq 0$. However, λ_r cannot be zero, because this would contradict the fact that $\mathbf{a}_1, \ldots, \mathbf{a}_k$ are linearly independent. Hence $\mathbf{a}_r$ can be written as a linear combination of $\mathbf{a}_1, \ldots, \mathbf{a}_k$.

EXAMPLES: (1). Consider the vectors of E^2, that is, two component vectors. Then any two non-null vectors **a**, **b** in E^2 are linearly dependent if $\lambda_1 \mathbf{a} + \lambda_2 \mathbf{b} = \mathbf{0}$, $\lambda_1, \lambda_2 \neq 0$ (why must both λ_1 and λ_2 differ from zero?) or $\mathbf{b} = \lambda \mathbf{a}$.

In E^2, two vectors are linearly dependent if one is a scalar multiple of the other. Geometrically, this means that two vectors are linearly dependent if they lie on the same line through the origin. For example, $(1, 0)$, $(3, 0)$ are linearly dependent since $(3, 0) = 3(1, 0)$. Both vectors lie along the x_1-axis.

Because only vectors lying along the same line are linearly dependent in E^2, it follows that any two vectors not lying along the same line in E^2 are linearly independent. Thus $(1, 2)$, $(2, 1)$ are linearly independent.

(2) It will be recalled that any vector in E^2 can be written as a linear combination of the two unit vectors $\mathbf{e}_1$, $\mathbf{e}_2$. Let $\mathbf{a}$, $\mathbf{b}$ be any two linearly independent vectors in E^2. We can write

$$\mathbf{a} = a_1\mathbf{e}_1 + a_2\mathbf{e}_2,$$

$$\mathbf{b} = b_1\mathbf{e}_1 + b_2\mathbf{e}_2.$$

Consider any other vector $\mathbf{x} = (x_1, x_2) = x_1\mathbf{e}_1 + x_2\mathbf{e}_2$ in E^2. Since $\mathbf{a}$, $\mathbf{b}$ are linearly independent, either a_1 or b_1 will differ from zero. Assume $a_1 \neq 0$. Then

$$\mathbf{e}_1 = \frac{1}{a_1}(\mathbf{a} - a_2\mathbf{e}_2),$$

$$\mathbf{b} = \frac{1}{a_1}[b_1\mathbf{a} + (a_1b_2 - a_2b_1)\mathbf{e}_2]. \tag{2-46}$$

However, $a_1b_2 - a_2b_1 \neq 0$, since $\mathbf{a}$, $\mathbf{b}$ are linearly independent. The vector $\mathbf{x}$ can be written

$$\mathbf{x} = \frac{1}{a_1}[x_1\mathbf{a} + (a_1x_2 - a_2x_1)\mathbf{e}_2]. \tag{2-47}$$

Solving for $\mathbf{e}_2$ in (2–46) and substituting into (2–47), we obtain

$$\mathbf{x} = \frac{1}{a_1}\left\{\left[x_1 - \frac{b_1(a_1x_2 - a_2x_1)}{a_1b_2 - a_2b_1}\right]\mathbf{a} + a_1\left[\frac{a_1x_2 - a_2x_1}{a_1b_2 - a_2b_1}\right]\mathbf{b}\right\}, \tag{2-48}$$

or

$$\mathbf{x} = \lambda_1\mathbf{a} + \lambda_2\mathbf{b}.$$

We have expressed $\mathbf{x}$ as a linear combination of $\mathbf{a}$, $\mathbf{b}$ and we see that there cannot be more than two linearly independent vectors in any collection of vectors from E^2. Any three vectors in E^2 are linearly dependent. The geometrical analogue of (2–48) is shown in Fig. 2–8.

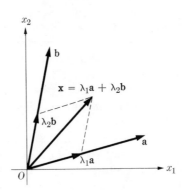

Figure 2–8

(3) The vectors $[1, 2, 4]$, $[2, 2, 8]$, $[1, 0, 4]$ are linearly dependent since

$$[1, 2, 4] - [2, 2, 8] + [1, 0, 4] = [0, 0, 0].$$

The vectors $\mathbf{e}_1 = [1, 0, 0]$, $\mathbf{e}_2 = [0, 1, 0]$, $\mathbf{e}_3 = [0, 0, 1]$ are linearly independent since

$$\lambda_1 \mathbf{e}_1 + \lambda_2 \mathbf{e}_2 + \lambda_3 \mathbf{e}_3 = [\lambda_1, \lambda_2, \lambda_3] = \mathbf{0}$$

implies the component equations $\lambda_1 = 0$, $\lambda_2 = 0$, $\lambda_3 = 0$.

The arguments outlined in the second example could be used also to demonstrate that any four vectors from E^3 are linearly dependent. Furthermore, if three vectors are linearly dependent, then they all lie in a plane which passes through the origin. Geometric reasoning will show that such a plane can be passed through any two vectors, and any linear combination of these vectors will lie in that plane (prove this). Hence, if any vector can be written as a linear combination of two vectors, it must lie in the plane determined by the two vectors. Conversely, any three vectors in E^3 which do not lie in a plane are linearly independent.

After having defined linear dependence and independence, we are faced with the question of how to determine whether any given set of vectors is linearly dependent. A complete answer will not be possible until after Chapters 3, 4, and 5 have been covered. However, we can indicate the nature of the problem. Suppose we have m n-component vectors $\mathbf{a}_1 = [a_{11}, \ldots, a_{n1}], \ldots, \mathbf{a}_m = [a_{1m}, \ldots, a_{nm}]$. Two subscripts are used on the components of the vectors. The first subscript refers to the relevant component of the vector, and the second subscript indicates the relevant vector. If the vectors are linearly dependent, then there exist λ_i not all zero such that

$$\sum_{i=1}^{m} \lambda_i \mathbf{a}_i = \mathbf{0}. \tag{2-49}$$

If (2–49) can be satisfied only with all $\lambda_i = 0$, the vectors are linearly independent. Let us write the n component equations for (2–49). They are:

$$a_{11}\lambda_1 + \cdots + a_{1m}\lambda_m = 0,$$
$$a_{21}\lambda_1 + \cdots + a_{2m}\lambda_m = 0, \qquad (2\text{–}50)$$
$$\vdots$$
$$a_{n1}\lambda_1 + \cdots + a_{nm}\lambda_m = 0.$$

Here we have a set of n simultaneous linear equations in m unknowns, and we are to solve for the λ_i. Equations (2–50) are called homogeneous since the right-hand side of each equation is zero. We have not studied the properties of solutions to a set of equations like those in (2–50). We shall do so in Chapter 5. The set of equations always has a solution, but there may or may not be a solution with one or more λ_i different from zero. If the vectors are linearly dependent, there will be a solution with not all $\lambda_i = 0$.

2–8 The concept of a basis. We have already seen that any vector in E^n can be written as a linear combination of the n unit vectors. Furthermore, we have noted that any vector in E^2 could be represented as a linear combination of any two linearly independent vectors. A set of vectors such that any vector in E^n can be represented as a linear combination of these vectors is of special interest. We make the following definition:

SPANNING SET: *A set of vectors* $\mathbf{a}_1, \ldots, \mathbf{a}_r$ *from* E^n *is said to span or generate* E^n *if every vector in* E^n *can be written as a linear combination of* $\mathbf{a}_1, \ldots, \mathbf{a}_r$.

However, we are not looking indiscriminately for any set of vectors which will span E^n; rather, we are seeking a set containing the smallest number of vectors which will span E^n. Any set of vectors spanning E^n which contains the smallest possible number of vectors must be linearly independent. If the vectors were linearly dependent, we could express some one of the vectors of the set as a linear combination of the others and hence eliminate it; in such a case, we should not have had the smallest set. However, if the set of vectors is linearly independent, then we have the smallest set, for if we dropped any vector, we could not express every vector in E^n as a linear combination of the vectors in the set (in particular, we could not express the vector dropped from the set as a linear combination of the remaining vectors). Any linearly independent set of vectors which spans E^n is called a basis for E^n:

BASIS: *A basis for* E^n *is a linearly independent subset of vectors from* E^n *which spans the entire space.*

A set of vectors which forms a basis has two properties. Its vectors must span E^n and must be linearly independent. A basis for E^n is by no means unique. As we shall see, there are an infinite number of bases for E^n.

To begin our discussion of bases, we prove that the n *unit vectors* $\mathbf{e}_1, \ldots, \mathbf{e}_n$ *form a basis for E^n*. The unit vectors are linearly independent since the expression

$$\lambda_1 \mathbf{e}_1 + \cdots + \lambda_n \mathbf{e}_n = [\lambda_1, \ldots, \lambda_n] = \mathbf{0} \tag{2-51}$$

implies that

$$\lambda_1 = 0, \lambda_2 = 0, \ldots, \lambda_n = 0.$$

As observed previously, any vector $\mathbf{x}$ in E^n can be written as a linear combination of the $\mathbf{e}_i$ in the following way:

$$\mathbf{x} = x_1 \mathbf{e}_1 + \cdots + x_n \mathbf{e}_n.$$

Thus the unit vectors for E^n yield a basis for E^n.

The representation of any vector in terms of a set of basis vectors is unique, that is, any vector in E^n can be written as a linear combination of a set of basis vectors in only one way. Let $\mathbf{b}$ be any vector in E^n, and $\mathbf{a}_1, \ldots, \mathbf{a}_r$ a set of basis vectors. Suppose that we can write $\mathbf{b}$ as a linear combination of the $\mathbf{a}_i$ in two different ways, namely:

$$\mathbf{b} = \lambda_1 \mathbf{a}_1 + \cdots + \lambda_r \mathbf{a}_r, \tag{2-52}$$

$$\mathbf{b} = \lambda_1' \mathbf{a}_1 + \cdots + \lambda_r' \mathbf{a}_r. \tag{2-53}$$

Subtracting (2–53) from (2–52), we obtain

$$(\lambda_1 - \lambda_1')\mathbf{a}_1 + \cdots + (\lambda_r - \lambda_r')\mathbf{a}_r = \mathbf{0}.$$

However, the $\mathbf{a}_i$ are linearly independent and thus

$$(\lambda_1 - \lambda_1') = 0, \ldots, (\lambda_r - \lambda_r') = 0. \tag{2-54}$$

Hence $\lambda_i = \lambda_i'$, and the linear combination is unique.

It is not at all true that the representation of a vector in terms of an *arbitrary* set of vectors is unique. Suppose we are given a set of m vectors $\mathbf{a}_1 = [a_{11}, \ldots, a_{n1}], \ldots, \mathbf{a}_m = [a_{1m}, \ldots, a_{nm}]$ and the vector $\mathbf{b} = [b_1, \ldots, b_n]$. We desire to write $\mathbf{b}$ as a linear combination of $\mathbf{a}_1, \ldots, \mathbf{a}_m$, that is,

$$\mathbf{b} = \lambda_1 \mathbf{a}_1 + \cdots + \lambda_m \mathbf{a}_m. \tag{2-55}$$

In component form, (2–55) becomes

$$\begin{aligned} b_1 &= a_{11}\lambda_1 + \cdots + a_{1m}\lambda_m, \\ &\vdots \\ b_n &= a_{n1}\lambda_1 + \cdots + a_{nm}\lambda_m. \end{aligned} \tag{2-56}$$

This is a set of n simultaneous linear equations in m unknowns (the λ_i). These equations may have no solution, a unique solution, or an infinite number of solutions. The conditions under which each of these possibilities can occur will be studied in Chapter 5. If the set of $\mathbf{a}_i$ forms a basis, then we know that the equations (2–56) have a unique solution. If the $\mathbf{a}_i$ do not form a basis, there may be a unique solution, no solution, or an infinite number of solutions.

EXAMPLE: The vector $\mathbf{a} = [2, 3, 4]$ can be written *uniquely* in terms of the vectors $\mathbf{e}_1 = [1, 0, 0]$, $\mathbf{e}_2 = [0, 1, 0]$, and $\mathbf{e}_3 = [0, 0, 1]$ which form a basis for E^3:

$$\mathbf{a} = 2\mathbf{e}_1 + 3\mathbf{e}_2 + 4\mathbf{e}_3.$$

Vector $\mathbf{a}$ can also be written uniquely in terms of the vectors $\mathbf{e}_1$ and $\mathbf{b} = [0, \tfrac{3}{4}, 1]$:

$$\mathbf{a} = 2\mathbf{e}_1 + 4\mathbf{b}.$$

However, it is not possible to express $\mathbf{a}$ as a linear combination of $\mathbf{e}_1$, $\mathbf{e}_2$ only. In this case, the set of equations (2–56) has *no solution*.

If we attempt to write $\mathbf{a}$ as a linear combination of the set of vectors $\mathbf{e}_1$, $\mathbf{e}_2$, $\mathbf{e}_3$, and $\mathbf{c} = [1, 0, 1]$, the set of equations (2–56) will provide an *infinite number* of solutions. Two possibilities are:

$$\mathbf{a} = 1.5\mathbf{e}_1 + 3\mathbf{e}_2 + 3.5\mathbf{e}_3 + 0.5\mathbf{c},$$

$$\mathbf{a} = \mathbf{e}_1 + 3\mathbf{e}_2 + 3\mathbf{e}_3 + \mathbf{c}.$$

2–9 Changing a single vector in a basis. We have mentioned earlier that there is no unique basis for E^n. We shall now investigate the conditions under which an arbitrary vector $\mathbf{b}$ from E^n can replace one of the vectors in a basis so that the new set of vectors is also a basis. The technique of replacing one vector in a basis by another such that the new set is also a basis is fundamental to the simplex technique for solving linear programming problems.

Given a set of basis vectors $\mathbf{a}_1, \ldots, \mathbf{a}_r$ *for* E^n *and any other vector* $\mathbf{b} \neq \mathbf{0}$ *from* E^n: *Then, if in the expression of* $\mathbf{b}$ *as a linear combination of the* $\mathbf{a}_i$,

$$\mathbf{b} = \sum_{i=1}^{r} \alpha_i \mathbf{a}_i, \tag{2-57}$$

any vector $\mathbf{a}_i$ *for which* $\alpha_i \neq 0$ *is removed from the set* $\mathbf{a}_1, \ldots, \mathbf{a}_r$, *and* $\mathbf{b}$ *is added to the set, the new collection of* r *vectors is also a basis for* E^n.

To prove this statement, note that, since $\mathbf{a}_1, \ldots, \mathbf{a}_r$ form a basis, they are linearly independent and

$$\sum_{i=1}^{r} \lambda_i \mathbf{a}_i = \mathbf{0} \quad \text{implies} \quad \lambda_i = 0, \quad i = 1, \ldots, r.$$

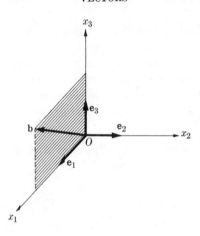

FIGURE 2–9

We would like to show that the new collection of vectors is also linearly independent. Without loss of generality, we can assume that in Eq. (2–57) $\alpha_r \neq 0$, replace $\mathbf{a}_r$ by $\mathbf{b}$, and obtain the new set $\mathbf{a}_1, \ldots, \mathbf{a}_{r-1}, \mathbf{b}$. To demonstrate that this collection is linearly independent, it must be shown that

$$\sum_{i=1}^{r-1} \delta_i \mathbf{a}_i + \delta \mathbf{b} = \mathbf{0} \quad \text{implies} \quad \delta_i = 0, \quad i = 1, \ldots, r-1,$$

and

$$\delta = 0. \tag{2–58}$$

If the set is linearly dependent, then δ cannot vanish since, by assumption, $\mathbf{a}_1, \ldots, \mathbf{a}_{r-1}$ are linearly independent.

Suppose $\delta \neq 0$. Using (2–57) in (2–58), we obtain

$$\sum_{i=1}^{r-1} (\delta_i + \alpha_i \delta) \mathbf{a}_i + \delta \alpha_r \mathbf{a}_r = \mathbf{0}.$$

But $\alpha_r \delta \neq 0$. This contradicts the assumption that $\mathbf{a}_1, \ldots, \mathbf{a}_r$ are linearly independent. Hence $\delta = 0$, and consequently $\delta_i = 0$, $i = 1, \ldots, r-1$, which implies that $\mathbf{a}_1, \ldots, \mathbf{a}_{r-1}, \mathbf{b}$ are linearly independent.

For $\mathbf{a}_1, \ldots, \mathbf{a}_{r-1}, \mathbf{b}$ to form a basis, it has to be shown that any vector $\mathbf{x}$ in E^n can be represented as a linear combination of this set. Vector $\mathbf{x}$ can be represented as a linear combination of $\mathbf{a}_1, \ldots, \mathbf{a}_r$ because these vectors form a basis, that is,

$$\mathbf{x} = \sum_{i=1}^{r} \gamma_i \mathbf{a}_i. \tag{2–59}$$

By assumption, $\alpha_r \neq 0$. From (2–57),

$$\mathbf{a}_r = \frac{1}{\alpha_r}\mathbf{b} - \sum_{i=1}^{r-1} \frac{\alpha_i}{\alpha_r}\mathbf{a}_i. \tag{2-60}$$

Substituting (2–60) into (2–59), we obtain

$$\mathbf{x} = \sum_{i=1}^{r-1}\left(\gamma_i - \frac{\alpha_i}{\alpha_r}\gamma_r\right)\mathbf{a}_i + \frac{\gamma_r}{\alpha_r}\mathbf{b}, \tag{2-61}$$

which expresses $\mathbf{x}$ as a linear combination of $\mathbf{a}_1, \ldots, \mathbf{a}_{r-1}, \mathbf{b}$. Thus $\mathbf{a}_1, \ldots, \mathbf{a}_{r-1}, \mathbf{b}$ form a basis for E^n.

If, in (2–57), $\mathbf{b}$ replaces a vector $\mathbf{a}_i$ for which $\alpha_i = 0$, then the new set of vectors is linearly dependent and does not form a basis for E^n. To prove this, take $\alpha_r = 0$ and replace $\mathbf{a}_r$ by $\mathbf{b}$. Then

$$\mathbf{b} = \sum_{i=1}^{r-1} \alpha_i \mathbf{a}_i,$$

or

$$\mathbf{b} - \sum_{i=1}^{r-1} \alpha_i \mathbf{a}_i = \mathbf{0}. \tag{2-62}$$

The vectors $\mathbf{a}_1, \ldots, \mathbf{a}_{r-1}, \mathbf{b}$ are linearly dependent.

EXAMPLE: Imagine that we want to insert the vector $\mathbf{b} = [3, 0, 4]$ into the basis $\mathbf{e}_1$, $\mathbf{e}_2$, $\mathbf{e}_3$ and remove one of the vectors in the basis. We have

$$\mathbf{b} = 3\mathbf{e}_1 + 0\mathbf{e}_2 + 4\mathbf{e}_3.$$

According to the preceding discussion, we can remove either $\mathbf{e}_1$ or $\mathbf{e}_3$ to obtain a new basis; that is, $\mathbf{e}_2$, $\mathbf{e}_3$, $\mathbf{b}$ or $\mathbf{e}_1$, $\mathbf{e}_2$, $\mathbf{b}$ will form a basis for E^3. We cannot remove the vector $\mathbf{e}_2$ and still maintain a basis. This is illustrated geometrically in Fig. 2–9. If $\mathbf{e}_2$ is removed, $\mathbf{e}_1$, $\mathbf{e}_3$, $\mathbf{b}$ all lie in the x_1x_3-plane and are linearly dependent. If either $\mathbf{e}_1$ or $\mathbf{e}_3$ is removed, the new set of vectors does not lie in a plane.

2–10 Number of vectors in a basis for E^n. The preceding theorems on bases have not depended on the actual number of vectors in the basis. However, our results have led us to suspect that every basis for E^n contains the same number of basis vectors, and that there are precisely n vectors in every basis for E^n. This is indeed correct, as we shall now show.

We shall begin by proving that any two bases for E^n have the same number of basis vectors. Let $\mathbf{a}_1, \ldots, \mathbf{a}_u$ be one set of basis vectors for E^n, and $\mathbf{b}_1, \ldots, \mathbf{b}_v$ be any other set of basis vectors. The theorem states that

$u = v$. To prove it, note that $\mathbf{b}_v$ can be expressed as a linear combination of $\mathbf{a}_1, \ldots, \mathbf{a}_u$:

$$\mathbf{b}_v = \sum_{i=1}^{u} \lambda_i \mathbf{a}_i,$$

and that at least one $\lambda_i \neq 0$ (since the null vector cannot be a member of a linearly independent set). Without loss of generality, we can set $\lambda_u \neq 0$. From the results of the preceding section we know that $\mathbf{a}_1, \ldots, \mathbf{a}_{u-1}, \mathbf{b}_v$ form a basis for E^n. Hence $\mathbf{b}_{v-1}$ can be written as a linear combination of this set of basis vectors,

$$\mathbf{b}_{v-1} = \sum_{i=1}^{u-1} \delta_i \mathbf{a}_i + \delta \mathbf{b}_v.$$

At least one $\delta_i \neq 0$; otherwise $\mathbf{b}_{v-1}$ would be merely a scalar multiple of $\mathbf{b}_v$, contradicting the fact that the $\mathbf{b}_j$ are linearly independent. We can assume that $\delta_{u-1} \neq 0$. Hence, the set $\mathbf{a}_1, \ldots, \mathbf{a}_{u-2}, \mathbf{b}_{v-1}, \mathbf{b}_v$ forms a basis for E^n. This process can be continued until a basis is found which must take one of the following forms:

$$\mathbf{a}_1, \ldots, \mathbf{a}_{u-v}, \mathbf{b}_1, \ldots, \mathbf{b}_v \quad \text{or} \quad \mathbf{b}_1, \ldots, \mathbf{b}_v. \tag{2-63}$$

There must be at least as many $\mathbf{a}_i$ as there are $\mathbf{b}_j$; otherwise a basis of the form $\mathbf{b}_{v-u+1}, \ldots, \mathbf{b}_v$ would be obtained, and the remaining $\mathbf{b}_j$ would be linearly dependent on $\mathbf{b}_{v-u+1}, \ldots, \mathbf{b}_v$, contradicting the fact that all the $\mathbf{b}_j$ are linearly independent. Thus we conclude that

$$u \geq v. \tag{2-64}$$

However, one can start with the $\mathbf{b}_j$ and insert the $\mathbf{a}_i$. This procedure would yield

$$v \geq u. \tag{2-65}$$

Therefore $u = v$, and the theorem is proved.

To determine the actual number of vectors in a basis for E^n, it is only necessary to find a basis for E^n and count the number of vectors in this basis. We have already shown that the n unit vectors $\mathbf{e}_1, \ldots, \mathbf{e}_n$ form a basis for E^n. *It immediately follows that every basis for E^n contains precisely n vectors.*

We can now see that *any set of n linearly independent vectors from E^n forms a basis for E^n.* Let $\mathbf{a}_1, \ldots, \mathbf{a}_n$ be such a set of linearly independent vectors. To prove this result, we only need to start with the basis $\mathbf{e}_1, \ldots, \mathbf{e}_n$ and insert the $\mathbf{a}_i$. After n insertions, which can always be made (for the same reasons used in proving that two bases always have the same number of elements), we obtain a new basis $\mathbf{a}_1, \ldots, \mathbf{a}_n$. Thus, it is obvious that *any $n + 1$ vectors from E^n are linearly dependent,* since any subset of n of these vectors is a basis, and the remaining vector can be written as a linear combination of the basis vectors.

Any $m < n$ *linearly independent vectors from E^n form part of a basis for E^n*, that is, if $\mathbf{a}_1, \ldots, \mathbf{a}_m$ from E^n are linearly independent, then $n - m$ additional vectors in E^n can be found such that $\mathbf{a}_1, \ldots, \mathbf{a}_m$ and the other $n - m$ vectors in E^n form a basis for E^n. Again, the proof depends on starting with a basis and inserting the $\mathbf{a}_i$ in precisely the same way as was done before. If we start with unit vectors as a basis, we see that the additional $n - m$ vectors used to fill out the basis can be unit vectors. As a specific example of this theorem, any non-null vector from E^n is part of a basis and, therefore, there is an infinite number of bases for E^n.

We have characterized the dimension n of a euclidean space by the number of components in the set of vectors which form the space, that is, E^n is the set of all n-component vectors. We have shown that the maximum number of linearly independent vectors in E^n is n. In some ways, it is more desirable to define the dimension of a space in terms of the maximum number of linearly independent vectors which can exist in that space. This definition avoids connecting the dimension of a space to the number of components in the vectors forming the space. The advantage of this approach will become clear in the discussion of subspaces.

*2–11 Orthogonal bases. It will be recalled that two n-component vectors $\mathbf{a}, \mathbf{b}$ are orthogonal if $\mathbf{a}'\mathbf{b} = 0$. Let us suppose that we have n vectors $\mathbf{v}_1, \ldots, \mathbf{v}_n$ from E^n which are mutually orthogonal and all different from $\mathbf{0}$ so that

$$\mathbf{v}'_i \mathbf{v}_j = 0 \quad (\text{all } i, j; i \neq j). \tag{2-66}$$

This set of vectors forms a basis for E^n. The proof follows immediately if we can show that the set $\mathbf{v}_1, \ldots, \mathbf{v}_n$ is linearly independent since any set of n linearly independent vectors from E^n forms a basis for E^n. Consider the problem of finding the λ_i which will satisfy

$$\lambda_1 \mathbf{v}_1 + \cdots + \lambda_n \mathbf{v}_n = \mathbf{0}. \tag{2-67}$$

Forming the scalar product of $\mathbf{v}_1$ and (2–67), we obtain

$$\lambda_1 |\mathbf{v}_1|^2 + \lambda_2 \mathbf{v}'_1 \mathbf{v}_2 + \cdots + \lambda_n \mathbf{v}'_1 \mathbf{v}_n = \mathbf{v}'_1 \mathbf{0} = 0, \tag{2-68}$$

and by (2–66)

$$\lambda_1 = 0,$$

since, by assumption, $|\mathbf{v}_1|^2 \neq 0$. If we form the scalar product of (2–67) and $\mathbf{v}_2$, we obtain $\lambda_2 = 0$, etc. Hence, each λ_i, $(i = 1, \ldots, n)$ is zero, so

---

* Starred sections can be omitted without loss of continuity.

the vectors are linearly independent and form a basis for E^n. Thus *any set* of n *mutually orthogonal nonzero vectors from E^n yields a basis for E^n.*

Let us divide each vector $\mathbf{v}_i$ by its length $|\mathbf{v}_i|$ and write

$$\mathbf{u}_i = \frac{\mathbf{v}_i}{|\mathbf{v}_i|}. \tag{2-69}$$

This can be done since $|\mathbf{v}_i| \neq 0$. The $\mathbf{u}_i$ represent vectors of unit length; thus

$$\mathbf{u}'_i \mathbf{u}_i = 1, \qquad \mathbf{u}'_i \mathbf{u}_j = 0 \quad (i \neq j). \tag{2-70}$$

ORTHONORMAL BASIS: *A set of n mutually orthogonal vectors of unit length from E^n forms what is called an orthonormal basis for E^n.*

Orthonormal bases are especially interesting because any vector $\mathbf{x}$ from E^n can be expressed very easily as a linear combination of the orthonormal basis vectors. If $\mathbf{u}_1, \ldots, \mathbf{u}_n$ form an orthonormal basis for E^n, and we want to write

$$\mathbf{x} = \lambda_1 \mathbf{u}_1 + \cdots + \lambda_n \mathbf{u}_n, \tag{2-71}$$

then the scalar product of $\mathbf{u}_i$ and (2-71) will yield

$$\lambda_i = \mathbf{u}'_i \mathbf{x} \qquad [\text{see } (2-70)]. \tag{2-72}$$

The scalar λ_i is found simply by forming the scalar product of $\mathbf{u}_i$ and $\mathbf{x}$. Note: the unit vectors $\mathbf{e}_1, \ldots, \mathbf{e}_n$ form an orthonormal basis for E^n.

Since any set of mutually orthogonal nonzero n-component vectors is linearly independent (proof?), it is not possible to have $n + 1$ mutually orthogonal nonzero vectors in E^n.

Any set of n given linearly independent vectors from E^n can be converted into an orthonormal basis by a procedure known as the Schmidt orthogonalization process. Let us suppose that $\mathbf{a}_1, \ldots, \mathbf{a}_n$ are n linearly independent vectors from E^n. We select any vector from this set, for example, $\mathbf{a}_1$. This vector defines a direction in space, and we build the orthonormal set around it. Let us define the vector of unit length $\mathbf{u}_1$ as

$$\mathbf{u}_1 = \frac{\mathbf{a}_1}{|\mathbf{a}_1|}. \tag{2-73}$$

To obtain a vector $\mathbf{v}_2$ orthogonal to $\mathbf{u}_1$, we subtract from $\mathbf{a}_2$ a scalar multiple of $\mathbf{u}_1$; that is, $\mathbf{v}_2$ is expressed as

$$\mathbf{v}_2 = \mathbf{a}_2 - \alpha_1 \mathbf{u}_1, \tag{2-74}$$

and α_1 is determined so that $\mathbf{u}'_1 \mathbf{v}_2 = 0$. Hence

$$\alpha_1 = \mathbf{u}'_1 \mathbf{a}_2,$$

which can be thought of as the component of $\mathbf{a}_2$ along $\mathbf{u}_1$. Also, the vector $\alpha_1 \mathbf{u}_1$ may be interpreted as the vector component of $\mathbf{a}_2$ along $\mathbf{u}_1$.

Thus,
$$\mathbf{v}_2 = \mathbf{a}_2 - (\mathbf{u}_1'\mathbf{a}_2)\mathbf{u}_1. \tag{2-75}$$

A second unit vector orthogonal to $\mathbf{u}_1$ is defined by
$$\mathbf{u}_2 = \frac{\mathbf{v}_2}{|\mathbf{v}_2|}; \tag{2-76}$$

this can be done since $|\mathbf{v}_2| \neq 0$ (why?). A vector orthogonal to $\mathbf{u}_1, \mathbf{u}_2$ is found by subtracting from $\mathbf{a}_3$ the vector components of $\mathbf{a}_3$ along $\mathbf{u}_1, \mathbf{u}_2$. This gives
$$\mathbf{v}_3 = \mathbf{a}_3 - (\mathbf{u}_1'\mathbf{a}_3)\mathbf{u}_1 - (\mathbf{u}_2'\mathbf{a}_3)\mathbf{u}_2. \tag{2-77}$$

The vector $\mathbf{v}_3$ is clearly orthogonal to $\mathbf{u}_1, \mathbf{u}_2$. The third unit vector which is orthogonal to $\mathbf{u}_1, \mathbf{u}_2$ is
$$\mathbf{u}_3 = \frac{\mathbf{v}_3}{|\mathbf{v}_3|}. \tag{2-78}$$

This procedure is continued until an orthonormal basis is obtained. In general,
$$\mathbf{v}_r = \mathbf{a}_r - \sum_{i=1}^{r-1} (\mathbf{u}_i'\mathbf{a}_r)\mathbf{u}_i, \tag{2-79}$$

$$\mathbf{u}_r = \frac{\mathbf{v}_r}{|\mathbf{v}_r|}. \tag{2-80}$$

EXAMPLE: Using the Schmidt process, construct an orthonormal basis from $\mathbf{a}_1 = [2, 3, 0]$, $\mathbf{a}_2 = [6, 1, 0]$, and $\mathbf{a}_3 = [0, 2, 4]$. If we plot the vectors $\mathbf{a}_1, \mathbf{a}_2, \mathbf{a}_3$, it is easily seen that they are linearly independent since they do not line in a plane. Then if

$$\mathbf{u}_1 = \frac{\mathbf{a}_1}{|\mathbf{a}_1|} = (13)^{-1/2}[2, 3, 0] = [0.554, 0.831, 0];$$

$$\mathbf{v}_2 = \mathbf{a}_2 - (\mathbf{u}_1'\mathbf{a}_2)\mathbf{u}_1;$$

$$(\mathbf{u}_1'\mathbf{a}_2) = 4.16, \qquad (\mathbf{u}_1'\mathbf{a}_2)\mathbf{u}_1 = [2.30, 3.45, 0];$$

$$\mathbf{v}_2 = [3.70, -2.45, 0];$$

$$\mathbf{u}_2 = \frac{\mathbf{v}_2}{|\mathbf{v}_2|} = [0.831, -0.554, 0];$$

$$\mathbf{v}_3 = \mathbf{a}_3 - (\mathbf{u}_1'\mathbf{a}_3)\mathbf{u}_1 - (\mathbf{u}_2'\mathbf{a}_3)\mathbf{u}_2;$$

$$\mathbf{u}_1'\mathbf{a}_3 = 1.664, \qquad \mathbf{u}_2'\mathbf{a}_3 = -1.106;$$

$$(\mathbf{u}_1'\mathbf{a}_3)\mathbf{u}_1 = [0.921, 1.386, 0], \qquad (\mathbf{u}_2'\mathbf{a}_3)\mathbf{u}_2 = [-0.921, 0.614, 0];$$

$$\mathbf{v}_3 = [0, 0, 4];$$

$$\mathbf{u}_3 = [0, 0, 1].$$

The fact that any set of $m < n$ linearly independent vectors from E^n constitutes part of a basis, and the Schmidt process demonstrate that any set of $m < n$ mutually orthogonal vectors of unit length in E^n is part of an orthonormal basis; that is, there exist $n - m$ vectors of unit length in E^n which, together with the m orthogonal vectors of unit length, form an orthonormal basis.

If two nonzero vectors $\mathbf{a}$, $\mathbf{b}$ *from* E^n *are both orthogonal to* $n - 1$ *mutually orthogonal vectors of unit length* $\mathbf{u}_1, \ldots, \mathbf{u}_{n-1}$ *from* E^n, *then* $\mathbf{b}$ *is a scalar multiple of* $\mathbf{a}$. This follows immediately since $\mathbf{a}, \mathbf{u}_1, \ldots, \mathbf{u}_{n-1}$ form a basis, and

$$\mathbf{b} = \lambda \mathbf{a} + \lambda_1 \mathbf{u}_1 + \cdots + \lambda_{n-1} \mathbf{u}_{n-1}.$$

However,

$$\mathbf{u}_i' \mathbf{b} = 0, \quad \mathbf{u}_i' \mathbf{a} = 0 \quad (i = 1, \ldots, n - 1), \quad \mathbf{u}_i' \mathbf{u}_j = 0 \quad (i \neq j);$$

hence

$$\lambda_i = 0 \quad (i = 1, \ldots, n - 1),$$

and

$$\mathbf{b} = \lambda \mathbf{a}.$$

*2–12 Generalized coordinate systems. When writing a vector $\mathbf{a} = [a_1, \ldots, a_n]$ as an ordered n-tuple of numbers, we have assumed that the components a_i lie along the coordinate axes defined by the unit vectors $\mathbf{e}_1, \ldots, \mathbf{e}_n$. Our preceding discussions were all implicitly based on a coordinate system defined by unit vectors. There is no reason, however, why we should be limited to one coordinate system since any set of basis vectors can define a system of coordinates.

Let us consider a set of basis vectors $\mathbf{v}_1, \ldots, \mathbf{v}_n$ for E^n. For convenience, we shall assume that each $\mathbf{v}_i$ is of unit length. If $\mathbf{x}$ is any point in E^n, then

$$\mathbf{x} = \alpha_1 \mathbf{v}_1 + \cdots + \alpha_n \mathbf{v}_n. \tag{2-81}$$

In (2–81), all vectors $\mathbf{x}$, $\mathbf{v}_i$ can be thought of as referred to the basic coordinate system defined by $\mathbf{e}_1, \ldots, \mathbf{e}_n$. However, $\mathbf{x}$ can be equally well characterized by the numbers $\alpha_1, \ldots, \alpha_n$. For example, the vector $\mathbf{x}_\mathbf{v} = [\alpha_1, \ldots, \alpha_n]$ can be considered to be the representation of $\mathbf{x}$ in the coordinate system defined by the basis vectors $\mathbf{v}_1, \ldots, \mathbf{v}_n$. A subscript is placed on $\mathbf{x}$ to indicate that $\mathbf{x}_\mathbf{v}$ is *not* referred to the coordinate system defined by $\mathbf{e}_1, \ldots, \mathbf{e}_n$, but instead to the coordinate system defined by the $\mathbf{v}_i$.

If λ is any scalar, multiplication of (2–81) by λ shows that

$$(\lambda \mathbf{x})_\mathbf{v} = \lambda \mathbf{x}_\mathbf{v} = [\lambda \alpha_1, \ldots, \lambda \alpha_n]. \tag{2-82}$$

Assuming **y** to be any other vector in E^n, we obtain

$$\mathbf{y} = \beta_1 \mathbf{v}_1 + \cdots + \beta_n \mathbf{v}_n, \tag{2-83}$$

and

$$\mathbf{y_v} = [\beta_1, \ldots, \beta_n]. \tag{2-84}$$

Adding (2–81) and (2–83), we see that

$$(\mathbf{x} + \mathbf{y})_\mathbf{v} = \mathbf{x_v} + \mathbf{y_v} = [\alpha_1 + \beta_1, \ldots, \alpha_n + \beta_n]. \tag{2-85}$$

Multiplication by a scalar and addition have precisely the same form in an arbitrary coordinate system as in the system defined by $\mathbf{e}_1, \ldots, \mathbf{e}_n$.

The scalar product, however, does not—in any coordinate system—have the form (2–31). Forming the scalar product of **y**, **x** by means of (2–81) and (2–83), we see that

$$\mathbf{y'x} = \sum_{i=1}^{n} \sum_{j=1}^{n} \alpha_i \beta_j \mathbf{v}'_i \mathbf{v}_j = \sum_{i=1}^{n} \sum_{j=1}^{n} \alpha_i \beta_j \cos \theta_{ij}$$

$$= \sum_{i=1}^{n} \alpha_i \beta_i + \sum_{\substack{i,j \\ i \neq j}} \alpha_i \beta_j \cos \theta_{ij}, \tag{2-86}$$

where θ_{ij} is the angle between $\mathbf{v}_i$ and $\mathbf{v}_j$. Equation (2–86) is not usually $\sum \alpha_i \beta_i$.

Let us suppose that $\mathbf{v}_1, \ldots, \mathbf{v}_n$ is an orthonormal basis. Equation (2–86) then becomes

$$\mathbf{y'x} = \sum_{i=1}^{n} \alpha_i \beta_i = \mathbf{y'_v x_v}. \tag{2-87}$$

Hence, in any orthogonal coordinate system, all the operations with vectors, including the scalar product, can be performed in the same way as in the coordinate system defined by $\mathbf{e}_1, \ldots, \mathbf{e}_n$. For this reason, orthogonal coordinate systems are particularly useful.

Many mathematics texts do not start out with defining a vector as an ordered n-tuple of numbers. Instead, they define vectors as algebraic elements for which the operations of addition and multiplication by a scalar are defined and assumed to possess the properties (2–23) through (2–26). The entire theory is then developed without referring to an n-tuple. This approach is abstract and tends to be somewhat difficult for students with little mathematical background. However, it does have the great advantage that it avoids the problem of coordinate systems since vectors are defined without reference to a coordinate system. Hence it follows immediately that the theory holds for any coordinate system. [The scalar product is also defined abstractly in terms of its properties, and then it is shown that, for orthogonal coordinate systems, it takes the concrete form of Eq. (2–31).]

2–13 Vector spaces and subspaces. For certain purposes, it is desirable to treat collections of vectors from a slightly more abstract point of view than we have as yet presented. We shall introduce now the concept of a vector space.

VECTOR SPACE: *A vector space is a collection of vectors which is closed under the operations of addition and multiplication by a scalar.*

The expression, "a set of vectors is closed under addition and multiplication by a scalar," means that if **a**, **b** are in the collection, the sum **a** + **b** is also in the collection, and if **a** is in the collection, λ**a** is in the collection for *any* scalar λ.

The definition of a vector space says nothing about a scalar product, length, or distance. These concepts do not need to be defined in a vector space. Only addition and multiplication by a scalar need to be defined. From the closure property, we see that if **a** is in a vector space, so is −**a**, since (−1)**a** is in the collection. Similarly, the null vector is always in a vector space, since λ**a** is in the space, and we can assume $\lambda = 0$. *The totality of all n-component vectors is called an n-dimensional vector space and is denoted by* V_n. The space V_n is identical with the n-dimensional euclidean space E^n if length is defined in V_n as in E^n. Although E^n is clearly a vector space, it does not follow that any V_n is an E^n.

Out of all the vectors in V_n, it is possible to find various subsets of vectors which are themselves vector spaces. For example, the set of all three component vectors forms a three-dimensional vector space. However, if we consider all three component vectors lying in the plane $x_2 = x_3$, that is, all vectors of the form $[x_1, x_2, x_2]$, then this collection is also a vector space. Although it is composed of three component vectors, we do not call it a three-dimensional vector space since all the vectors lie in a plane. Such a vector space is referred to as a subspace of V_3.

SUBSPACE: *A subspace S_n of the n-dimensional vector space V_n is defined to be a subset of V_n which is itself a vector space.*

The subscript n on S_n means that the vectors have n components, i.e., are elements of V_n. In assigning a dimension to a subspace, we shall find it convenient to define the dimension of a space as the maximum number of linearly independent vectors in the space, rather than to refer to the number of components in the vectors generating the space. *We define the dimension of a subspace S_n as the maximum number of linearly independent vectors in the subspace.* In our example, any vector $[x_1, x_2, x_2]$ can be written

$$[x_1, x_2, x_2] = x_1\mathbf{e}_1 + x_2[0, 1, 1].$$

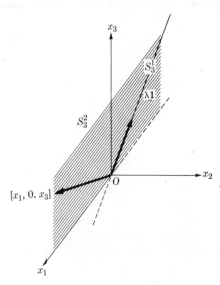

FIGURE 2-10

The maximum number of linearly independent vectors in this subspace is two; hence its dimension is two. This is precisely what we want intuitively because we know that all the vectors lie in a plane. To simplify later discussions, we consider $S_n = V_n$ to be a subspace of V_n having dimension n, i.e., the subspace is all of V_n.

A clear geometrical interpretation of subspaces in E^3 can be given. Any subspace of E^3 is either E^3 itself, a plane through the origin, a line through the origin, or just the origin itself (this is a subspace of a single element and has dimension 0).

EXAMPLE: The collection of all vectors $\lambda[1, 1, 1]$ for any scalar λ form a subspace of E^3 since

$$\lambda_1 \mathbf{1} + \lambda_2 \mathbf{1} = (\lambda_1 + \lambda_2)\mathbf{1},$$

which is also an element of the collection. This collection of vectors lies on the line which passes through the origin and the point $[1, 1, 1]$ (see line S_3^1 in Fig. 2-10).

The collection of all vectors $[x_1, 0, x_3]$ is a subspace of E^3 and represents the $x_1 x_3$-plane (see S_3^2 in Fig. 2-10).

The collection of all linear combinations of m n-component vectors $\mathbf{a}_1, \ldots, \mathbf{a}_m$ is a subspace of V_n. To prove this, note that any vectors $\mathbf{a}$, $\mathbf{b}$ in the collection can be written

$$\mathbf{a} = \sum_{i=1}^m \alpha_i \mathbf{a}_i \qquad \mathbf{b} = \sum_{i=1}^m \alpha_i' \mathbf{a}_i.$$

Then
$$\lambda \mathbf{a} = \sum_{i=1}^{m} (\lambda \alpha_i) \mathbf{a}_i$$
is also a linear combination of the $\mathbf{a}_i$ and is in the collection. Furthermore,
$$\mathbf{a} + \mathbf{b} = \sum_{i=1}^{m} (\alpha_i + \alpha'_i) \mathbf{a}_i$$
is a linear combination of the $\mathbf{a}_i$ and is in the collection.

A set of vectors from a subspace S_n is said to span or generate S_n if every vector in S_n can be written as a linear combination of this set of vectors. A linearly independent set of vectors from S_n which spans S_n is called a basis for S_n. If S_n has dimension k, then a basis for S_n will contain precisely k vectors. The arguments used in Section 2-10 show that if S_n has dimension k, any k linearly independent vectors from S_n are a basis for S_n.

If S_n is a subspace of E^n having dimension k, then, except for notation, S_n is the same as E^k. The vectors in S_n will have n components rather than k components. However, it is possible to establish a unique correspondence between the vectors in S_n and the vectors in E^k. For example, the subspace of E^3 (discussed above) which is the collection of all vectors $[x_1, x_2, x_2]$ has dimension 2. Corresponding to any vector in this subspace there is one and only one vector $[x_1, x_2]$ in E^2, and for any vector $[x_1, x_2]$ in E^2 there is one and only one vector $[x_1, x_2, x_2]$ in the subspace. We say that a subspace of E^n of dimension k is isomorphic to E^k.

References

1. G. Birkhoff and S. A. MacLane, *A Survey of Modern Algebra*. New York: Macmillan, 1941.

   Chapter 7 of this well-known text gives a good treatment of the subject of vector spaces from the more abstract point of view. This chapter is not too easy to read for one not used to the abstract approach.

2. P. R. Halmos, *Finite Dimensional Vector Spaces*. 2nd ed. Princeton: Van Nostrand, 1958.

   This text presents a complete tratement of vector spaces from the more abstract, axiomatic point of view. It includes a great deal of material not covered in this chapter and not treated in the other references listed here.

3. S. Perlis, *Theory of Matrices*. Reading, Mass.: Addison-Wesley, 1952.

   Chapter 2 and parts of Chapter 10 give a simple and clear discussion of the more elementary topics in the theory of vector spaces. The approach is not abstract; it begins with n-component vectors, as we have done.

4. R. M. Thrall and L. Tornheim, *Vector Spaces and Matrices*. New York: Wiley, 1958.

Chapter 1 and part of Chapter 7 give a reasonably clear and complete treatment of vector spaces. The intuitive foundations are not emphasized to any extent. An abstract approach is used.

Problems

2-1. Given the vectors $\mathbf{a} = (2, 1)$, $\mathbf{b} = (1, 3)$, plot the vectors $-2\mathbf{a}$, $\mathbf{a} + \mathbf{b}$, and illustrate the use of the parallelogram rule to find $\mathbf{a} + \mathbf{b}$, $2\mathbf{a} + \frac{1}{2}\mathbf{b}$.

2-2. Given $\mathbf{a} = [a_1, \ldots, a_n]$, $\mathbf{b} = [b_1, \ldots, b_n]$, $\mathbf{c} = [c_1, \ldots, c_n]$, show by direct computation that

$$(\mathbf{a} + \mathbf{b}) + \mathbf{c} = \mathbf{a} + (\mathbf{b} + \mathbf{c}) = \mathbf{a} + \mathbf{b} + \mathbf{c},$$

$$\mathbf{a}'\mathbf{b} = \mathbf{b}'\mathbf{a},$$

$$|\mathbf{a} - \mathbf{b}| = |(\mathbf{a} - \mathbf{c}) - (\mathbf{b} - \mathbf{c})|.$$

2-3. If $\mathbf{a} = [3, 2, 1]$, $\mathbf{b} = [1, 5, 6]$, solve the following vector equation for $\mathbf{x}$:

$$3\mathbf{a} + 2\mathbf{x} = 5\mathbf{b}.$$

2-4. If $\mathbf{a}$ is an n-component vector, show by direct computation that

$$\mathbf{a} + \mathbf{0} = \mathbf{0} + \mathbf{a} = \mathbf{a}, \qquad 3\mathbf{a} = \mathbf{a} + \mathbf{a} + \mathbf{a}, \qquad \mathbf{a}'\mathbf{0} = 0.$$

Show that the equation $\mathbf{a} + \mathbf{x} = \mathbf{0}$ has the unique solution $\mathbf{x} = -\mathbf{a}$.

2-5. Show that the vectors $[2, 3, 1]$, $[1, 0, 4]$, $[2, 4, 1]$, $[0, 3, 2]$ are linearly dependent.

2-6. Express $\mathbf{x} = [4, 5]$ as a linear combination of $\mathbf{a} = [1, 3]$, $\mathbf{b} = [2, 2]$.

2-7. Given three linearly independent vectors $\mathbf{a}_1$, $\mathbf{a}_2$, $\mathbf{a}_3$ from E^3, obtain a formula for any vector $\mathbf{x}$ in E^3 as a linear combination of $\mathbf{a}_1$, $\mathbf{a}_2$, $\mathbf{a}_3$, that is, evaluate the λ_i in terms of the components of $\mathbf{x}$, $\mathbf{a}_1$, $\mathbf{a}_2$, $\mathbf{a}_3$ when

$$\mathbf{x} = \lambda_1 \mathbf{a}_1 + \lambda_2 \mathbf{a}_2 + \lambda_3 \mathbf{a}_3.$$

2-8. Given the basis vectors $\mathbf{e}_1 = [1, 0, 0]$, $[0, 1, 1]$, $\mathbf{e}_3 = [0, 0, 1]$ for E^3. Which vectors can be removed from the basis and be replaced by $\mathbf{b} = [4, 3, 6]$, while still maintaining a basis? Illustrate this geometrically.

2-9. Given the basis vectors $\mathbf{e}_1$, $[0, 1, 1]$, $\mathbf{e}_2$ for E^3. Which vectors can be removed from the basis and be replaced by $\mathbf{b} = [4, 3, 3]$, while still maintaining a basis? Illustrate this geometrically.

2-10. Let $\mathbf{b}_1, \ldots, \mathbf{b}_n$ be a basis for E^n. Suppose that when a given vector $\mathbf{x}$ in E^n is written as

$$\mathbf{x} = \lambda_1 \mathbf{b}_1 + \cdots + \lambda_n \mathbf{b}_n,$$

it has all $\lambda_i > 0$. Assume that we have a vector $\mathbf{a}$ and want to remove some $\mathbf{b}_i$

from the basis and replace it by **a** to form a new basis:

$$\mathbf{a} = \alpha_1 \mathbf{b}_1 + \cdots + \alpha_n \mathbf{b}_n.$$

Under what conditions can **a** replace $\mathbf{b}_i$? Assuming **a** can replace $\mathbf{b}_i$, write **x** as a linear combination of the new basis vectors. In addition, let us require that the scalar coefficients in the linear combination which expresses **x** in terms of the new basis vectors be non-negative. Show that a necessary condition for the new scalar multipliers to be non-negative when $\mathbf{b}_i$ is removed is that $\alpha_i > 0$. It does not follow, however, that simply because $\alpha_i > 0$, all the new scalar coefficients will be non-negative. If several $\alpha_i > 0$, show how to determine which vector $\mathbf{b}_i$ is to be removed so that the new scalar coefficients will be non-negative. Hint:

$$\mathbf{x} = \sum \lambda_i \mathbf{b}_i = \sum \lambda_i \mathbf{b}_i - \theta \mathbf{a} + \theta \mathbf{a},$$

and

$$\theta \mathbf{a} = \theta \sum \alpha_i \mathbf{b}_i, \; \theta \geq 0.$$

Note that $\lambda_i - \theta \alpha_i$ must be zero for one $\mathbf{b}_i$, and $\lambda_i - \theta \alpha_i \geq 0$ for all other $\mathbf{b}_i$.

2–11. Suppose that in Problem 2–10, we have defined a function $z_0 = \sum c_i \lambda_i$, where the c_i are scalars, and the λ_i are as defined there. Corresponding to each $\mathbf{b}_i$ there is a number c_i which we shall call its price. Let the price of **a** be c. Now **a** is inserted into the basis and some $\mathbf{b}_i$ is removed, the $\mathbf{b}_i$ being chosen by the method developed in Problem 2–10, so that the new scalar coefficients are non-negative. Find an equation for the new value of z_0 when **a** is introduced into the basis. Denote $\sum c_i \alpha_i$ by z. Imagine next that several different vectors $\mathbf{a}_j$ can be inserted into the basis, but that only one of them will be actually introduced. Develop a criterion for selecting the $\mathbf{a}_j$ which will produce the maximum increase for z_0. What happens when there is no increase in z_0 after insertion of any of the $\mathbf{a}_j$? How is this expressed mathematically?

2–12. Compute the angle between the vectors

$$\mathbf{a} = [4, 7, 9, 1, 3], \quad \mathbf{b} = [2, 1, 1, 6, 8].$$

2–13. If the n-component vectors **a**, **b**, **c** are linearly independent, show that $\mathbf{a} + \mathbf{b}$, $\mathbf{b} + \mathbf{c}$, $\mathbf{a} + \mathbf{c}$ are also linearly independent. Is this true of $\mathbf{a} - \mathbf{b}$, $\mathbf{b} + \mathbf{c}$, $\mathbf{a} + \mathbf{c}$?

2–14. Show how the change of basis technique (discussed in Section 2–9) can be used to express a vector in terms of some arbitrary basis by starting with the unit vectors as a basis and then proceeding to insert the vectors of the arbitrary basis. Illustrate by expressing $\mathbf{b} = [4, 6, 1]$ in terms of $\mathbf{a}_1 = [3, 1, 2]$, $\mathbf{a}_2 = [1, 3, 9]$, $\mathbf{a}_3 = [2, 8, 5]$.

2–15. Consider the set of n-tuples $\mathbf{x} = [x_1, x_2, \ldots, x_n]$ for which $x_1, \ldots, x_m$ are completely unrestricted, while

$$x_{m+i} = \sum_{j=1}^{m} a_{ij} x_j, \quad i = 1, \ldots, n - m.$$

Show that this set of vectors forms a subspace of V_n. What is the dimension of the subspace? Is the set of vectors

$$\mathbf{a}_1 = [1, 0, 0, \ldots, 0, a_{11}, a_{21}, \ldots, a_{n-m 1}], \ldots,$$

$$\mathbf{a}_m = [0, 0, \ldots, 1, a_{1m}, \ldots, a_{n-m,m}]$$

a basis for the subspace?

2–16. Do the following vectors form a basis for E^3?

(a) $\mathbf{a}_1 = [3, 0, 2]$, $\quad \mathbf{a}_2 = [7, 0, 9]$, $\quad \mathbf{a}_3 = [4, 1, 2]$;

(b) $\mathbf{a}_1 = [1, 1, 0]$, $\quad \mathbf{a}_2 = [3, 0, 1]$, $\quad \mathbf{a}_3 = [5, 2, 1]$;

(c) $\mathbf{a}_1 = [1, 5, 7]$, $\quad \mathbf{a}_2 = [4, 0, 6]$, $\quad \mathbf{a}_3 = [1, 0, 0]$.

2–17. Consider two subspaces S'_n and S''_n of V_n. Define the sum of the subspaces, written symbolically $S'_n + S''_n$, as the collection of all vectors $\mathbf{a} + \mathbf{b}$, where $\mathbf{a}$ is any vector in S'_n, and $\mathbf{b}$ is any vector in S''_n. Prove that $S'_n + S''_n$ is also a subspace of V_n. If S'_n is the collection of vectors $\lambda[2, 1]$, and S''_n is the collection of vectors $\alpha[1, 3]$, show that $S'_n + S''_n$ is V_2. Illustrate geometrically.

2–18. Referring to Problem 2–17, prove that if $\mathbf{0}$ is the only vector which is common to S'_n and S''_n, the dimension of $(S'_n + S''_n)$ equals dimension of S'_n + dimension of S''_n. Hint: Let $\mathbf{a}_1, \ldots, \mathbf{a}_n$ be a basis for S'_n, and $\mathbf{b}_1, \ldots, \mathbf{b}_v$ a basis for S''_n. Show that any vector in $S'_n + S''_n$ can be written as a linear combination of $\mathbf{a}_1, \ldots, \mathbf{a}_u, \mathbf{b}_1, \ldots, \mathbf{b}_v$. To prove that $\mathbf{a}_1, \ldots, \mathbf{a}_u, \mathbf{b}_1, \ldots, \mathbf{b}_v$ are linearly independent, suppose

$$\sum \lambda_i \mathbf{a}_i + \sum \delta_i \mathbf{b}_i = 0 \quad \text{or} \quad \sum \lambda_i \mathbf{a}_i = -\sum \delta_i \mathbf{b}_i.$$

Note that $\sum \lambda_i \mathbf{a}_i$ is in S'_n and $\sum \delta_i \mathbf{b}_i$ is in S''_n.

2–19. The intersection of two subspaces, written $S'_n \cap S''_n$, is the collection of all vectors which are in both S'_n and S''_n. Prove that the collection of vectors $S'_n \cap S''_n$ is a subspace of V_n. Let S'_2 be the collection of vectors $\lambda \mathbf{e}_1$ and $S''_2 = V_2$. What is $S'_2 \cap S''_2$? Illustrate geometrically.

2–20. Prove that the dimension of $S'_n \cap S''_n \leq$ minimum [dimension of S'_n and dimension of S''_n].

2–21. Prove that dimension $(S'_n + S''_n)$ + dimension $(S'_n \cap S''_n)$ = dimension S'_n + dimension S''_n. Hint: Let $\mathbf{u}_1, \ldots, \mathbf{u}_r$ be a basis for $S'_n \cap S''_n$, $\mathbf{a}_1, \ldots, \mathbf{a}_s$, $\mathbf{u}_1, \ldots, \mathbf{u}_r$ a basis for S'_n, and $\mathbf{b}_1, \ldots, \mathbf{b}_v, \mathbf{u}_1, \ldots, \mathbf{u}_r$ a basis for S''_n.

2–22. Let $\mathbf{a}_1, \ldots, \mathbf{a}_m$ be a set of linearly independent vectors from E^n and $\mathbf{a}$ any other vector in E^n. Show that if $\mathbf{a}$ is linearly independent of the set $\mathbf{a}_1, \ldots, \mathbf{a}_m$, then $\mathbf{a}, \mathbf{a}_1, \ldots, \mathbf{a}_m$ is a linearly independent set of vectors.

2–23. In the text it was shown under what circumstances a vector could be removed from a basis and replaced by another vector in such a way that the new set of vectors was also a basis. Suppose now that we wish to form a new basis by removing two vectors from a basis and replacing them by two other vectors. Try to derive the conditions under which the new set of vectors will be a basis. What are the difficulties? Is it easier to insert the two vectors one at a time or both together?

2-24. Prove that $|\mathbf{a} - \mathbf{b}| > 0$ unless $\mathbf{a} = \mathbf{b}$. Prove that $|\mathbf{a} - \mathbf{b}| = |\mathbf{b} - \mathbf{a}|$.

2-25. Prove the Schwarz inequality $|\mathbf{a}'\mathbf{b}| \leq |\mathbf{a}||\mathbf{b}|$. Hint: $|\lambda\mathbf{a} + \mathbf{b}|^2 = \lambda^2|\mathbf{a}|^2 + 2\lambda\mathbf{a}'\mathbf{b} + |\mathbf{b}|^2 \geq 0$ for all λ. Thus the quadratic equation $|\lambda\mathbf{a} + \mathbf{b}|^2 = 0$ cannot have two different real roots.

2-26. Prove the triangle inequality $|\mathbf{a} - \mathbf{b}| + |\mathbf{b} - \mathbf{c}| \geq |\mathbf{a} - \mathbf{c}|$. Hint: $|\mathbf{a} - \mathbf{c}|^2 \leq |\mathbf{a} - \mathbf{b}|^2 + 2|(\mathbf{a} - \mathbf{b})'(\mathbf{b} - \mathbf{c})| + |\mathbf{b} - \mathbf{c}|^2$. Use the Schwarz inequality.

2-27. By direct computation, verify the Schwarz inequality for the special case

$$\mathbf{a} = [4, 1, 2], \quad \mathbf{b} = [3, 7, 9].$$

2-28. Verify by direct computation the triangle inequality for the special case

$$\mathbf{a} = [3, 7, 1], \quad \mathbf{b} = [9, 1, 4], \quad \mathbf{c} = [3, 0, 2].$$

2-29. Verify the triangle inequality when

$$\mathbf{a} = [1, 1, 1], \quad \mathbf{b} = [3, 3, 3], \quad \mathbf{c} = [-2, -2, -2].$$

Illustrate geometrically. Repeat with $\mathbf{c} = [2, 2, 2]$.

*2-30. Consider the basis $\mathbf{a}_1 = 5^{-1/2}[1, 2]$, $\mathbf{a}_2 = 10^{-1/2}[-1, 3]$ for E^2. Using $\mathbf{a}_1, \mathbf{a}_2$ to define a coordinate system, find α_1, α_2 in

$$\mathbf{x} = \alpha_1\mathbf{a}_1 + \alpha_2\mathbf{a}_2,$$

where $\mathbf{x} = [x_1, x_2]$ when $\mathbf{e}_1, \mathbf{e}_2$ define the coordinate system. Illustrate this geometrically.

*2-31. Using the Schmidt orthogonalization process, construct an orthonormal basis for E^3 from the following set of basis vectors:

$$\mathbf{a}_1 = [2, 6, 3], \quad \mathbf{a}_2 = [9, 1, 0], \quad \mathbf{a}_3 = [1, 2, 7].$$

Illustrate this geometrically.

*2-32. Consider any subspace S_n of V_n. The set of vectors orthogonal to every vector in S_n is called the orthogonal complement of S_n [written $O(S_n)$]. Prove that $O(S_n)$ is a subspace of V_n. Let S_3 be a set of vectors of the form $[x_1, x_2, 0]$. What is $O(S_3)$?

*2-33. Prove that any vector $\mathbf{x}$ of V_n is in $S_n + O(S_n)$, that is, prove $S_n + O(S_n) = V_n$. Hint: Choose an orthonormal basis $\mathbf{a}_1, \ldots, \mathbf{a}_u$ for S_n and an orthonormal basis $\mathbf{b}_1, \ldots, \mathbf{b}_v$ for $O(S_n)$. No vector in V_n can be orthogonal to these basis vectors for S_n and $O(S_n)$. If $\mathbf{x}$ is not orthogonal to the set $\mathbf{a}_1, \ldots, \mathbf{a}_u$ or the set $\mathbf{b}_1, \ldots, \mathbf{b}_v$, then $\mathbf{x} - \sum_i (\mathbf{a}_i'\mathbf{x})\mathbf{a}_i$ is orthogonal to the set $\mathbf{a}_1, \ldots, \mathbf{a}_u$. Illustrate this theorem for the case where $S_n = S_3$ is the set of vectors $[x_1, x_2, 0]$.

*2-34. Prove that $O[O(S_n)] = S_n$. Illustrate this geometrically.

---

* Starred problems refer to starred sections in the text.

The following problems deal with generalizations to vectors whose components are complex numbers.

2-35. The solutions to a quadratic equation, such as $\alpha x^2 + \beta x + \gamma = 0$ (α, β, γ real), cannot always be written as real numbers. Depending on the values of α, β, γ, the solutions may take the form $x = a \pm ib$, where $i^2 = -1$ or $i = \sqrt{-1}$, and a, b are real numbers. To see this, solve $x^2 + 2x + 5 = 0$. A number of the form $z = a + ib$ (a, b real) is called a complex number. Furthermore, a is called the real part of z, and b the imaginary part of z (note that there is nothing "imaginary" about b; it is a real number). We write $z = 0$ ($0 = 0 + i0$) if and only if $a = b = 0$. Given two complex numbers $z_1 = a_1 + ib_1$, $z_2 = a_2 + ib_2$, $z_1 = z_2$ if and only if $a_1 = a_2$, and $b_1 = b_2$. If the complex number z_3 denotes the sum of z_1 and z_2, it is defined to be

$$z_3 = z_1 + z_2 = (a_1 + a_2) + i(b_1 + b_2).$$

If z_3 represents the product of z_1 and z_2, it is defined to be

$$z_3 = z_1 z_2 = (a_1 + ib_1)(a_2 + ib_2) = (a_1 a_2 - b_1 b_2) + i(a_1 b_2 + a_2 b_1).$$

Note that we can obtain this expression by using the ordinary laws of multiplication and $i^2 = -1$. If the complex number z_3 is the quotient of z_1 and z_2, it is defined to be

$$z_3 = \frac{z_1}{z_2} = \frac{1}{a_2^2 + b_2^2}[(a_1 a_2 + b_1 b_2) + i(a_2 b_1 - a_1 b_2)].$$

Problem 2-38 will illustrate an easy method of obtaining the above expression. Note that $z_1 = z_2 z_3$.

Show that the definitions for equality, addition, multiplication, and division are consistent with the theorems on solutions of quadratic equations. How should subtraction be defined? Demonstrate that addition and multiplication of complex numbers satisfy the associative and commutative laws [for multiplication this means that $z_1 z_2 = z_2 z_1$ and $z_1(z_2 z_3) = (z_1 z_2)z_3 = z_1 z_2 z_3$]. Observe that real numbers are included in the set of all complex numbers since b in $z = a + ib$ may be zero.

2-36. The complex conjugate of a complex number $z = a + ib$ is defined to be $z^* = a - ib$. Show that if z is the solution to a quadratic equation, so is z^*. Prove that

$$(z_1 + z_2)^* = z_1^* + z_2^*, \qquad (z_1 z_2)^* = z_1^* z_2^*, \qquad \left(\frac{z_1}{z_2}\right)^* = \frac{z_1^*}{z_2^*}.$$

2-37. The absolute value of a complex number $z = a + ib$, written $|z|$, is defined to be $|z| = \sqrt{a^2 + b^2}$. Show that $zz^* = z^*z = |z|^2$. Prove that $|z^*| = |z|$, $|z_1 z_2| = |z_1||z_2|$, $|z_1/z_2| = |z_1|/|z_2|$. To write z_1/z_2 as the complex number $z_3 = a_3 + ib_3$, it is necessary to eliminate the i in the denominator, that is, we multiply the numerator and denominator by z_2^* and obtain $z_3 = z_1 z_2^*/|z_2|^2$; this is the result given above for the quotient of two complex numbers.

2-38. Let $z_1 = 3 + 4i$, $z_2 = 2 + i$. Compute $z_1 z_2$, $z_2 z_1$, $z_1 + z_2$, z_1/z_2, z_2/z_1, z_1^*, z_2^*, $|z_1|$, $|z_2|$, $(z_1 z_2)^*$, $z_1^* z_2^*$, $|z_1/z_2|$, and verify the general rules developed above.

2-39. Most of the theory developed in this chapter can be extended directly to vectors whose components are complex numbers. Given two vectors $\mathbf{z} = [z_1, \ldots, z_n]$ and $\mathbf{w} = [w_1, \ldots, w_n]$ with complex components, $\mathbf{z} = \mathbf{w}$ if and only if $z_i = w_i$, $i = 1, \ldots, n$. Then $\mathbf{z} + \mathbf{w} = [z_1 + w_1, \ldots, z_n + w_n]$, and $\gamma \mathbf{z} = [\gamma z_1, \ldots, \gamma z_n]$ for any complex number γ. Show that addition is associative and commutative. The scalar product is defined somewhat differently when the vectors have complex components. The scalar product of $\mathbf{z}$ and $\mathbf{w}$, written $\bar{\mathbf{z}} \mathbf{w}$, is defined to be

$$\bar{\mathbf{z}} \mathbf{w} = \sum_{i=1}^{n} z_i^* w_i,$$

where the * denotes the complex conjugate. This is often referred to as the Hermitian scalar product which, if the components are real, reduces to the definition of the scalar product given in the text. Show that $\bar{\mathbf{z}} \mathbf{w} = (\bar{\mathbf{w}} \mathbf{z})^*$ and $\bar{\mathbf{z}} \mathbf{z} = \sum_{i=1}^{n} |z_i|^2$; $\bar{\mathbf{z}} \mathbf{z}$ is called the square of the magnitude of $\mathbf{z}$ and is written $|\mathbf{z}|^2$. Observe that $|\mathbf{z}|$ is a real number. One important difference between vectors with real components and those with complex components is the loss of the simple geometrical interpretation in the latter case, although it is still possible to think of n-dimensional spaces with complex coordinates.

2-40. Let $\mathbf{z} = [3 + 5i, 2 + i, 1 + 9i]$, $\mathbf{w} = [2i, 4, 5 + 6i]$, $\gamma = 2 + 4i$. Compute $\mathbf{z} + \mathbf{w}$, $\gamma \mathbf{z}$, $\gamma \mathbf{w}$, $\bar{\mathbf{z}} \mathbf{w}$, $\bar{\mathbf{w}} \mathbf{z}$, $|\mathbf{z}|$, $|\mathbf{w}|$.

2-41. A set of k vectors $\mathbf{z}_1, \ldots, \mathbf{z}_k$ with complex components is linearly independent if the only set of complex numbers α_i for which $\sum_{i=1}^{k} \alpha_i \mathbf{z}_i = \mathbf{0}$ is $\alpha_i = 0$, $i = 1, \ldots, k$. Otherwise the set is linearly dependent. Thus we see that the same definition applies formally to both types of vectors, i.e., to those with complex components and to those whose components are real numbers. Under what conditions are the two vectors $\mathbf{z} = [z_1, z_2]$ and $\mathbf{w} = [w_1, w_2]$ linearly dependent? Reduce the condition to one involving real numbers only. Consider the set of n-component vectors $\hat{\mathbf{e}}_i$, $i = 1, \ldots, n$; all components of $\hat{\mathbf{e}}_i$ are zero except the ith component which is $1 + i$. Show that these n vectors are linearly independent. How can any n-component vector (with real or complex components) be expressed as a linear combination of these vectors such that $\mathbf{z} = \sum_{i=1}^{n} \lambda_i \hat{\mathbf{e}}_i$ (the λ_i being complex numbers). Hint: If $w = a + ib$, $w(1 + i) = (a - b) + i(a + b)$. Express $\mathbf{z} = [i, 4 + 5i, 2 + 3i]$ as a linear combination of $\hat{\mathbf{e}}_1$, $\hat{\mathbf{e}}_2$, $\hat{\mathbf{e}}_3$.

2-42. The set of all n-component vectors $\mathbf{z} = [z_1, \ldots, z_n]$, with the z_i taken from the set of complex numbers, forms an n-dimensional vector space which may be denoted by $V_n(C)$, the C indicating that the components belong to the set of complex numbers. A basis for this space is a set of linearly independent vectors which spans the space. Show that every basis contains precisely n vectors and that any set of n linearly independent vectors from $V_n(C)$ is a basis for $V_n(C)$. Demonstrate that any $n + 1$ vectors in $V_n(C)$ are linearly dependent.

Show that the definition of a subspace can be applied to $V_n(C)$ without any modification.

2-43. Two vectors with complex components are orthogonal if their Hermitian scalar product is zero. Generalize the Schmidt orthogonalization procedure to vectors with complex components. Construct an orthogonal basis from $z_1 = [i, 4 + 2i]$, $z_2 = [5 + 6i, 1]$.

2-44. Show that the unit vectors $e_1, \ldots, e_n$ are a basis for $V_n(C)$ as well as E^n.

CHAPTER 3

MATRICES AND DETERMINANTS

*"like the glaze in a
katydid wing
subdivided by sun
till the nettings are legion."*

Marianne Moore.

3–1 Matrices. Matrices provide a very powerful tool for dealing with linear models. This chapter will discuss some of the elementary properties of matrices and investigate the closely related theory of determinants.

MATRIX: *A matrix is defined to be a rectangular array of numbers arranged into rows and columns. It is written as follows:*

$$\begin{bmatrix} a_{11} & a_{12} & \cdots & a_{1n} \\ a_{21} & a_{22} & \cdots & a_{2n} \\ \vdots & & & \vdots \\ a_{m1} & a_{m2} & \cdots & a_{mn} \end{bmatrix}. \tag{3-1}$$

The above array is called an m by n matrix (written m × n) since it has m rows and n columns. As a rule, brackets [ ], parentheses ( ), or the form ‖ ‖ is used to enclose the rectangular array of numbers. It should be noted right at the beginning that a matrix has no numerical value. It is simply a convenient way of representing arrays (tables) of numbers.

Since a matrix is in general a two-dimensional array of numbers, a double subscript must be used to represent any one of its elements (*entries, matrix elements*). By convention, the first subscript refers to the row, the second to the column. Thus a_{23} refers to the element in the second row, third column; and a_{ij} refers to the element in the ith row, jth column. No relation at all needs to exist between the number of rows and the number of columns. A matrix, for example, can have 100 rows and 10 columns or 1 row and 1000 columns. *Any matrix which has the same number of rows as columns is called a square matrix. A square matrix with n rows and n columns is often called an nth-order matrix.* Any nth-order matrix is by definition a square matrix.

EXAMPLES: The following are matrices:

$$\begin{bmatrix} 1 & 0 \\ 0 & 1 \end{bmatrix}, \quad \begin{bmatrix} a & b \\ c & e \end{bmatrix}, \quad \begin{bmatrix} 1 & 2 & 3 \\ 0 & 1 & 0 \end{bmatrix}, \quad (3, 1, 7).$$

But

$$\begin{bmatrix} 1 & 2 \\ 3 & \end{bmatrix}, \quad \begin{bmatrix} & 4 & \\ 5 & & 6 \end{bmatrix}, \quad \begin{bmatrix} 11 \\ 7 \\ 8 \quad 10 \\ 9 \end{bmatrix}$$

are not matrices since they are not rectangular arrays arranged into rows and columns.

Matrices will usually be denoted by upper-case boldface roman letters (**A**, **B**, etc.), and elements by italicized lower case letters (a_{ij}, b_{ij}, etc.) unless specific numbers are used. We can write:

$$\mathbf{A} = \|a_{ij}\| = \begin{bmatrix} a_{11} & \cdots & a_{1n} \\ \vdots & & \vdots \\ a_{m1} & \cdots & a_{mn} \end{bmatrix}. \quad (3\text{–}2)$$

Observe that the expressions **A**, $\|a_{ij}\|$ do not indicate how many rows or columns the matrix has. This must be known from other sources. When only the typical element a_{ij} of the matrix is shown we use $\|a_{ij}\|$ rather than (a_{ij}) or $[a_{ij}]$. The latter notations do not indicate clearly whether a matrix is implied or whether parentheses (brackets) have been placed around a single element. We shall adopt the convention that brackets will be used to enclose matrices having at least two rows and two columns. Parentheses will be used to enclose a matrix of a single row. To simplify the printing in the text of matrices having a single column, they will be printed as a row and will be enclosed by brackets. The same convention is used in printing column vectors. In equations and examples, a matrix with a single column will sometimes be printed as a column for added clarity.

In this chapter and in the remainder of the text, the elements of matrices will be real numbers. However, the results obtained also hold for matrices with complex elements (see Problems 3–78 through 3–86).

3–2 Matrix operations. It is the proper definition of matrix operations that determines their usefulness, since a matrix *per se* is merely a table of numbers. We shall see that the definitions which seem intuitively obvious for operating with tables of numbers are also the most useful ones.

EQUALITY: *Two matrices* **A** *and* **B** *are said to be equal, written* **A** = **B**, *if they are identical, that is, if the corresponding elements are equal. Thus,* **A** = **B** *if and only if* $a_{ij} = b_{ij}$ *for every* i, j. *If* **A** *is not equal to* **B**, *we write* **A** ≠ **B**.

EXAMPLES:

(1) $\mathbf{A} = \begin{bmatrix} 1 & 0 \\ 0 & 1 \end{bmatrix}$, $\mathbf{B} = \begin{bmatrix} 1 & 1 \\ 0 & 0 \end{bmatrix}$;

A ≠ **B** since $a_{12} \neq b_{12}, a_{22} \neq b_{22}$.

(2) $\mathbf{A} = \begin{bmatrix} 1 & 0 & 0 \\ 0 & 1 & 0 \end{bmatrix}$, $\mathbf{B} = \begin{bmatrix} 1 & 0 \\ 0 & 1 \end{bmatrix}$; **A** ≠ **B**.

(3) $\mathbf{A} = \begin{bmatrix} 1 & 2 \\ 3 & 4 \end{bmatrix}$, $\mathbf{B} = \begin{bmatrix} 1 & 2 \\ 3 & 4 \end{bmatrix}$; **A** = **B**.

Clearly, **A** cannot be equal to **B** unless the number of rows and columns in **A** is the same as the number of rows and columns in **B**. If **A** = **B**, then **B** = **A**.

MULTIPLICATION BY A SCALAR: *Given a matrix* **A** *and a scalar* λ, *the product of* λ *and* **A**, *written* λ**A**, *is defined to be*

$$\lambda \mathbf{A} = \begin{bmatrix} \lambda a_{11} & \cdots & \lambda a_{1n} \\ \lambda a_{21} & \cdots & \lambda a_{2n} \\ \vdots & & \vdots \\ \lambda a_{m1} & \cdots & \lambda a_{mn} \end{bmatrix}. \tag{3-3}$$

Each element of **A** is multiplied by the scalar λ. The product λ**A** is then another matrix having m rows and n columns if **A** has m rows and n columns. We note:

$$\lambda \mathbf{A} = \|\lambda a_{ij}\| = \|a_{ij}\lambda\| = \mathbf{A}\lambda.$$

EXAMPLES:

(1) $\lambda = 4$, $\mathbf{A} = \begin{bmatrix} 1 & 2 \\ 0 & 1 \end{bmatrix}$; $\lambda \mathbf{A} = \begin{bmatrix} 4 & 8 \\ 0 & 4 \end{bmatrix}$.

(2) $\lambda = -2$, $\mathbf{A} = \begin{bmatrix} 1 & 4 & 1 \\ 2 & 0 & 3 \end{bmatrix}$; $\lambda \mathbf{A} = \begin{bmatrix} -2 & -8 & -2 \\ -4 & 0 & -6 \end{bmatrix}$.

ADDITION: *The sum* **C** *of a matrix* **A** *having m rows and n columns and a* matrix **B** *having m rows and n columns is a matrix having m rows and n columns whose elements are given by*

$$c_{ij} = a_{ij} + b_{ij} \quad (\text{all } i, j). \tag{3-4}$$

Written out in detail, (3–4) becomes

$$\mathbf{C} = \begin{bmatrix} c_{11} & \cdots & c_{1n} \\ \vdots & & \vdots \\ c_{m1} & \cdots & c_{mn} \end{bmatrix} = \begin{bmatrix} a_{11} & \cdots & a_{1n} \\ \vdots & & \vdots \\ a_{m1} & \cdots & a_{mn} \end{bmatrix} + \begin{bmatrix} b_{11} & \cdots & b_{1n} \\ \vdots & & \vdots \\ b_{m1} & \cdots & b_{mn} \end{bmatrix}$$

$$= \begin{bmatrix} a_{11} + b_{11} & \cdots & a_{1n} + b_{1n} \\ \vdots & & \vdots \\ a_{m1} + b_{m1} & \cdots & a_{mn} + b_{mn} \end{bmatrix}. \tag{3-5}$$

This expression can be abbreviated to

$$\mathbf{C} = \mathbf{A} + \mathbf{B}. \tag{3-6}$$

Two matrices are added by adding the corresponding elements.

EXAMPLES:

(1) $\mathbf{A} = \begin{bmatrix} 1 & 2 \\ 3 & 4 \end{bmatrix}, \quad \mathbf{B} = \begin{bmatrix} 1 & 0 \\ 0 & 1 \end{bmatrix};$

$\mathbf{C} = \mathbf{A} + \mathbf{B} = \begin{bmatrix} 1+1 & 2+0 \\ 3+0 & 4+1 \end{bmatrix} = \begin{bmatrix} 2 & 2 \\ 3 & 5 \end{bmatrix}.$

(2) $\mathbf{A} = \begin{bmatrix} 4 & 5 & 6 \\ 3 & 1 & 2 \end{bmatrix}, \quad \mathbf{B} = \begin{bmatrix} 1 & 0 & 1 \\ 0 & 1 & 0 \end{bmatrix};$

$\mathbf{C} = \mathbf{A} + \mathbf{B} = \begin{bmatrix} 5 & 5 & 7 \\ 3 & 2 & 2 \end{bmatrix}.$

(3) $\mathbf{A} + \mathbf{A} = \|a_{ij}\| + \|a_{ij}\| = \|a_{ij} + a_{ij}\| = \|2a_{ij}\| = 2\mathbf{A},$

and

$$\mathbf{A} + \mathbf{A} + \mathbf{A} = 3\mathbf{A}, \text{ etc.}$$

Note that addition of matrices **A**, **B** *is defined only when* **B** *has the same number of rows and the same number of columns as* **A**. Addition is not defined in the following example:

(4) $\mathbf{A} = \begin{bmatrix} 1 & 2 \\ 5 & 2 \end{bmatrix}, \quad \mathbf{B} = \begin{bmatrix} 2 & 4 & 6 \\ 1 & 3 & 5 \end{bmatrix} \quad \text{(addition not defined)}.$

It is obvious from the definition that matrix addition follows the associative and commutative laws, that is,

$$\mathbf{A} + \mathbf{B} = \mathbf{B} + \mathbf{A}, \tag{3-7}$$

$$\mathbf{A} + (\mathbf{B} + \mathbf{C}) = (\mathbf{A} + \mathbf{B}) + \mathbf{C} = \mathbf{A} + \mathbf{B} + \mathbf{C}, \tag{3-8}$$

since addition proceeds by elements and the laws hold for real numbers. *Subtraction is defined in terms of operations already considered:*

$$\mathbf{A} - \mathbf{B} = \mathbf{A} + (-1)\mathbf{B}. \tag{3-9}$$

Thus, we subtract two matrices by subtracting the corresponding elements. The restrictions on number of rows and columns are the same for subtraction and addition.

EXAMPLE:

$$\mathbf{A} = \begin{bmatrix} 2 & 5 \\ 3 & 6 \\ 4 & 8 \end{bmatrix}, \quad \mathbf{B} = \begin{bmatrix} 4 & 2 \\ 1 & 0 \\ 2 & 0 \end{bmatrix}; \quad \mathbf{A} - \mathbf{B} = \begin{bmatrix} -2 & 3 \\ 2 & 6 \\ 2 & 8 \end{bmatrix}.$$

3–3 Matrix multiplication—introduction. It was fairly obvious how equality, multiplication by a scalar, and addition should be defined for matrices. It is not nearly so obvious, however, how matrix multiplication should be defined. At this point, the concept of matrices as mere tables of data must be abandoned since this approach does not give a clue to a proper definition. Our first thought would be to multiply the corresponding elements to obtain the product. However, this definition turns out to be of little value in the majority of problems requiring application of matrices (although recently, in some data processing operations, it has been found useful).

Instead, we shall approach the subject in a different way. The value of matrices stems mostly from the fact that they permit us to deal efficiently with systems of simultaneous linear equations and related subjects. We have already encountered systems of simultaneous linear equations in Section 2–8 on the linear dependence of vectors. Let us consider a set of m simultaneous linear equations in n unknowns, the x_j, which can be written

$$\begin{aligned} a_{11}x_1 + a_{12}x_2 + \cdots + a_{1n}x_n &= d_1, \\ &\vdots \\ a_{m1}x_1 + a_{m2}x_2 + \cdots + a_{mn}x_n &= d_m. \end{aligned} \tag{3-10}$$

Note, first of all, that the coefficients a_{ij} can be arranged into an $m \times n$

matrix $\mathbf{A} = \|a_{ij}\|$. Furthermore, the variables x_j may be arranged into a matrix of a single column and n rows which will be denoted by $\mathbf{X} = [x_1, \ldots, x_n]$. Similarly, the constants d_i may be arranged into a matrix of one column and m rows which can be written $\mathbf{D} = [d_1, \ldots, d_m]$. The idea suggests itself to write the entire set of equations (3–10) in the abbreviated form*

$$\mathbf{AX} = \mathbf{D}. \tag{3-11}$$

Equation (3–11) will be interpreted as meaning that the matrix $\mathbf{D}$ is obtained by multiplying matrix $\mathbf{A}$ by matrix $\mathbf{X}$. This interpretation will be correct only if element d_i of $\mathbf{D}$ is computed from the elements in $\mathbf{A}$, $\mathbf{X}$ according to (3–10):

$$d_i = \sum_{j=1}^{n} a_{ij} x_j, \qquad i = 1, \ldots, m. \tag{3-12}$$

Here we have a lead toward a useful definition of matrix multiplication. However, Eqs. (3–11) and (3–12) suggest a way only for the multiplication of a matrix by another matrix of a single column. It is immediately apparent, however, that if our definition is to make sense, the number of columns in $\mathbf{A}$ must be the same as the number of rows in $\mathbf{X}$; that is, in (3–10), there is one column in $\mathbf{A}$ for each variable and one row (element) in $\mathbf{X}$ for each variable.

To generalize the definition of matrix multiplication, let us imagine that the variables x_j are defined in terms of another set of variables y_k by the following relations:

$$\begin{aligned} x_1 &= b_{11} y_1 + \cdots + b_{1r} y_r, \\ &\vdots \\ x_n &= b_{n1} y_1 + \cdots + b_{nr} y_r, \end{aligned} \tag{3-13}$$

or

$$x_j = \sum_{k=1}^{r} b_{jk} y_k, \qquad j = 1, \ldots, n.$$

Each original variable x_j is a linear combination of the new variables y_k.

---

* The reader can, no doubt, think of many other possible forms beside that given in (3–11). For example, we might write (3–10) as $\mathbf{XA} = \mathbf{D}$. Furthermore, we could define $\mathbf{X}$ as a matrix with one row and n columns and $\mathbf{D}$ as a matrix with one row and m columns, and with this notation, write $\mathbf{AX} = \mathbf{D}$ or $\mathbf{XA} = \mathbf{D}$. There are many other feasible combinations. The one we have selected leads to a useful definition of matrix multiplication, while many of the others do not. Instead of investigating all the possibilities, we shall begin with the one which leads directly to our goal.

Substitution of (3–13) into (3–12) gives

$$\sum_{j=1}^{n} a_{ij} \sum_{k=1}^{r} b_{jk} y_k = d_i. \qquad (3\text{–}14)$$

Rearranging the summation signs, we have

$$\sum_{k=1}^{r} \left(\sum_{j=1}^{n} a_{ij} b_{jk} \right) y_k = d_i, \quad i = 1, \ldots, m. \qquad (3\text{–}15)$$

Let us write

$$c_{ik} = \sum_{j=1}^{n} a_{ij} b_{jk}, \quad i = 1, \ldots, m, \quad k = 1, \ldots, r. \qquad (3\text{–}16)$$

Then (3–15) becomes

$$\sum_{k=1}^{r} c_{ik} y_k = d_i, \quad i = 1, \ldots, m. \qquad (3\text{–}17)$$

This is a set of m simultaneous linear equations in r variables, the y_k.

Suppose we write $\mathbf{B} = \|b_{jk}\|$ and $\mathbf{Y} = [y_1, \ldots, y_r]$ so that, using our simplified notation (3–11), Eq. (3–13) can be written $\mathbf{X} = \mathbf{BY}$. Furthermore, if $\mathbf{C} = \|c_{ik}\|$, (3–17) becomes $\mathbf{CY} = \mathbf{D}$. If we replace $\mathbf{X}$ by $\mathbf{BY}$ in (3–11), we obtain $\mathbf{ABY} = \mathbf{D}$. Thus it appears that $\mathbf{C}$ should be the product of $\mathbf{A}$ and $\mathbf{B}$, that is:

$$\mathbf{C} = \mathbf{AB}. \qquad (3\text{–}18)$$

For this interpretation to be valid, we must define multiplication in such a way that each element c_{ik} of $\mathbf{C}$ is computed from the elements of $\mathbf{A}, \mathbf{B}$ by means of (3–16).

The matrix $\mathbf{C}$ is $m \times r$, while $\mathbf{A}$ is $m \times n$, and $\mathbf{B}$ is $n \times r$. To define the product $\mathbf{AB}$ [see (3–18)], it is necessary that the number of columns in $\mathbf{A}$ be the same as the number of rows in $\mathbf{B}$ since there is one column in $\mathbf{A}$ and one row in $\mathbf{B}$ for each variable x_j. However, there is no restriction whatever on the number of rows (the number of equations) in $\mathbf{A}$ or the number of columns (the number of variables y_k) in $\mathbf{B}$. Given any two matrices $\mathbf{A}, \mathbf{B}$, the preceding discussion suggests a way to form the product $\mathbf{C} = \mathbf{AB}$ provided the number of columns in $\mathbf{A}$ is equal to the number of rows in $\mathbf{B}$, and we are now in a position to give the general definition of matrix multiplication.

MATRIX MULTIPLICATION: *Given an $m \times n$ matrix $\mathbf{A}$ and an $n \times r$ matrix $\mathbf{B}$, the product $\mathbf{AB}$ is defined to be an $m \times r$ matrix $\mathbf{C}$, whose elements are computed from the elements of $\mathbf{A}, \mathbf{B}$ according to*

$$c_{ij} = \sum_{k=1}^{n} a_{ik}b_{kj}, \quad i = 1, \ldots, m, \quad j = 1, \ldots, r. \quad (3\text{-}19)$$

In the matrix product **AB**, **A** is called the premultiplier, and **B** the postmultiplier. The product **AB** is defined only when the number of columns in **A** is equal to the number of rows in **B**, so that we arrive at the following rule:

RULE. *A and B are conformable for multiplication to yield the product AB if and only if the number of columns in A is the same as the number of rows in B.*

As long as the number of columns in **A** is the same as the number of rows in **B**, **AB** is defined; it is immaterial how many rows there are in **A** or how many columns in **B**. The product **C** = **AB** has the same number of rows as **A** and the same number of columns as **B**.

Note: The reader who is not familiar with double sums and with interchanging the summation signs in going from (3–14) to (3–15) should consider the following example: Imagine the $m \times n$ matrix

$$\begin{bmatrix} a_1b_{11} & a_1b_{12} & \cdots & a_1b_{1n} \\ a_2b_{21} & a_2b_{22} & \cdots & a_2b_{2n} \\ \vdots & & & \vdots \\ a_mb_{m1} & a_mb_{m2} & \cdots & a_mb_{mn} \end{bmatrix}.$$

Then

$$\sum_{k=1}^{m}\left[\sum_{j=1}^{n} a_k b_{kj}\right] = \sum_{k=1}^{m}\left[a_k \sum_{j=1}^{n} b_{kj}\right] \quad (3\text{-}20)$$

is the sum of all the elements $a_k b_{kj}$ in the matrix obtained by first adding all the elements in one row and then summing over the rows. Similarly,

$$\sum_{j=1}^{n}\left[\sum_{k=1}^{m} a_k b_{kj}\right] \quad (3\text{-}21)$$

is the sum of all the elements in the matrix where we first sum all the elements in one column and then sum over the columns. Since (3–20) and (3–21) are two expressions for the same thing, they are equal. Identification of the a_k, b_{kj} of (3–20) and (3–21) with the a_{ij}, $b_{jk}y_k$ of (3–14), respectively, yields the desired result. Often it is useful to represent either (3–20) or (3–21) by the simplified notation

$$\sum_{j,k} a_k b_{jk}.$$

3–4 Matrix multiplication—further development.

The consideration of linear equations and the change of variables in linear equations led us to the most useful definition of matrix multiplication. In fact, it was the search for a simplified notation applicable to systems of linear equations which inspired Cayley in the 1850's to introduce matrix notation. We shall now clarify the definition of matrix multiplication and develop some of the properties of the matrix product.

At first glance, Eq. (3–19) looks rather strange; quite a few subscripts appear. Note first of all that the element c_{ij} depends on all the elements in row i of **A** and on all the elements in column j of **B**. The rule for matrix multiplication is quite similar to the definition of the scalar product of vectors. Indeed, if we think of the ith row of **A** and the jth column of **B** as vectors, then the element in the ith row and the jth column of **C** is, in fact, the scalar product of the ith row of **A** and the jth column of **B**. Diagrammatically,

$$\begin{bmatrix} c_{11} & \cdots & & & c_{1r} \\ \vdots & & & & \vdots \\ c_{i1} & \cdots & \boxed{c_{ij}} & \cdots & c_{ir} \\ \vdots & & & & \vdots \\ c_{m1} & \cdots & & & c_{mr} \end{bmatrix} = \begin{bmatrix} a_{11} & \cdots & a_{1n} \\ \vdots & & \vdots \\ \boxed{a_{i1} \cdots a_{in}} \\ \vdots & & \vdots \\ a_{m1} & \cdots & a_{mn} \end{bmatrix} \begin{bmatrix} b_{11} & \cdots & \boxed{b_{1j}} & \cdots & b_{1r} \\ \vdots & & \vdots & & \vdots \\ b_{n1} & \cdots & \boxed{b_{nj}} & \cdots & b_{nr} \end{bmatrix}. \quad (3\text{–}22)$$

Before we proceed any further, it will be helpful to examine some examples of multiplication.

EXAMPLES:

(1) $\mathbf{A} = \begin{bmatrix} 1 & 3 \\ 2 & 4 \end{bmatrix}, \quad \mathbf{B} = \begin{bmatrix} 2 & 1 \\ 3 & 5 \end{bmatrix};$

$\mathbf{C} = \mathbf{AB} = \begin{bmatrix} [1(2) + 3(3)] & [1(1) + 3(5)] \\ [2(2) + 4(3)] & [2(1) + 4(5)] \end{bmatrix} = \begin{bmatrix} 11 & 16 \\ 16 & 22 \end{bmatrix}.$

(2) $\mathbf{A} = \begin{bmatrix} 3 & 2 \\ 6 & 1 \end{bmatrix}, \quad \mathbf{B} = \begin{bmatrix} 4 \\ 5 \end{bmatrix};$

$\mathbf{C} = \mathbf{AB} = \begin{bmatrix} 3 & 2 \\ 6 & 1 \end{bmatrix} \begin{bmatrix} 4 \\ 5 \end{bmatrix} = \begin{bmatrix} [3(4) + 2(5)] \\ [6(4) + 1(5)] \end{bmatrix} = \begin{bmatrix} 22 \\ 29 \end{bmatrix}.$

(3) $\mathbf{A} = \begin{bmatrix} a_{11} & a_{12} \\ a_{21} & a_{22} \end{bmatrix}, \quad \mathbf{B} = \begin{bmatrix} b_{11} & b_{12} \\ b_{21} & b_{22} \end{bmatrix};$

$\mathbf{C} = \begin{bmatrix} [a_{11}b_{11} + a_{12}b_{21}] & [a_{11}b_{12} + a_{12}b_{22}] \\ [a_{21}b_{11} + a_{22}b_{21}] & [a_{21}b_{12} + a_{22}b_{22}] \end{bmatrix}.$

(4) $\mathbf{A} = \begin{bmatrix} 3 & 0 \\ 1 & 1 \\ 5 & 2 \end{bmatrix}$, $\quad \mathbf{B} = \begin{bmatrix} 4 & 7 \\ 6 & 8 \end{bmatrix}$;

$$\mathbf{C} = \mathbf{AB} = \begin{bmatrix} 3 & 0 \\ 1 & 1 \\ 5 & 2 \end{bmatrix} \begin{bmatrix} 4 & 7 \\ 6 & 8 \end{bmatrix}$$

$$= \begin{bmatrix} [3(4) + 0(6)] & [3(7) + 0(8)] \\ [1(4) + 1(6)] & [1(7) + 1(8)] \\ [5(4) + 2(6)] & [5(7) + 2(8)] \end{bmatrix} = \begin{bmatrix} 12 & 21 \\ 10 & 15 \\ 32 & 51 \end{bmatrix}.$$

In order to get a feeling for the size of the matrix resulting from multiplication of two matrices, it may be helpful to represent the matrices as rectangular blocks (see Fig. 3–1).

Matrix multiplication does not obey all the rules for the multiplication of ordinary numbers. One of the most important differences is the fact that, in general, matrix multiplication is not commutative, that is, **AB** and **BA** are not the same thing. In fact, it does not even need to be true that **AB** and **BA** are both defined. Thus, as a rule, $\mathbf{AB} \neq \mathbf{BA}$. In the special case where $\mathbf{AB} = \mathbf{BA}$, the matrices are said to commute.

EXAMPLES:

(1) $\mathbf{A} = \begin{bmatrix} 2 & 1 \\ 3 & 1 \end{bmatrix}$; $\quad \mathbf{B} = \begin{bmatrix} 2 \\ 1 \end{bmatrix}$; $\quad \mathbf{AB} = \begin{bmatrix} 5 \\ 7 \end{bmatrix}$.

However, **BA** is not defined since the number of columns in **B** is not the same as the number of rows in **A**.

(2) $\mathbf{A} = \begin{bmatrix} 1 \\ 2 \\ 0 \\ 1 \end{bmatrix}$, $\mathbf{B} = (3, 4, 1, 5)$; $\mathbf{AB} = \begin{bmatrix} 1 \\ 2 \\ 0 \\ 1 \end{bmatrix} (3, 4, 1, 5) = \begin{bmatrix} 3 & 4 & 1 & 5 \\ 6 & 8 & 2 & 10 \\ 0 & 0 & 0 & 0 \\ 3 & 4 & 1 & 5 \end{bmatrix}$,

$$\mathbf{BA} = (3, 4, 1, 5) \begin{bmatrix} 1 \\ 2 \\ 0 \\ 1 \end{bmatrix} = (3 + 8 + 5) = (16).$$

In this example, both **AB** and **BA** are defined, but the products are completely different.

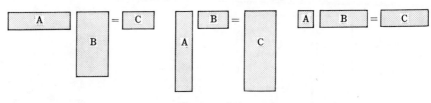

FIGURE 3–1

Multiplication does not have to be commutative even for two $n \times n$ matrices where both **AB** and **BA** are defined and are $n \times n$ matrices.

EXAMPLE:
$$\mathbf{A} = \begin{bmatrix} 1 & 1 \\ 0 & 1 \end{bmatrix}, \quad \mathbf{B} = \begin{bmatrix} 0 & 1 \\ 1 & 0 \end{bmatrix};$$

$$\mathbf{AB} = \begin{bmatrix} 1 & 1 \\ 0 & 1 \end{bmatrix} \begin{bmatrix} 0 & 1 \\ 1 & 0 \end{bmatrix} = \begin{bmatrix} 1 & 1 \\ 1 & 0 \end{bmatrix},$$

$$\mathbf{BA} = \begin{bmatrix} 0 & 1 \\ 1 & 0 \end{bmatrix} \begin{bmatrix} 1 & 1 \\ 0 & 1 \end{bmatrix} = \begin{bmatrix} 0 & 1 \\ 1 & 1 \end{bmatrix},$$

and
$$\mathbf{AB} \neq \mathbf{BA}.$$

Although matrix multiplication is not in general commutative, the associative and distributive laws do hold (when the appropriate operations are defined):

$$(\mathbf{AB})\mathbf{C} = \mathbf{A}(\mathbf{BC}) = \mathbf{ABC} \quad \text{(associative law)} \quad (3\text{--}23)$$

$$\mathbf{A}(\mathbf{B} + \mathbf{C}) = \mathbf{AB} + \mathbf{AC} \quad \text{(distributive law).} \quad (3\text{--}24)$$

To prove the associative law, we only need to note that

$$\sum_r \left(\sum_k a_{ik} b_{kr} \right) c_{rj} = \sum_k a_{ik} \left(\sum_r b_{kr} c_{rj} \right) = \sum_{k,r} a_{ik} b_{kr} c_{rj}, \quad (3\text{--}25)$$

since, as discussed before, summation signs are interchangeable. Similarly, for the distributive law,

$$\sum_k a_{ik}(b_{kj} + c_{kj}) = \sum_k a_{ik} b_{kj} + \sum_k a_{ik} c_{kj}. \quad (3\text{--}26)$$

According to the associative law, **ABC** is unique; we can compute this product by either computing **AB** and postmultiplying by **C**, or by computing **BC** and premultiplying by **A**.

EXAMPLES:

(1) $\mathbf{A} = (1, 2)$, $\mathbf{B} = \begin{bmatrix} 3 & 4 \\ 2 & 1 \end{bmatrix}$, $\mathbf{C} = \begin{bmatrix} 3 & 0 & 2 \\ 5 & 1 & 0 \end{bmatrix}$;

$$\mathbf{AB} = (1, 2) \begin{bmatrix} 3 & 4 \\ 2 & 1 \end{bmatrix} = (7, 6),$$

$$(\mathbf{AB})\mathbf{C} = (7, 6) \begin{bmatrix} 3 & 0 & 2 \\ 5 & 1 & 0 \end{bmatrix} = (51, 6, 14),$$

$$\mathbf{BC} = \begin{bmatrix} 3 & 4 \\ 2 & 1 \end{bmatrix} \begin{bmatrix} 3 & 0 & 2 \\ 5 & 1 & 0 \end{bmatrix} = \begin{bmatrix} 29 & 4 & 6 \\ 11 & 1 & 4 \end{bmatrix},$$

$$\mathbf{A}(\mathbf{BC}) = (1, 2) \begin{bmatrix} 29 & 4 & 6 \\ 11 & 1 & 4 \end{bmatrix} = (51, 6, 14).$$

(2) $\mathbf{A} = (1, 2)$, $\mathbf{B} = \begin{bmatrix} 3 & 4 \\ 0 & 5 \end{bmatrix}$, $\mathbf{C} = \begin{bmatrix} 4 & 2 \\ 1 & 7 \end{bmatrix}$;

$$\mathbf{B} + \mathbf{C} = \begin{bmatrix} 7 & 6 \\ 1 & 12 \end{bmatrix}, \quad \mathbf{A}(\mathbf{B} + \mathbf{C}) = (1, 2) \begin{bmatrix} 7 & 6 \\ 1 & 12 \end{bmatrix} = (9, 30),$$

$$\mathbf{AB} = (1, 2) \begin{bmatrix} 3 & 4 \\ 0 & 5 \end{bmatrix} = (3, 14),$$

$$\mathbf{AC} = (1, 2) \begin{bmatrix} 4 & 2 \\ 1 & 7 \end{bmatrix} = (6, 16),$$

$$\mathbf{AB} + \mathbf{AC} = (9, 30).$$

3–5 Vectors and matrices. The reader has no doubt noticed a pronounced similarity between matrices of one row or column and vectors. Indeed, even the notation is the same. We shall now point out the complete equivalence between vectors and matrices having a single row or column. First, if one examines the definitions of equality, addition, and multiplication by a scalar for both vectors and matrices, it will be observed that they are equivalent when a matrix has a single row or column. Furthermore, the scalar product of a row vector and a column vector corresponds precisely to our definition of matrix multiplication, when

the premultiplier is a row matrix, and the postmultiplier a column matrix.*
We shall see later that the notation indicating the scalar product of two
column vectors or two row vectors will again be identical with the appropriate matrix notation. Hence, there is a complete equivalence between
matrices of one row or one column and vectors. We shall not distinguish
between them. Consequently, we shall continue to use lower case boldface letters for vectors or, equivalently, for matrices of one row or column
rather than the upper case boldface type denoting matrices in general.
This distinction makes it easy to differentiate between vectors and
matrices having more than a single row and a single column. Following
this convention, we shall henceforth write a system of simultaneous linear
equations as $\mathbf{Ax} = \mathbf{d}$ and abstain from the notation $\mathbf{AX} = \mathbf{D}$ used in
Section 3–3.

It has already been suggested that the element c_{ij} of the product $\mathbf{AB}$
is found by forming the scalar product of the ith row of matrix $\mathbf{A}$ with
the jth column of $\mathbf{B}$. On many occasions, it is convenient to represent the
rows or columns of a matrix as vectors. Suppose that $\mathbf{A}$ is an $m \times n$
matrix. If we define the column vectors as

$$\mathbf{a}_j = [a_{1j}, \ldots, a_{mj}], \tag{3-27}$$

then $\mathbf{A}$ can be written as a row of column vectors,

$$\mathbf{A} = (\mathbf{a}_1, \ldots, \mathbf{a}_n). \tag{3-28}$$

Similarly, if we define the row vectors as

$$\mathbf{a}^i = (a_{i1}, \ldots, a_{in}), \tag{3-29}$$

then $\mathbf{A}$ can be written as a column of row vectors,

$$\mathbf{A} = [\mathbf{a}^1, \ldots, \mathbf{a}^m]. \tag{3-30}$$

(To refresh the memory: Square brackets mean that a column is printed
as a row.)

Let us consider the product $\mathbf{AB}$ where $\mathbf{A}$ is $m \times n$ and $\mathbf{B}$ is $n \times r$.
Represent $\mathbf{A}$ as a column of row vectors and $\mathbf{B}$ as a row of column vectors,
that is,

$$\mathbf{A} = [\mathbf{a}^1, \ldots, \mathbf{a}^m], \quad \mathbf{B} = (\mathbf{b}_1, \ldots, \mathbf{b}_r).$$

---

* The careful reader will note that the scalar product of two vectors is a
scalar, while the matrix product of a row and column is a matrix having one
element. There is a complete equivalence between numbers and 1×1 matrices.
The details of demonstrating this equivalence are left to Problem 3–56.

Thus

$$\mathbf{AB} = [\mathbf{a}^1, \ldots, \mathbf{a}^m](\mathbf{b}_1, \ldots, \mathbf{b}_r) = \begin{bmatrix} \mathbf{a}^1\mathbf{b}_1 & \mathbf{a}^1\mathbf{b}_2 & \cdots & \mathbf{a}^1\mathbf{b}_r \\ \mathbf{a}^2\mathbf{b}_1 & \mathbf{a}^2\mathbf{b}_2 & \cdots & \mathbf{a}^2\mathbf{b}_r \\ \vdots & & & \vdots \\ \mathbf{a}^m\mathbf{b}_1 & \mathbf{a}^m\mathbf{b}_2 & \cdots & \mathbf{a}^m\mathbf{b}_r \end{bmatrix}. \quad (3\text{–}31)$$

Here, we can see clearly that each element of the product matrix is a scalar product of a row from **A** and a column from **B**. As noted above, matrices will be frequently represented as a row of column vectors or as a column of row vectors. Column vectors will always be denoted by subscripts ($\mathbf{a}_j$), rows by superscripts ($\mathbf{a}^i$).

EXAMPLE: Let **A** be an arbitrary $m \times n$ matrix. Then

$$\mathbf{A}\mathbf{e}_j = \mathbf{a}_j. \quad (3\text{–}32)$$

This is easily seen by direct multiplication; it can also be shown to be true by writing

$$\mathbf{A}\mathbf{e}_j = [\mathbf{a}^1\mathbf{e}_j, \ldots, \mathbf{a}^m\mathbf{e}_j] = [a_{1j}, \ldots, a_{mj}] = \mathbf{a}_j.$$

3–6 Identity, scalar, diagonal, and null matrices. The number one in the real number system has the property that for any other number a, $1a = a1 = a$; also $1(1) = 1$. A matrix with similar properties is called the identity or unit matrix.

IDENTITY MATRIX: *The identity matrix of order n, written $\mathbf{I}$ or $\mathbf{I}_n$, is a square matrix having ones along the main diagonal (the diagonal running from upper left to lower right) and zeros elsewhere.*

$$\mathbf{I} = \begin{bmatrix} 1 & 0 & 0 & \cdots & 0 \\ 0 & 1 & 0 & \cdots & 0 \\ 0 & 0 & 1 & \cdots & 0 \\ \vdots & & & & \vdots \\ 0 & 0 & 0 & \cdots & 1 \end{bmatrix}. \quad (3\text{–}33)$$

If we write $\mathbf{I} = \|\delta_{ij}\|$, then

$$\delta_{ij} = \begin{cases} 1, & i = j, \\ 0, & i \neq j. \end{cases} \quad (3\text{–}34)$$

The symbol δ_{ij}, defined by (3–34), is called the Kronecker delta. The symbol will always refer to the Kronecker delta unless otherwise specified. It is tacitly assumed that in (3–34) the indices i, j run from 1 to n.

Occasionally, we shall also find it convenient to write an identity matrix as a row of column vectors. The column vectors are the unit vectors $\mathbf{e}_i$. Thus

$$\mathbf{I} = (\mathbf{e}_1, \mathbf{e}_2, \ldots, \mathbf{e}_n). \tag{3-35}$$

Frequently, we must deal with several identity matrices of different sizes. Differentiation between such matrices is facilitated by writing $\mathbf{I}_n$, where the subscript indicates the size of $\mathbf{I}$.

EXAMPLE: The identity matrix of order 2 is

$$\mathbf{I} = \mathbf{I}_2 = \begin{bmatrix} 1 & 0 \\ 0 & 1 \end{bmatrix},$$

and of order 3,

$$\mathbf{I} = \mathbf{I}_3 = \begin{bmatrix} 1 & 0 & 0 \\ 0 & 1 & 0 \\ 0 & 0 & 1 \end{bmatrix}.$$

If $\mathbf{A}$ is a square matrix of order n and $\mathbf{I}$ the identity matrix of order n, $\mathbf{I}$ commutes with $\mathbf{A}$:

$$\mathbf{IA} = \mathbf{AI} = \mathbf{A}. \tag{3-36}$$

To prove this, write $\mathbf{B} = \mathbf{IA}$, and note that by (3-34)

$$b_{ij} = \sum_{k=1}^{n} \delta_{ik} a_{kj} = a_{ij}.$$

Similarly, $\mathbf{AI} = \mathbf{A}$. If we take $\mathbf{A} = \mathbf{I}$,

$$\mathbf{II} = \mathbf{I}.$$

It seems natural to write $\mathbf{II}$ as $\mathbf{I}^2$. Thus

$$\mathbf{I}^2 = \mathbf{I} \quad \text{and hence} \quad \mathbf{I}^3 = \mathbf{I}, \text{ etc.} \tag{3-37}$$

If $\mathbf{A}$ is not square, then our result cannot be true, since either $\mathbf{IA}$ or $\mathbf{AI}$ will not be defined. However, if $\mathbf{A}$ is $m \times n$, then $\mathbf{I}_m \mathbf{A} = \mathbf{A}$ and $\mathbf{AI}_n = \mathbf{A}$. Hence, we can always replace any $m \times n$ matrix $\mathbf{A}$ by $\mathbf{I}_m \mathbf{A}$ or $\mathbf{AI}_n$, without changing the expression. This substitution is frequently useful.

Let us consider again a square matrix of order n. Then

$$\mathbf{AA} = \mathbf{A}^2; \quad \mathbf{AAA} = \mathbf{A}^3; \quad \mathbf{A}^k = \mathbf{AA}^{k-1}, \quad k = 4, 5, \ldots,$$

are also square matrices of order n. By analogy to the kth-degree polynomial in the real number x, we can construct the matrix polynomial

$$\lambda_k \mathbf{A}^k + \lambda_{k-1}\mathbf{A}^{k-1} + \cdots + \lambda_0 \mathbf{I}, \qquad (3\text{--}38)$$

where the λ_i are, of course, scalars. It is important to note that λ_0 must multiply $\mathbf{I}$, not the number 1. The matrix polynomial is an $n \times n$ matrix.

SCALAR MATRIX: *For any scalar λ, the square matrix*

$$\mathbf{S} = \|\lambda\ \delta_{ij}\| = \lambda \mathbf{I} \qquad (3\text{--}39)$$

is called a scalar matrix.

Problem 3–55 will ask you to show that there is an exact equivalence between the scalar matrices $\lambda \mathbf{I}$ and the real numbers λ.

DIAGONAL MATRIX: *A square matrix*

$$\mathbf{D} = \|\lambda_i\ \delta_{ij}\| \qquad (3\text{--}40)$$

is called a diagonal matrix.

Note that the λ_i may vary with i.

EXAMPLES:

(1) $\begin{bmatrix} 2 & 0 \\ 0 & 2 \end{bmatrix} = 2\begin{bmatrix} 1 & 0 \\ 0 & 1 \end{bmatrix}$ is a scalar matrix.

(2) $\begin{bmatrix} 2 & 0 & 0 \\ 0 & 1 & 0 \\ 0 & 0 & 3 \end{bmatrix}$ is a diagonal matrix.

In the real number system, zero has the property $a0 = 0a = 0$ and $a + 0 = a$. A matrix with similar properties is called the null matrix or zero matrix.

NULL MATRIX: *A matrix whose elements are all zero is called a null or zero matrix and is denoted by $\mathbf{0}$.*

A null matrix does not need to be square.

EXAMPLE: The following are null matrices:

$$\mathbf{0} = \begin{bmatrix} 0 & 0 \\ 0 & 0 \end{bmatrix}, \quad \mathbf{0} = \begin{bmatrix} 0 & 0 & 0 \\ 0 & 0 & 0 \end{bmatrix}, \quad \mathbf{0} = \begin{bmatrix} 0 \\ 0 \\ 0 \end{bmatrix}.$$

When the operations are defined, we have

$$\mathbf{A} + \mathbf{0} = \mathbf{A} = \mathbf{0} + \mathbf{A}, \qquad (3\text{--}41)$$

$$\mathbf{A} - \mathbf{A} = \mathbf{0}, \qquad (3\text{--}42)$$

$$\mathbf{A}\mathbf{0} = \mathbf{0}, \qquad (3\text{--}43)$$

$$\mathbf{0}\mathbf{A} = \mathbf{0}. \qquad (3\text{--}44)$$

When $\mathbf{A}$, $\mathbf{0}$ are square matrices of order n,

$$\mathbf{A}\mathbf{0} = \mathbf{0}\mathbf{A} = \mathbf{0}. \qquad (3\text{--}45)$$

If a, b are real numbers, then $ab = 0$ means that $a = 0$, or $b = 0$, or a and $b = 0$. The matrix equation

$$\mathbf{AB} = \mathbf{0}, \qquad (3\text{--}46)$$

however, does not imply that $\mathbf{A} = \mathbf{0}$ or $\mathbf{B} = \mathbf{0}$. It is easy to find non-null matrices $\mathbf{A}$, $\mathbf{B}$ whose product $\mathbf{AB} = \mathbf{0}$. This is another case where matrix operations do not follow the behavior of real numbers.

EXAMPLE:

$$\begin{bmatrix} 1 & 4 \\ 0 & 0 \end{bmatrix} \begin{bmatrix} 4 & 0 \\ -1 & 0 \end{bmatrix} = \begin{bmatrix} 0 & 0 \\ 0 & 0 \end{bmatrix}.$$

This product of two matrices is a null matrix, although neither of the factors is a null matrix.

3–7 The transpose. Sometimes it is of interest to interchange the rows and columns of a matrix. This new form is called the transpose of the original matrix.

TRANSPOSE: *The transpose of a matrix* $\mathbf{A} = \|a_{ij}\|$ *is a matrix formed from* $\mathbf{A}$ *by interchanging rows and columns such that row i of* $\mathbf{A}$ *becomes column i of the transposed matrix. The transpose is denoted by* $\mathbf{A}'$ *and*

$$\mathbf{A}' = \|a_{ji}\| \qquad \text{when } \mathbf{A} = \|a_{ij}\|. \qquad (3\text{--}47)$$

If a'_{ij} is the ijth element of $\mathbf{A}'$, then $a'_{ij} = a_{ji}$.

EXAMPLES:

(1) $\mathbf{A} = \begin{bmatrix} 1 & 3 \\ 2 & 5 \end{bmatrix}$; $\mathbf{A}' = \begin{bmatrix} 1 & 2 \\ 3 & 5 \end{bmatrix}$. (2) $\mathbf{A} = \begin{bmatrix} 1 & 3 & 4 \\ 0 & 1 & 0 \end{bmatrix}$; $\mathbf{A}' = \begin{bmatrix} 1 & 0 \\ 3 & 1 \\ 4 & 0 \end{bmatrix}$.

It will be observed that if $\mathbf{A}$ is $m \times n$, $\mathbf{A}'$ is $n \times m$.

If the sum $\mathbf{C} = \mathbf{A} + \mathbf{B}$ is defined, then

$$\mathbf{C}' = \mathbf{A}' + \mathbf{B}'. \tag{3-48}$$

We prove this by noting that

$$c'_{ij} = c_{ji} = a_{ji} + b_{ji} = a'_{ij} + b'_{ij}.$$

It is interesting to consider the transpose of the product of two matrices. If $\mathbf{AB}$ is defined, then

$$(\mathbf{AB})' = \mathbf{B}'\mathbf{A}', \tag{3-49}$$

that is, *the transpose of the product is the product of the transposes in reverse order*. To prove this, let $\mathbf{C} = \mathbf{AB}$. Thus,

$$c'_{ij} = c_{ji} = \sum_k a_{jk} b_{ki} = \sum_k a'_{kj} b'_{ik} = \sum_k b'_{ik} a'_{kj}.$$

Hence we have shown that the ijth element of $\mathbf{C}'$ is the ijth element of $\mathbf{B}'\mathbf{A}'$, and (3–49) follows. Suppose $\mathbf{A}$ is $m \times n$ and $\mathbf{B}$ is $n \times r$ so that $\mathbf{AB}$ is defined. Then $\mathbf{A}'$ is $n \times m$ and $\mathbf{B}'$ is $r \times n$. Thus if $\mathbf{AB}$ is defined, then $\mathbf{B}'\mathbf{A}'$ is defined also. The same result as in (3–49) holds for any finite number of factors, i.e.,

$$(\mathbf{A}_1 \mathbf{A}_2 \ldots \mathbf{A}_n)' = \mathbf{A}'_n \ldots \mathbf{A}'_2 \mathbf{A}'_1. \tag{3-50}$$

This is easily proved by induction; the details are left to be worked out in Problem 3–25.

EXAMPLE:

$$\mathbf{A} = \begin{bmatrix} 1 & 3 \\ 0 & 5 \end{bmatrix}, \quad \mathbf{B} = \begin{bmatrix} 2 & 4 \\ 1 & 2 \end{bmatrix}; \quad \mathbf{AB} = \begin{bmatrix} 5 & 10 \\ 5 & 10 \end{bmatrix}, \quad (\mathbf{AB})' = \begin{bmatrix} 5 & 5 \\ 10 & 10 \end{bmatrix}.$$

$$\mathbf{A}' = \begin{bmatrix} 1 & 0 \\ 3 & 5 \end{bmatrix}, \quad \mathbf{B}' = \begin{bmatrix} 2 & 1 \\ 4 & 2 \end{bmatrix}; \quad \mathbf{B}'\mathbf{A}' = \begin{bmatrix} 5 & 5 \\ 10 & 10 \end{bmatrix} = (\mathbf{AB})'.$$

It will be noted that

$$\mathbf{I}' = \mathbf{I}. \tag{3-51}$$

Also,

$$(\mathbf{A}')' = \mathbf{A}, \tag{3-52}$$

since

$$(a'_{ij})' = a'_{ji} = a_{ij}.$$

It should now be clear why we used $\mathbf{a}'\mathbf{b}$ for the scalar product of two column vectors and $\mathbf{a}\mathbf{b}'$ for the scalar product of two row vectors. The transpose symbol was introduced so that the product, expressed in matrix terms, would be defined.

3–8 Symmetric and skew-symmetric matrices.

SYMMETRIC MATRIX: *A symmetric matrix is a matrix* **A** *for which*

$$\mathbf{A} = \mathbf{A}'. \tag{3-53}$$

Clearly, a symmetric matrix must be square; it is symmetric about the main diagonal, that is, a reflection in the main diagonal leaves the matrix unchanged. An nth-order symmetric matrix does not have n^2 arbitrary elements since $a_{ij} = a_{ji}$ both below and above the main diagonal. The number of elements above the main diagonal is $(n^2 - n)/2$. The diagonal elements are also arbitrary. Thus the total number of arbitrary elements in an nth-order symmetric matrix is

$$\frac{n^2 - n}{2} + n = \frac{n(n+1)}{2}.$$

EXAMPLE: The following is a symmetric matrix:

$$\begin{bmatrix} 2 & 0 & 7 \\ 0 & 3 & 5 \\ 7 & 5 & 1 \end{bmatrix}.$$

SKEW-SYMMETRIC MATRIX: *A skew-symmetric matrix is a matrix* **A** *for which*

$$\mathbf{A} = -\mathbf{A}'. \tag{3-54}$$

A skew-symmetric matrix is also a square matrix, and

$$a_{ij} = -a_{ji}.$$

Hence, the diagonal elements are zero, $a_{ii} = 0$, and the number of arbitrary elements in an nth-order skew-symmetric matrix is

$$\frac{n(n-1)}{2}.$$

EXAMPLE: The following matrix is skew-symmetric:

$$\begin{bmatrix} 0 & 1 & 2 \\ -1 & 0 & -3 \\ -2 & 3 & 0 \end{bmatrix}.$$

Any square matrix can be written as the sum of a symmetric and a skew-symmetric matrix:

$$\mathbf{A} = \mathbf{A} + \frac{\mathbf{A}'}{2} - \frac{\mathbf{A}'}{2} = \frac{\mathbf{A} + \mathbf{A}'}{2} + \frac{\mathbf{A} - \mathbf{A}'}{2}. \quad (3\text{-}55)$$

Now

$$\left(\frac{\mathbf{A} + \mathbf{A}'}{2}\right)' = \frac{\mathbf{A} + \mathbf{A}'}{2}, \quad (3\text{-}56)$$

and

$$\left(\frac{\mathbf{A} - \mathbf{A}'}{2}\right)' = -\frac{\mathbf{A} - \mathbf{A}'}{2}. \quad (3\text{-}57)$$

If we write

$$\mathbf{A}_s = \frac{\mathbf{A} + \mathbf{A}'}{2} \quad \text{and} \quad \mathbf{A}_a = \frac{\mathbf{A} - \mathbf{A}'}{2},$$

then $\mathbf{A}_s$ is symmetric, and $\mathbf{A}_a$ is skew-symmetric. We have thus expressed the square matrix $\mathbf{A}$ as the sum of a symmetric and a skew-symmetric matrix.

EXAMPLE:

$$\mathbf{A} = \begin{bmatrix} 2 & -4 & 7 \\ 3 & 1 & 9 \\ 8 & 6 & 9 \end{bmatrix}, \quad \frac{\mathbf{A} + \mathbf{A}'}{2} = \begin{bmatrix} 2 & -\frac{1}{2} & \frac{15}{2} \\ -\frac{1}{2} & 1 & \frac{15}{2} \\ \frac{15}{2} & \frac{15}{2} & 9 \end{bmatrix}.$$

$$\frac{\mathbf{A} - \mathbf{A}'}{2} = \begin{bmatrix} 0 & -\frac{7}{2} & -\frac{1}{2} \\ \frac{7}{2} & 0 & \frac{3}{2} \\ \frac{1}{2} & -\frac{3}{2} & 0 \end{bmatrix}, \quad \mathbf{A} = \frac{\mathbf{A} + \mathbf{A}'}{2} + \frac{\mathbf{A} - \mathbf{A}'}{2}.$$

3–9 Partitioning of matrices. It is often necessary to study some subset of elements in a matrix which form a so-called submatrix.

SUBMATRIX: *If we cross out all but k rows and s columns of an $m \times n$ matrix $\mathbf{A}$, the resulting $k \times s$ matrix is called a submatrix of $\mathbf{A}$.*

EXAMPLE: If we cross out rows 1, 3 and columns 2, 6 of the 4×6 matrix $\mathbf{A} = \|a_{ij}\|$, we are left with the 2×4 submatrix

$$\begin{bmatrix} a_{21} & a_{23} & a_{24} & a_{25} \\ a_{41} & a_{43} & a_{44} & a_{45} \end{bmatrix}.$$

For a number of reasons we wish to introduce here the notion of partitioning matrices into submatrices. Some of these reasons are:
(1) The partitioning may simplify the writing or printing of $\mathbf{A}$.
(2) It exhibits some particular structure of $\mathbf{A}$ which is of interest.
(3) It simplifies computation.

We have already introduced one special case of partitioning when we wrote a matrix as a column of row vectors or a row of column vectors. We have seen that an $m \times n$ matrix $\mathbf{B}$ could be written

$$\mathbf{B} = \|b_{ij}\| = (\mathbf{b}_1, \ldots, \mathbf{b}_n), \qquad \mathbf{b}_j = [b_{1j}, \ldots, b_{mj}];$$

$\mathbf{B}$ has been partitioned into n submatrices, the column vectors $\mathbf{b}_j$. Suppose that we have an $r \times m$ matrix $\mathbf{A}$; the product $\mathbf{C} = \mathbf{AB}$ is then defined. The matrix $\mathbf{C}$ will be $r \times n$ and can be written

$$\mathbf{C} = (\mathbf{c}_1, \ldots, \mathbf{c}_n), \qquad \mathbf{c}_j = [c_{1j}, \ldots, c_{rj}].$$

However,
$$c_{ij} = \sum_k a_{ik} b_{kj},$$

or
$$\mathbf{c}_j = [\textstyle\sum a_{1k} b_{kj}, \sum a_{2k} b_{kj}, \ldots, \sum a_{rk} b_{kj}].$$

This can be expressed in matrix form as follows:

$$\mathbf{c}_j = \mathbf{A}\mathbf{b}_j. \tag{3-58}$$

Hence, we can write the product $\mathbf{AB}$ as

$$\mathbf{C} = \mathbf{AB} = (\mathbf{A}\mathbf{b}_1, \mathbf{A}\mathbf{b}_2, \ldots, \mathbf{A}\mathbf{b}_n). \tag{3-59}$$

Each column of $\mathbf{C}$ is computed by multiplying the corresponding column of $\mathbf{B}$ by $\mathbf{A}$.

We shall now consider partitioning from a more general point of view: We have the matrix

$$\mathbf{A} = \begin{bmatrix} a_{11} & a_{12} & a_{13} & a_{14} \\ a_{21} & a_{22} & a_{23} & a_{24} \\ \hdashline a_{31} & a_{32} & a_{33} & a_{34} \\ a_{41} & a_{42} & a_{43} & a_{44} \\ a_{51} & a_{52} & a_{53} & a_{54} \end{bmatrix}. \tag{3-60}$$

Imagine $\mathbf{A}$ to be divided up by dotted lines as shown. Now, if we write

$$\mathbf{A}_{11} = \begin{bmatrix} a_{11} & a_{12} & a_{13} \\ a_{21} & a_{22} & a_{23} \end{bmatrix}, \qquad \mathbf{A}_{12} = \begin{bmatrix} a_{14} \\ a_{24} \end{bmatrix},$$

$$\mathbf{A}_{21} = \begin{bmatrix} a_{31} & a_{32} & a_{33} \\ a_{41} & a_{42} & a_{43} \\ a_{51} & a_{52} & a_{53} \end{bmatrix}, \qquad \mathbf{A}_{22} = \begin{bmatrix} a_{34} \\ a_{44} \\ a_{54} \end{bmatrix}, \tag{3-61}$$

then $\mathbf{A}$ can be written

$$\mathbf{A} = \begin{bmatrix} \mathbf{A}_{11} & \mathbf{A}_{12} \\ \mathbf{A}_{21} & \mathbf{A}_{22} \end{bmatrix}. \tag{3-62}$$

Matrix **A** has been partitioned into four submatrices, the $\mathbf{A}_{ij}$. The submatrices will be denoted by upper-case boldface roman letters; sometimes, if they are vectors, by lower-case boldface type.

If partitioning is to be of any use for computations, we must be able to perform the usual operations of addition and multiplication in partitioned form. Clearly, if

$$\mathbf{A} = \begin{bmatrix} \mathbf{A}_{11} & \mathbf{A}_{12} \\ \mathbf{A}_{21} & \mathbf{A}_{22} \end{bmatrix}, \quad \mathbf{B} = \begin{bmatrix} \mathbf{B}_{11} & \mathbf{B}_{12} \\ \mathbf{B}_{21} & \mathbf{B}_{22} \end{bmatrix}, \tag{3-63}$$

then

$$\mathbf{A} + \mathbf{B} = \begin{bmatrix} \mathbf{A}_{11} + \mathbf{B}_{11} & \mathbf{A}_{12} + \mathbf{B}_{12} \\ \mathbf{A}_{21} + \mathbf{B}_{21} & \mathbf{A}_{22} + \mathbf{B}_{22} \end{bmatrix}, \tag{3-64}$$

provided that for every $\mathbf{A}_{ij}$, the corresponding $\mathbf{B}_{ij}$ has the same number of rows and the same number of columns as $\mathbf{A}_{ij}$. *Hence the rule for addition of partitioned matrices is the same as the rule for addition of ordinary matrices if the submatrices are conformable for addition.* In other words, the matrices can be added "by blocks." However, **A**, **B** must be partitioned "in the same way" if (3-64) is to hold.

Next, we shall consider multiplication of partitioned matrices. We would like the formula for block multiplication to follow the pattern of the usual formula

$$\mathbf{C}_{ij} = \sum_k \mathbf{A}_{ik} \mathbf{B}_{kj}, \tag{3-65}$$

where the elements in (3-19) are now submatrices in (3-65). We note at once that this rule cannot hold unless $\mathbf{A}_{ik}$ and $\mathbf{B}_{kj}$ are conformable for multiplication, that is, the number of columns in submatrix $\mathbf{A}_{ik}$ must be the same as the number of rows in submatrix $\mathbf{B}_{kj}$. This will be true only if the columns of **A** are partitioned in the same way as the rows of **B**.

We wish to show that multiplication by blocks does follow the ordinary rules for multiplication:

$$\mathbf{A} = \begin{bmatrix} \mathbf{A}_{11} & \mathbf{A}_{12} \\ \mathbf{A}_{21} & \mathbf{A}_{22} \end{bmatrix}, \quad \mathbf{B} = \begin{bmatrix} \mathbf{B}_{11} & \mathbf{B}_{12} \\ \mathbf{B}_{21} & \mathbf{B}_{22} \end{bmatrix};$$

$$\mathbf{C} = \mathbf{AB} = \begin{bmatrix} \mathbf{A}_{11}\mathbf{B}_{11} + \mathbf{A}_{12}\mathbf{B}_{21} & \mathbf{A}_{11}\mathbf{B}_{12} + \mathbf{A}_{12}\mathbf{B}_{22} \\ \mathbf{A}_{21}\mathbf{B}_{11} + \mathbf{A}_{22}\mathbf{B}_{21} & \mathbf{A}_{21}\mathbf{B}_{12} + \mathbf{A}_{22}\mathbf{B}_{22} \end{bmatrix} = \begin{bmatrix} \mathbf{C}_{11} & \mathbf{C}_{12} \\ \mathbf{C}_{21} & \mathbf{C}_{22} \end{bmatrix}, \tag{3-66}$$

provided that the submatrices $\mathbf{A}_{ik}$, $\mathbf{B}_{kj}$ are conformable for multiplication. As usual, if $\mathbf{A}_{ik}$ is $m \times n$ and $\mathbf{B}_{kj}$ is $n \times r$, then $\mathbf{C}_{ij}$ is $m \times r$.

It is not hard to see that multiplication by blocks is correct. Suppose, for example, that we have a 3 × 3 matrix **A** and partition it as

$$\mathbf{A} = \begin{bmatrix} a_{11} & a_{12} & a_{13} \\ a_{21} & a_{22} & a_{23} \\ \hline a_{31} & a_{32} & a_{33} \end{bmatrix} = \begin{bmatrix} \mathbf{A}_{11} & \mathbf{A}_{12} \\ \mathbf{A}_{21} & \mathbf{A}_{22} \end{bmatrix}, \qquad (3\text{-}67)$$

and a 3 × 2 matrix **B** which we partition as

$$\mathbf{B} = \begin{bmatrix} b_{11} & b_{12} \\ b_{21} & b_{22} \\ \hline b_{31} & b_{32} \end{bmatrix} = \begin{bmatrix} \mathbf{B}_{11} \\ \mathbf{B}_{21} \end{bmatrix}. \qquad (3\text{-}68)$$

Then any element of $\mathbf{C} = \|c_{ij}\| = \mathbf{AB}$ is

$$c_{ij} = (a_{i1}b_{1j} + a_{i2}b_{2j}) + a_{i3}b_{3j}. \qquad (3\text{-}69)$$

The quantity $a_{i1}b_{1j} + a_{i2}b_{2j}$ is simply the ijth element of $\mathbf{A}_{11}\mathbf{B}_{11}$ if $i \leq 2$, and the ijth element of $\mathbf{A}_{21}\mathbf{B}_{11}$ if $i = 3$. Similarly $a_{i3}b_{3j}$ is the ijth element of $\mathbf{A}_{12}\mathbf{B}_{21}$ if $i \leq 2$, and of $\mathbf{A}_{22}\mathbf{B}_{21}$ if $i = 3$. Thus **C** can be written

$$\mathbf{C} = \begin{bmatrix} \mathbf{C}_{11} \\ \mathbf{C}_{12} \end{bmatrix} = \begin{bmatrix} \mathbf{A}_{11}\mathbf{B}_{11} + \mathbf{A}_{12}\mathbf{B}_{21} \\ \mathbf{A}_{21}\mathbf{B}_{11} + \mathbf{A}_{22}\mathbf{B}_{21} \end{bmatrix}. \qquad (3\text{-}70)$$

It is important to note that in partitioning matrices for multiplication, the partition line between rows s and $s + 1$ is drawn all the way across the matrix so that there is a partition between rows s and $s + 1$ in every column. These partitions are never jagged lines like that shown in the following matrix:

$$\begin{bmatrix} a_{11} & a_{12} & a_{13} & a_{14} \\ \hline a_{21} & a_{22} & a_{23} & a_{24} \\ a_{31} & a_{32} & a_{33} & a_{34} \\ a_{41} & a_{42} & a_{43} & a_{44} \end{bmatrix}.$$

Similarly, when we start drawing a partition line between column r and $r + 1$, we draw it all the way down the matrix. Only by drawing partitions "all the way across" and "all the way down" can we be sure that the expression $\sum \mathbf{A}_{ik}\mathbf{B}_{kj}$ is defined in terms of addition and multiplication. For example, the number of rows in each $\mathbf{A}_{ik}$ is the same for each k and a given i only because every column is partitioned between rows s and $s + 1$.

Using the concept of partitioning, we can write the system of equations $\mathbf{Ax} = \mathbf{b}$ in another useful way. If we write **A** as a row of column vectors

$(\mathbf{a}_1, \ldots, \mathbf{a}_n)$, then

$$\mathbf{A}\mathbf{x} = \mathbf{a}_1 x_1 + \cdots + \mathbf{a}_n x_n = x_1 \mathbf{a}_1 + \cdots + x_n \mathbf{a}_n = \mathbf{b}. \quad (3\text{-}71)$$

Similarly, if we write $\mathbf{A}$ as a column of row vectors

$$[\mathbf{a}^1, \ldots, \mathbf{a}^m],$$

then,

$$[\mathbf{a}^1 \mathbf{x}, \ldots, \mathbf{a}^m \mathbf{x}] = \mathbf{b}, \quad \text{or} \quad \mathbf{a}^i \mathbf{x} = b_i, \quad i = 1, \ldots, m. \quad (3\text{-}72)$$

EXAMPLE: We are given the matrix $\mathbf{A}$ and partition it as follows:

$$\mathbf{A} = \begin{bmatrix} 1 & 3 & 2 \\ 2 & 5 & 0 \\ \hline 4 & 1 & 7 \end{bmatrix} = \begin{bmatrix} \mathbf{A}_{11} & \mathbf{A}_{12} \\ \mathbf{A}_{21} & \mathbf{A}_{22} \end{bmatrix};$$

$$\mathbf{A}_{11} = \begin{bmatrix} 1 & 3 \\ 2 & 5 \end{bmatrix}, \quad \mathbf{A}_{12} = \begin{bmatrix} 2 \\ 0 \end{bmatrix}, \quad \mathbf{A}_{21} = (4, 1), \quad \mathbf{A}_{22} = (7).$$

We also have matrix $\mathbf{B}$; this will be partitioned so that block multiplication of $\mathbf{AB}$ is defined, that is,

$$\mathbf{B} = \begin{bmatrix} 0 & 1 & 2 \\ 2 & 4 & 5 \\ \hline 6 & 0 & 1 \end{bmatrix} = \begin{bmatrix} \mathbf{B}_{11} & \mathbf{B}_{12} \\ \mathbf{B}_{21} & \mathbf{B}_{22} \end{bmatrix};$$

$$\mathbf{B}_{11} = \begin{bmatrix} 0 \\ 2 \end{bmatrix}, \quad \mathbf{B}_{12} = \begin{bmatrix} 1 & 2 \\ 4 & 5 \end{bmatrix}, \quad \mathbf{B}_{21} = (6), \quad \mathbf{B}_{22} = (0, 1).$$

First, we shall compute $\mathbf{C} = \mathbf{AB}$ by multiplying the matrices directly without partitioning. This gives

$$\mathbf{C} = \begin{bmatrix} 1 & 3 & 2 \\ 2 & 5 & 0 \\ 4 & 1 & 7 \end{bmatrix} \begin{bmatrix} 0 & 1 & 2 \\ 2 & 4 & 5 \\ 6 & 0 & 1 \end{bmatrix} = \begin{bmatrix} 18 & 13 & 19 \\ 10 & 22 & 29 \\ 44 & 8 & 20 \end{bmatrix}.$$

If we use block multiplication, $\mathbf{C}$ should be

$$\mathbf{C} = \begin{bmatrix} \mathbf{A}_{11}\mathbf{B}_{11} + \mathbf{A}_{12}\mathbf{B}_{21} & \mathbf{A}_{11}\mathbf{B}_{12} + \mathbf{A}_{12}\mathbf{B}_{22} \\ \mathbf{A}_{21}\mathbf{B}_{11} + \mathbf{A}_{22}\mathbf{B}_{21} & \mathbf{A}_{21}\mathbf{B}_{12} + \mathbf{A}_{22}\mathbf{B}_{22} \end{bmatrix}.$$

However,

$$\mathbf{A}_{11}\mathbf{B}_{11} + \mathbf{A}_{12}\mathbf{B}_{21} = \begin{bmatrix} 1 & 3 \\ 2 & 5 \end{bmatrix} \begin{bmatrix} 0 \\ 2 \end{bmatrix} + \begin{bmatrix} 2 \\ 0 \end{bmatrix}(6) = \begin{bmatrix} 6 \\ 10 \end{bmatrix} + \begin{bmatrix} 12 \\ 0 \end{bmatrix} = \begin{bmatrix} 18 \\ 10 \end{bmatrix},$$

$$\mathbf{A}_{11}\mathbf{B}_{12} + \mathbf{A}_{12}\mathbf{B}_{22} = \begin{bmatrix} 1 & 3 \\ 2 & 5 \end{bmatrix} \begin{bmatrix} 1 & 2 \\ 4 & 5 \end{bmatrix} + \begin{bmatrix} 2 \\ 0 \end{bmatrix}(0, 1) = \begin{bmatrix} 13 & 17 \\ 22 & 29 \end{bmatrix} + \begin{bmatrix} 0 & 2 \\ 0 & 0 \end{bmatrix}$$

$$= \begin{bmatrix} 13 & 19 \\ 22 & 29 \end{bmatrix},$$

$$\mathbf{A}_{21}\mathbf{B}_{11} + \mathbf{A}_{22}\mathbf{B}_{21} = (4, 1) \begin{bmatrix} 0 \\ 2 \end{bmatrix} + (7)(6) = (44),$$

$$\mathbf{A}_{21}\mathbf{B}_{12} + \mathbf{A}_{22}\mathbf{B}_{22} = (4, 1) \begin{bmatrix} 1 & 2 \\ 4 & 5 \end{bmatrix} + (7)(0, 1) = (8, 13) + (0, 7)$$

$$= (8, 20).$$

Thus, by block multiplication,

$$\mathbf{C} = \begin{bmatrix} \begin{bmatrix} 18 \\ 10 \end{bmatrix} & \begin{bmatrix} 13 & 19 \\ 22 & 29 \end{bmatrix} \\ (44) & (8\ 20) \end{bmatrix}.$$

The same result was obtained by direct multiplication.

We have seen that multiplication of matrices by blocks requires that the partitioning of the columns in the premultiplier be the same as the partitioning of the rows in the postmultiplier. It makes no difference at all how the rows in the premultiplier or the columns in the postmultiplier are partitioned. If we are considering the product $\mathbf{AB}$, then the number of columns in $\mathbf{A}_{ik}$ must be the same as the number of rows in $\mathbf{B}_{kj}$.

The results of this section have shown that when matrices are appropriately partitioned for addition or multiplication, the submatrices behave as if they were ordinary elements of the matrix. It should be emphasized, however, that submatrices behave like ordinary elements only when the matrices have been partitioned properly so that the operations to be performed are defined.

We shall now interrupt our study of matrices for a while in order to develop some aspects of the theory of determinants which will be needed in our further investigation of the properties of matrices.

3–10 Basic notion of a determinant.

The concept of a second- or third-order determinant should be familiar to everyone. A second-order determinant is defined to be

$$\begin{vmatrix} a_{11} & a_{12} \\ a_{21} & a_{22} \end{vmatrix} = a_{11}a_{22} - a_{12}a_{21}. \tag{3-73}$$

It is a number computed from the four elements of the square array according to (3–73). Determinants occur naturally in the solution of simultaneous linear equations. In elementary algebra, we learn that the solution to two simultaneous linear equations in two unknowns,

$$\begin{aligned} a_{11}x_1 + a_{12}x_2 &= b_1, \\ a_{21}x_1 + a_{22}x_2 &= b_2, \end{aligned} \tag{3-74}$$

can be written (when the denominator does not vanish):

$$x_1 = \frac{\begin{vmatrix} b_1 & a_{12} \\ b_2 & a_{22} \end{vmatrix}}{\begin{vmatrix} a_{11} & a_{12} \\ a_{21} & a_{22} \end{vmatrix}}, \quad x_2 = \frac{\begin{vmatrix} a_{11} & b_1 \\ a_{21} & b_2 \end{vmatrix}}{\begin{vmatrix} a_{11} & a_{12} \\ a_{21} & a_{22} \end{vmatrix}}. \tag{3-75}$$

A third-order determinant is defined to be

$$\begin{vmatrix} a_{11} & a_{12} & a_{13} \\ a_{21} & a_{22} & a_{23} \\ a_{31} & a_{32} & a_{33} \end{vmatrix} = \begin{aligned} & a_{11}a_{22}a_{33} - a_{12}a_{21}a_{33} + a_{12}a_{23}a_{31} \\ & - a_{13}a_{22}a_{31} + a_{13}a_{21}a_{32} - a_{11}a_{23}a_{32}. \end{aligned} \tag{3-76}$$

It is a number computed from the elements of a 3×3 array according to formula (3–76). Third-order determinants arise in solving three simultaneous linear equations in three unknowns.

It is of interest to generalize the notion of a determinant to $n \times n$ arrays. In the 2×2 and 3×3 cases, it will be observed that the determinant is the sum of terms such that each term contains one and only one element from each row and each column in the square array. Furthermore, the number of elements in each term is the same as the number of rows in the array, that is, no element is repeated. We notice also some alternation in the sign of the terms. We expect that a more general definition of a determinant will include these features.

3–11 General definition of a determinant.

First, we wish to develop some properties of permutations of numbers. A set of integers $1, \ldots, n$ are in "natural order" when they appear in the order $1, 2, 3, \ldots, n$. If two integers are out of natural order in a set of n integers, then a larger integer will precede a smaller one. For example, the natural order of the first five integers, beginning with 1, is $(1, 2, 3, 4, 5)$. When the integers 2 and 4 are interchanged, we obtain (1, 4, 3, 2, 5). The set is now out of natural order because 4 precedes 3, 2, and 3 precedes 2. Any rearrangement of the natural order of n integers is called a *permutation* of these integers. The interchange of two integers, such as 2 and 4 in the above example, is called a *transposition*. The number of *inversions* in a permutation of n integers is the number of pairs of elements (not necessarily adjacent) in which a larger integer precedes a smaller one. In our example, there are three inversions: (4, 3), (4, 2), and (3, 2). It should be noted that the number of inversions in any permutation of n integers from their natural order is unique and can be counted directly and systematically. A permutation is *even* when the number of inversions is even, and *odd* when the number of inversions is odd.

Consider now the nth-order matrix

$$\mathbf{A} = \begin{bmatrix} a_{11} & \cdots & a_{1n} \\ \vdots & & \vdots \\ a_{n1} & \cdots & a_{nn} \end{bmatrix}. \quad (3\text{-}77)$$

If one element is selected from each row and column of $\mathbf{A}$, n elements are obtained. The product of these elements can be written

$$a_{1i}a_{2j}a_{3k}\cdots a_{nr}, \quad (3\text{-}78)$$

where the set of second subscripts $(i, j, k, \ldots, r)$ is a permutation of the set of integers $(1, 2, \ldots, n)$. To determine the number of possible different products of this type, we note that there are n choices for i; given i, there are $n - 1$ choices for j, and given j, there are $n - 2$ choices for k, etc. Thus in all there are

$$n(n-1)(n-2)\ldots 1 = n! \quad (3\text{-}79)$$

products of type (3–78) which can be formed from the n^2 elements of $\mathbf{A}$. The symbol $n!$ is read n-factorial and is defined by (3–79). All possible products (3–78) are obtained by using all possible permutations of the integers $1, 2, \ldots, n$ as the set of second subscripts on the elements.

EXAMPLE: Let us find all the different products of three elements in a 3×3 matrix $\mathbf{A}$ such that any one product contains one and only one element from each row and column of $\mathbf{A}$. These products can be obtained from $a_{1i}a_{2j}a_{3k}$ by substituting all permutations of the set of numbers 1, 2, 3 for the set of subscripts i, j, k.

We obtain $a_{12}a_{21}a_{33}$, $a_{13}a_{22}a_{31}$, $a_{11}a_{23}a_{32}$ which represent odd permutations of the second subscripts from the natural order 1, 2, 3, since they have an odd number of inversions (1 for the first, 3 for the second, 1 for the third). Then there are the terms $a_{13}a_{21}a_{32}$, $a_{12}a_{23}a_{31}$ which are even permutations. Finally, we have $a_{11}a_{22}a_{33}$ which is an even permutation (no inversions); this is the identity permutation—the subscripts are in their natural order.

Hence, six terms can be obtained from the nine elements a_{ij}. It will also be noted that the above terms are precisely those which appeared in the definition of a third-order determinant. If these terms are added together, with a plus sign attached to terms representing even permutations, and a minus sign to those representing odd permutations, we have arrived at the expansion of a third-order determinant. This is the key to the general definition of a determinant.

DETERMINANT: *The determinant of an nth-order matrix* $\mathbf{A} = \|a_{ij}\|$, *written* $|\mathbf{A}|$, *is defined to be the number computed from the following sum involving the* n^2 *elements in* $\mathbf{A}$:

$$|\mathbf{A}| = \sum(\pm) a_{1i} a_{2j} \ldots a_{nr}, \qquad (3\text{--}80)$$

the sum being taken over all permutations of the second subscripts. A term is assigned a plus sign if $(i, j, \ldots, r)$ *is an even permutation of* $(1, 2, \ldots, n)$, *and a minus sign if it is an odd permutation.*

We shall find it convenient to refer to the determinant of an nth-order matrix as an nth-order or $n \times n$ determinant. If we examine our definitions of second- and third-order determinants, we see that they follow directly from the above definition. The third-order case was worked out in the preceding example. We have shown that there are $n!$ terms in the summation. One and only one element from each row and column of $\mathbf{A}$ appears in every term of the summation.

The definition of a determinant implies that only square matrices have determinants associated with them. Often we shall write $|\mathbf{A}|$ as

$$|\mathbf{A}| = \begin{vmatrix} a_{11} & a_{12} \cdots a_{1n} \\ a_{21} & a_{22} \cdots a_{2n} \\ \vdots & \vdots \\ a_{n1} & a_{n2} \cdots a_{nn} \end{vmatrix}, \qquad (3\text{--}81)$$

with straight lines denoting the determinant (instead of brackets). The notational form of (3–80) was used by early writers, such as Jacobi and Cauchy. Cayley introduced the notation of (3–81). Originally, determinants were called eliminants. This name indicated more clearly that they arose on elimination of variables in solving simultaneous linear equations. Perhaps it should be emphasized again that, while a matrix has *no numerical value*, a determinant is *a number*.

3–12 Some properties of determinants. Before turning to the properties of determinants, we shall find it helpful to establish the fact that a single transposition of two elements in any permutation will change an odd permutation to an even one, and vice versa. To prove this, consider the set of numbers $\alpha_1, \alpha_2, \ldots, \alpha_n$ representing some permutation of $1, 2, \ldots, n$. Suppose now that α_i and α_j are interchanged $(j > i)$. In doing this, α_j passes over $\alpha_{j-1}, \alpha_{j-2}, \ldots, \alpha_i$, or $j - i$ subscripts. On the other hand α_i passes over $\alpha_{i+1}, \ldots, \alpha_{j-1}$, or $j - i - 1$ subscripts. Wherever we pass over any given index, we either introduce an inversion or remove one. The total number of inversion changes on a single transposition is then $2(j - i) - 1$ which is always an odd number. Hence, if the original permutation was odd, the new permutation is even, and if the original was even, the new one is odd.

EXAMPLE: Consider $(1, 3, 2, 6, 4, 5)$. This permutation is odd and the inversions are $(3, 2)$, $(6, 4)$, $(6, 5)$. Suppose that we interchange 5 and 1 to yield $(5, 3, 2, 6, 4, 1)$. This permutation is even, and the inversions are: $(5, 3)$, $(5, 2)$, $(5, 4)$, $(5, 1)$, $(3, 2)$, $(3, 1)$, $(2, 1)$, $(6, 4)$, $(6, 1)$, $(4, 1)$.

Using the result just obtained, we can immediately show that *an interchange of two columns in an nth-order matrix* **A** *changes the sign of* $|\mathbf{A}|$. The interchange of two columns interchanges two second subscripts in each term of (3–80) and changes the sign of each term since in the new determinant an originally odd permutation becomes an even permutation, and vice versa.

EXAMPLE:
$$|\mathbf{A}| = \begin{vmatrix} 1 & 2 \\ 3 & 4 \end{vmatrix} = -2.$$

If the two columns are interchanged, the new determinant is
$$|\mathbf{B}| = \begin{vmatrix} 2 & 1 \\ 4 & 3 \end{vmatrix} = 2,$$

and the sign of the determinant is changed.

Next we would like to show that *for a square matrix* **A**, $|\mathbf{A}| = |\mathbf{A}'|$; *that is, the determinant of the transposed matrix is the same as the determinant of the matrix itself.* If $\mathbf{B} = \mathbf{A}'$, then a typical term in the expansion of $|\mathbf{B}|$ is

$$b_{1i} b_{2j} \cdots b_{nr} = a_{i1} a_{j2} \cdots a_{rn}. \tag{3-82}$$

Hence, for every term in $|\mathbf{B}|$ there is precisely the same term in $|\mathbf{A}|$. It only remains to be shown that the signs are the same. Clearly, the number of inversions of the second subscripts of the left-hand side of (3–82) is

the same as the number of inversions of the first subscripts on the right-hand side. If the elements are rearranged on the right-hand side so that the first subscripts are in their natural order, then, by symmetry, the second subscripts have the same number of inversions that the first subscripts had at the outset. Thus the signs of the corresponding terms are the same, and
$$|\mathbf{A}'| = |\mathbf{A}|. \tag{3-83}$$

EXAMPLE:
$$|\mathbf{A}| = \begin{vmatrix} 4 & 1 \\ 3 & 2 \end{vmatrix} = 5, \quad |\mathbf{A}'| = \begin{vmatrix} 4 & 3 \\ 1 & 2 \end{vmatrix} = 5.$$

Equation (3-83) demonstrates clearly that, since an interchange of two columns in **A** changes the sign of $|\mathbf{A}|$, *an interchange of two rows in **A** must also change the sign of* $|\mathbf{A}|$.

The result of the interchange of rows and columns points to another useful property. If a matrix has two rows or two columns which are identical, then the value of its determinant is zero. Suppose that $a_{ik} = a_{jk}$ (all k) or $a_{ik} = a_{ij}$ (all i); if we interchange the two rows or columns which are the same, **A** remains unchanged. However, the sign of the determinant changes:
$$|\mathbf{A}| = -|\mathbf{A}| \quad \text{or} \quad |\mathbf{A}| + |\mathbf{A}| = 0.$$

The only real number for which this equation holds is $|\mathbf{A}| = 0$.

EXAMPLE:
$$\begin{vmatrix} 1 & 1 \\ 2 & 2 \end{vmatrix} = 0.$$

Multiplication of every element of the ith row or the jth column of **A** by a number λ multiplies $|\mathbf{A}|$ by λ. We only have to keep in mind that an element from each row and column appears once and only once in each term of (3-80), so that if, for example, a_{ik} is replaced by λa_{ik} for a given i and all k, then the whole determinant is multiplied by λ. Consequently, if every element of an nth-order matrix **A** is multiplied by λ, then
$$|\lambda \mathbf{A}| = \lambda^n |\mathbf{A}|. \tag{3-84}$$

It is important to note that it is not true that $|\lambda \mathbf{A}| = \lambda |\mathbf{A}|$. If $\lambda = -1$, then
$$|-\mathbf{A}| = (-1)^n |\mathbf{A}|. \tag{3-85}$$

The determinant of a square matrix having a row or column whose elements are all zero clearly is zero. This follows directly from both the definition and the result of (3–84) if λ, the factor multiplying a row or column, is zero.

EXAMPLE:

$$|\mathbf{A}| = \begin{vmatrix} 2 & 3 \\ 1 & 4 \end{vmatrix} = 5, \qquad |\mathbf{B}| = \begin{vmatrix} 4 & 3 \\ 2 & 4 \end{vmatrix} = 10, \qquad |\mathbf{B}| = 2|\mathbf{A}|;$$

$$|2\mathbf{A}| = \begin{vmatrix} 4 & 6 \\ 2 & 8 \end{vmatrix} = 20, \qquad |2\mathbf{A}| = 4|\mathbf{A}|.$$

3–13 Expansion by cofactors. It is not easy to evaluate numerically a determinant by the definition (3–80) if n is large. The task of finding all the permutations and assigning the proper sign is a difficult one. Hence, we shall develop another method of evaluating a determinant, which is of considerable theoretical importance and simplifies the procedure. However, it is by no means efficient. A more adequate numerical procedure is developed in Problem 4–21.

In (3–80) an element from row i of $\mathbf{A}$ appears in every term of the summation. If the elements a_{ij} are factored out, the determinant of $\mathbf{A}$ can be written*

$$|\mathbf{A}| = \sum_{j=1}^{n} a_{ij} A_{ij}, \qquad (3\text{–}86)$$

where i can be any row. Each A_{ij} is the sum of $(n-1)!$ terms involving the products of $n-1$ elements from $\mathbf{A}$, the sum being taken over all possible permutations of the second subscripts. In addition, there is a plus or minus sign attached to each term, depending on whether a_{ij} times that term yields an even or odd permutation of $(1, 2, \ldots, n)$ as the set of second subscripts $(q, k, \ldots, r)$. In A_{ij} there are no elements from row i or column j of $\mathbf{A}$ since these two subscripts appear in a_{ij}, and only one element from each row and column of $\mathbf{A}$ can occur in any term.

EXAMPLE: In expanding a third-order determinant in the form of (3–86), we see from (3–76) that if $i = 1$ (row 1),

$$A_{11} = a_{22}a_{33} - a_{23}a_{32}, \qquad A_{12} = -(a_{21}a_{33} - a_{23}a_{31}),$$

$$A_{13} = a_{21}a_{32} - a_{22}a_{31};$$

---

* It is not easy to follow the material in this section in the abstract. We suggest therefore that the reader simultaneously work through an example such as Problem 3–18. Note that in discussing the second subscripts, we always assume that the first subscripts are in their natural order. Also, when the second subscript j is fixed to position i, then the question whether the permutation of the second subscripts is odd or even depends only on the positions of the $n-1$ second subscripts other than j.

and
$$|\mathbf{A}| = a_{11}A_{11} + a_{12}A_{12} + a_{13}A_{13}.$$

If $i = 3$ (row 3), then

$$A_{31} = a_{12}a_{23} - a_{13}a_{22}, \qquad A_{32} = -(a_{11}a_{23} - a_{13}a_{21}),$$

$$A_{33} = a_{11}a_{22} - a_{12}a_{21},$$

and
$$|\mathbf{A}| = a_{31}A_{31} + a_{32}A_{32} + a_{33}A_{33}.$$

The description of the A_{ij} indicates that A_{ij} can be considered as a determinant of order $n - 1$. All elements of $\mathbf{A}$ appear in A_{ij} except those in row i and column j. In fact, except for a possible difference in sign, A_{ij} is the determinant of the submatrix formed from $\mathbf{A}$ by crossing out row i and column j. If the reader examines the A_{ij} in our example, he will see that in the 3×3 case this is indeed true.

Let us now determine the sign that should be assigned to the determinant of the submatrix obtained by crossing out row i and column j of $\mathbf{A}$ in order to convert it to A_{ij}. We do this most simply by moving row i to row 1 by $i - 1$ interchanges of rows. The other rows, while retaining their original order, will be moved down one. Then, by $j - 1$ interchanges, column j will be moved to the position of column 1. We shall call this new matrix $\mathbf{B}$. Then

$$|\mathbf{B}| = (-1)^{i+j-2}|\mathbf{A}| = (-1)^{i+j}|\mathbf{A}|. \qquad (3\text{–}87)$$

However, the product of the elements on the main diagonal of $\mathbf{B}$ have a plus sign when $|\mathbf{B}|$ is written in the form of (3–80). Now $b_{11} = a_{ij}$, and the remaining elements on the main diagonal of $\mathbf{B}$ are the same as those appearing on the main diagonal of the submatrix whose determinant is A_{ij}. But, from (3–87), $(-1)^{i+j}$ times this term in the expansion of $|\mathbf{A}|$ must be positive. Therefore, A_{ij} is $(-1)^{i+j}$ times the determinant of the submatrix formed from $\mathbf{A}$ by deleting row i and column j. The number A_{ij} is called the cofactor of a_{ij}.

COFACTOR: *The cofactor A_{ij} of the element a_{ij} of any square matrix $\mathbf{A}$ is $(-1)^{i+j}$ times the determinant of the submatrix obtained from $\mathbf{A}$ by deleting row i and column j.*

We have now provided a way to evaluate the A_{ij} in expansion (3–86) of $|\mathbf{A}|$. This important method of evaluating determinants is called *expansion by cofactors*. Thus, for example, expression (3–86) is said to be an expansion by row i of $\mathbf{A}$. To make (3–86) consistent with (3–80) for the case of $n = 1$, the cofactor of the element in a first-order matrix must be defined to be 1 since (3–80) requires that $|\mathbf{A}| = a_{11}$ for the matrix $\mathbf{A} = (a_{11})$.

The use of the expansion by cofactors [Eq. (3–86)] reduces the problem of evaluating an nth-order determinant to that of evaluating n determinants of order $n-1$, which are the cofactors A_{ij} when $j = 1, \ldots, n$. Thus, proceeding by steps, we arrive at an evaluation of $|\mathbf{A}|$. Application of the cofactor-expansion method to the cofactors reduces the task of evaluating each cofactor of order $n-1$ to that of evaluating $n-1$ determinants of order $n-2$, etc. The expansion by cofactors is not a very efficient numerical procedure for evaluating determinants of high order. It is, however, of considerable theoretical value.

We find it useful to give a name to the kth-order determinants, $k = 1, \ldots, n-1$, which appear in the step-by-step evaluation of $|\mathbf{A}|$ by cofactor expansion. They will be referred to as minors of $\mathbf{A}$. More precisely:

MINOR OF ORDER k: *For any $m \times n$ matrix $\mathbf{A}$ consider the kth-order submatrix $\mathbf{R}$ obtained by deleting all but some k rows and k columns of $\mathbf{A}$. Then $|\mathbf{R}|$ is called a kth-order minor of $\mathbf{A}$.*

Note: In this definition $\mathbf{A}$ is not required to be a square matrix.

Instead of writing $|\mathbf{A}|$ in the form (3–86), we could have shown equally well that

$$|\mathbf{A}| = \sum_{i=1}^{n} a_{ij} A_{ij}, \qquad (3\text{–}88)$$

where A_{ij} is again the cofactor of a_{ij}. This is called an expansion of $|\mathbf{A}|$ by column j of $\mathbf{A}$.

EXAMPLES: (1) Let us expand

$$|\mathbf{A}| = \begin{vmatrix} a_{11} & a_{12} & a_{13} \\ a_{21} & a_{22} & a_{23} \\ a_{31} & a_{32} & a_{33} \end{vmatrix}$$

in cofactors by the second row. This expansion should read

$$|\mathbf{A}| = a_{21} A_{21} + a_{22} A_{22} + a_{23} A_{23}.$$

The cofactor A_{ij} is found by crossing out row i and column j and by multiplying the resulting determinant by $(-1)^{i+j}$. Hence

$$A_{21} = (-1) \begin{vmatrix} a_{12} & a_{13} \\ a_{32} & a_{33} \end{vmatrix} = a_{32} a_{13} - a_{12} a_{33},$$

$$A_{22} = \begin{vmatrix} a_{11} & a_{13} \\ a_{31} & a_{33} \end{vmatrix} = a_{11} a_{33} - a_{13} a_{31},$$

$$A_{23} = (-1) \begin{vmatrix} a_{11} & a_{12} \\ a_{31} & a_{32} \end{vmatrix} = a_{12} a_{31} - a_{11} a_{32}.$$

Thus $|\mathbf{A}| = a_{21}(a_{32}a_{13} - a_{12}a_{33}) + a_{22}(a_{11}a_{33} - a_{13}a_{31})$
$+ a_{23}(a_{12}a_{31} - a_{11}a_{32}).$

If the definition of a third-order determinant (3–76) is examined, it will be seen that the above result is identical with (3–76); hence, the correct expansion of $|\mathbf{A}|$ has been obtained.

(2) The method of expansion by cofactors facilitates the evaluation of $|\mathbf{I}_n|$. We expand by the first row. This gives

$$|\mathbf{I}_n| = (1)|\mathbf{I}_{n-1}|.$$

Expanding $|\mathbf{I}_{n-1}|$, etc., in the same way, we finally obtain

$$|\mathbf{I}_n| = (1)(1)\ldots(1) = 1.$$

Thus the determinant of any identity matrix is 1.

3–14 Additional properties of determinants. Expansion by cofactors can be used to prove some additional properties of determinants. Expanding by the first column, we can immediately see that

$$\begin{vmatrix} (\lambda_1 a_{11} + \lambda_2 b_{11} + \lambda_3 c_{11}) & a_{12} & \cdots & a_{1n} \\ (\lambda_1 a_{21} + \lambda_2 b_{21} + \lambda_3 c_{21}) & a_{22} & \cdots & a_{2n} \\ \vdots & & & \vdots \\ (\lambda_1 a_{n1} + \lambda_2 b_{n1} + \lambda_3 c_{n1}) & a_{n2} & \cdots & a_{nn} \end{vmatrix} = \lambda_1 \begin{vmatrix} a_{11} & a_{12} & \cdots & a_{1n} \\ a_{21} & a_{22} & \cdots & a_{2n} \\ \vdots & & & \vdots \\ a_{n1} & a_{n2} & \cdots & a_{nn} \end{vmatrix}$$

$$+ \lambda_2 \begin{vmatrix} b_{11} & a_{12} & \cdots & a_{1n} \\ b_{21} & a_{22} & \cdots & a_{2n} \\ \vdots & & & \vdots \\ b_{n1} & a_{n2} & \cdots & a_{nn} \end{vmatrix} + \lambda_3 \begin{vmatrix} c_{11} & a_{12} & \cdots & a_{1n} \\ c_{21} & a_{22} & \cdots & a_{2n} \\ \vdots & & & \vdots \\ c_{n1} & a_{n2} & \cdots & a_{nn} \end{vmatrix}, \quad (3\text{–}89)$$

since

$$\sum(\lambda_1 a_{i1} + \lambda_2 b_{i1} + \lambda_3 c_{i1})A_{i1} = \lambda_1 \sum a_{i1} A_{i1} + \lambda_2 \sum b_{i1} A_{i1} + \lambda_3 \sum c_{i1} A_{i1}.$$

The same sort of result holds when the ith row or jth column of $\mathbf{A}$ is written as a sum of terms.

EXAMPLE:

$$\begin{vmatrix} 6 & 2 \\ 5 & 1 \end{vmatrix} = \begin{vmatrix} (3+3) & 2 \\ (4+1) & 1 \end{vmatrix} = \begin{vmatrix} 3 & 2 \\ 4 & 1 \end{vmatrix} + \begin{vmatrix} 3 & 2 \\ 1 & 1 \end{vmatrix};$$

$-4 = -5 + 1 = -4.$

$$\begin{vmatrix} 6 & 2 \\ 5 & 1 \end{vmatrix} = \begin{vmatrix} (4+2) & (2+0) \\ 5 & 1 \end{vmatrix} = \begin{vmatrix} 4 & 2 \\ 5 & 1 \end{vmatrix} + \begin{vmatrix} 2 & 0 \\ 5 & 1 \end{vmatrix};$$

$-4 = -6 + 2 = -4.$

It should be observed that if **A**, **B** are nth-order matrices, it is definitely not true that $|\mathbf{A} + \mathbf{B}|$ is the same as $|\mathbf{A}| + |\mathbf{B}|$ in all cases.

$$|\mathbf{A} + \mathbf{B}| \neq |\mathbf{A}| + |\mathbf{B}| \quad \text{(in general)}. \tag{3-90}$$

From result (3–89) and the fact that a determinant vanishes if any two rows or columns are the same, it is easy to see that *adding a multiple of row k to row i ($i \neq k$) or a multiple of column k to column i ($i \neq k$) of* **A** *does not change the value of* $|\mathbf{A}|$.

This can be easily proved: On expanding by row or column i we get $|\mathbf{A}|$ plus a constant times the determinant of a matrix with two identical rows or two identical columns. This determinant vanishes, and hence $|\mathbf{A}|$ is left unchanged. Consequently, if one column (or row) of a square matrix **A** is a linear combination of the other columns (or rows), the value of $|\mathbf{A}|$ is zero.

EXAMPLES:

(1) $\begin{vmatrix} 6 & 2 \\ 5 & 1 \end{vmatrix} = \begin{vmatrix} (6+2\lambda) & 2 \\ (5+\lambda) & 1 \end{vmatrix} = \begin{vmatrix} 6 & 2 \\ 5 & 1 \end{vmatrix} + \lambda \begin{vmatrix} 2 & 2 \\ 1 & 1 \end{vmatrix}.$

(2) $|\mathbf{A}| = \begin{vmatrix} a_{11} a_{12} \cdots a_{1n} \\ a_{21} a_{22} \cdots a_{2n} \\ \vdots \quad \vdots \quad \quad \vdots \\ a_{n1} a_{n2} \cdots a_{nn} \end{vmatrix} = \begin{vmatrix} (a_{11} + 5a_{n1}) & \cdots & (a_{1n} + 5a_{nn}) \\ a_{21} & \cdots & a_{2n} \\ \vdots & & \vdots \\ a_{n1} & \cdots & a_{nn} \end{vmatrix}.$

The fact that a determinant $|\mathbf{A}|$ vanishes if any two rows or columns of **A** are the same, leads us to a very important result concerning the expansion by cofactors. We have shown that

$$|\mathbf{A}| = \sum_j a_{ij} A_{ij} = \sum_i a_{ij} A_{ij}. \tag{3-91}$$

Now consider the expression

$$\sum_j a_{kj} A_{ij} \quad (k \neq i).$$

We are using the cofactors for row i and the elements of row k. This is exactly the expansion of a determinant where rows i and k are the same; hence, the above expression vanishes, that is,

$$\sum_j a_{kj} A_{ij} = \sum_j a_{ji} A_{jk} = 0, \quad (i \neq k). \tag{3-92}$$

Equations (3–91) and (3–92) can be combined:

$$\sum_j a_{ij} A_{kj} = \sum_j a_{ji} A_{jk} = |\mathbf{A}|\, \delta_{ki}, \tag{3-93}$$

where δ_{ki} is the Kronecker delta defined earlier. We have shown that *an expansion by a row or column i in terms of the cofactors of row or column k vanishes when $i \neq k$, and is equal to $|\mathbf{A}|$ when $i = k$.* This result will be used a number of times in the future.

EXAMPLE:

$$|\mathbf{A}| = \begin{vmatrix} a_{11} & a_{12} & a_{13} \\ a_{21} & a_{22} & a_{23} \\ a_{31} & a_{32} & a_{33} \end{vmatrix} ; \quad A_{11} = \begin{vmatrix} a_{22} & a_{23} \\ a_{32} & a_{33} \end{vmatrix}, \quad A_{12} = - \begin{vmatrix} a_{21} & a_{23} \\ a_{31} & a_{33} \end{vmatrix},$$

$$A_{13} = \begin{vmatrix} a_{21} & a_{22} \\ a_{31} & a_{32} \end{vmatrix}.$$

We expand by row 2, using the cofactors of row 1, and obtain, as expected:

$$a_{21}(a_{22}a_{33} - a_{23}a_{32}) - a_{22}(a_{21}a_{33} - a_{23}a_{31}) + a_{23}(a_{21}a_{32} - a_{22}a_{31}) = 0.$$

It will be noted that (3–93) looks very much like a matrix product. If we set $a_{ij}^+ = A_{ji}$, then (3–93) becomes

$$\sum_j a_{ij} a_{jk}^+ = \sum_j a_{kj}^+ a_{ji} = |\mathbf{A}| \, \delta_{ki}. \qquad (3\text{--}94)$$

Thus if

$$\mathbf{A}^+ = \|a_{ij}^+\| = \begin{bmatrix} A_{11} & A_{21} & \cdots & A_{n1} \\ A_{12} & A_{22} & \cdots & A_{n2} \\ \vdots & & & \vdots \\ A_{1n} & A_{2n} & \cdots & A_{nn} \end{bmatrix}, \qquad (3\text{--}95)$$

then Eq. (3–94) can be written in matrix form:

$$\mathbf{A}\mathbf{A}^+ = \mathbf{A}^+\mathbf{A} = |\mathbf{A}|\mathbf{I}_n. \qquad (3\text{--}96)$$

The matrix $\mathbf{A}^+$ is called the adjoint of matrix $\mathbf{A}$. In fact, $\mathbf{A}^+$ is the transpose of the matrix obtained from $\mathbf{A}$ by replacing each element a_{ij} by its cofactor A_{ij}.

*3–15 **Laplace expansion.*** In this section, we shall consider another, more general, technique of expanding a determinant known as the Laplace expansion method, which includes, as a special case, the expansion by cofactors. Instead of expanding by a single row or column, we now expand by several rows or columns. The determinant $|\mathbf{A}|$ is written as the sum of terms, each of which is the product of two determinants.

---

* Starred sections can be omitted without loss of continuity.

We begin by considering the first m rows ($m < n$) and columns of **A**. If we collect the terms containing $a_{11}a_{22} \ldots a_{mm}$ in the expansion of $|\mathbf{A}|$, and factor out this quantity, we are left with

$$\sum (\pm 1) a_{m+1\,i} a_{m+2\,j} \cdots a_{nr}, \qquad (3\text{-}97)$$

the sum being taken over all permutations of the second subscripts. This sum, however, represents the determinant of the submatrix formed from **A** by crossing out the first m rows and columns. Note that the sum (3-97) will be obtained also if we collect all terms containing $a_{1u} a_{2v} \ldots a_{mw}$ [$(u, v, \ldots, w)$ represents a permutation of $(1, 2, \ldots, m)$] and factor out $a_{1u} a_{2v} \ldots a_{mw}$. The sign will alternate, depending on whether the permutation $(u, v, \ldots, w)$ is even or odd. Thus, the following terms appear in the expansion of $|\mathbf{A}|$:

$$\sum (\pm 1) a_{1u} a_{2v} \cdots a_{mw} \sum (\pm 1) a_{m+1\,i} a_{m+2\,j} \cdots a_{nr}, \qquad (3\text{-}98)$$

where

$$\sum (\pm 1) a_{1u} a_{2v} \cdots a_{mw}$$

is the determinant of the submatrix formed from the first m rows and columns of **A**. Expression (3-98) is thus the product of the determinant of the submatrix formed from the first m rows and columns and the determinant of the submatrix formed by crossing out the first m rows and columns. We have the correct sign, since in the expansion of $|\mathbf{A}|$ the term $a_{11} a_{22} \cdots a_{nn}$ has a plus sign.

Next, we shall consider the $m \times m$ submatrix formed from rows $i_1, i_2, \ldots, i_m$ and columns $j_1, j_2, \ldots, j_m$. Except for the sign, the expansion of $|\mathbf{A}|$ will contain the product of the determinant of this submatrix and the determinant of the submatrix formed by crossing out rows $i_1, i_2, \ldots, i_m$ and columns $j_1, j_2, \ldots, j_m$. The sign of the product is determined by the method used in the expansion by cofactors. The $m \times m$ submatrix is moved so that it occupies the first m rows and columns. Let us assume quite logically that $i_1 < i_2 < \cdots < i_m$, and $j_1 < j_2 < \cdots < j_m$. Then after $(i_1 - 1) + (i_2 - 2) + \cdots + (i_m - m)$ interchanges of rows and $(j_1 - 1) + (j_2 - 2) + \cdots + (j_m - m)$ interchanges of columns, the $m \times m$ submatrix formed from rows $i_1, \ldots, i_m$ and columns $j_1, \ldots, j_m$ lies in rows $1, \ldots, m$ and columns $1, \ldots, m$. Furthermore, the order of the remaining columns has not been changed. Once this rearrangement is completed, we are back at the case already considered. The sign depends on

$$\sum_{k=1}^{m} (i_k + j_k) - 2(1 + 2 + \cdots + m).$$

However,

$$1 + 2 + \cdots + m = \tfrac{1}{2} m(m+1).$$

Note: To prove $S = 1 + 2 + \cdots + m = \frac{1}{2}m(m+1)$, simply write

$$S = 1 + 2 + \cdots + m;$$
$$S = m + m - 1 + \cdots + 1;$$

addition yields $2S = m(m+1)$.

Since $m(m+1)$ is always even, the sign to be attached to the product of the two determinants is

$$(-1)^{\sum_{k=1}^{m}(i_k + j_k)}.$$

Two definitions will be helpful.

COMPLEMENTARY MINOR: *Given the nth-order matrix* **A**. *The determinant of the* $(n-m)$*th-order submatrix* **P** *formed by crossing out rows* $i_1, \ldots, i_m$ *and columns* $j_1, \ldots, j_m$ *is called the complementary minor of the mth-order submatrix* **N** *formed from rows* $i_1, \ldots, i_m$ *and columns* $j_1, \ldots, j_m$.

COMPLEMENTARY COFACTOR: *With* **N**, **P** *as defined above, the determinant*

$$|\mathbf{M}| = (-1)^{\sum_{k=1}^{m}(i_k + j_k)} |\mathbf{P}| \qquad (3\text{-}99)$$

is called the complementary cofactor of **N** *in* **A**.

EXAMPLE: Consider

$$\begin{vmatrix} a_{11} & a_{12} & a_{13} & a_{14} & a_{15} \\ a_{21} & a_{22} & a_{23} & a_{24} & a_{25} \\ a_{31} & a_{32} & a_{33} & a_{34} & a_{35} \\ a_{41} & a_{42} & a_{43} & a_{44} & a_{45} \\ a_{51} & a_{52} & a_{53} & a_{54} & a_{55} \end{vmatrix}.$$

The determinant of the submatrix **N** formed from columns 2 and 5 and rows 1 and 3 is

$$|\mathbf{N}| = \begin{vmatrix} a_{12} & a_{15} \\ a_{32} & a_{35} \end{vmatrix}.$$

The complementary minor is

$$|\mathbf{P}| = \begin{vmatrix} a_{21} & a_{23} & a_{24} \\ a_{41} & a_{43} & a_{44} \\ a_{51} & a_{53} & a_{54} \end{vmatrix}.$$

Hence $\sum(i_k + j_k) = 2 + 5 + 1 + 3 = 11$, and the complementary cofactor of **N** is

$$|\mathbf{M}| = -|\mathbf{P}|.$$

From these results we can immediately derive a new way of expanding the determinant $|\mathbf{A}|$. Select any m rows of $\mathbf{A}$. From these m rows we can form $n!/m!(n-m)!$ different mth-order submatrices, where $n!/m!(n-m)!$ defines the number of combinations of n columns of $\mathbf{A}$ taken m at a time. In choosing these submatrices $\mathbf{N}$, we always keep the order of the columns in $\mathbf{A}$ unchanged. For each submatrix $\mathbf{N}$ we find $|\mathbf{N}|$ and the corresponding complimentary cofactor $|\mathbf{M}|$; then we form the product $|\mathbf{N}|\,|\mathbf{M}|$. From $|\mathbf{N}|$, we obtain $m!$ terms and from $|\mathbf{M}|$, $(n-m)!$ terms. Hence each product $|\mathbf{N}|\,|\mathbf{M}|$ yields, with the correct sign, $m!(n-m)!$ terms of the expansion of $|\mathbf{A}|$. In all, there are $n!/m!(n-m)!$ products $|\mathbf{N}|\,|\mathbf{M}|$ which yield a total of

$$\frac{n!}{m!(n-m)!}\,m!(n-m)! = n!\text{ terms in the expansion of }|\mathbf{A}|.$$

Thus, we have obtained all terms in $|\mathbf{A}|$ since our method of selecting the mth-order submatrices from the m rows eliminates any possible repetition of terms. Hence we can write

$$|\mathbf{A}| = \sum_{j_1 < j_2 < \cdots < j_m} |\mathbf{N}(i_1,\ldots,i_m\,|\,j_1,\ldots,j_m)|\,|\mathbf{M}|, \qquad (3\text{–}100)$$

where $|\mathbf{N}(i_1,\ldots,i_m|j_1,\ldots,j_m)|$ is the mth-order determinant of the submatrix formed from rows $i_1,\ldots,i_m$ and columns $j_1,\ldots,j_m$. The sum is taken over the $n!/m!(n-m)!$ choices for $j_1,\ldots,j_m$. The notation $j_1 < j_2 < \cdots < j_m$ indicates that the sum is taken over all choices of the columns such that the column order is maintained. This technique of expansion is called the Laplace method: We select any m rows from $\mathbf{A}$ (note that they do not need to be adjacent). From these m rows we form the $n!/m!(n-m)!$ possible mth-order determinants and find their individual complimentary cofactors. We then multiply the determinant by its complimentary cofactor and add the $n!/m!(n-m)!$ terms to obtain $|\mathbf{A}|$. This is an example of expansion by rows $i_1,\ldots,i_m$.

We can, of course, expand by any m columns and obtain

$$|\mathbf{A}| = \sum_{i_1 < i_2 < \cdots < i_m} |\mathbf{N}(i_1,\ldots,i_m\,|\,j_1,\ldots,j_m)|\,|\mathbf{M}|. \qquad (3\text{–}101)$$

EXAMPLE: Let us expand the determinant

$$\begin{vmatrix} a_{11} & a_{12} & a_{13} & a_{14} \\ a_{21} & a_{22} & a_{23} & a_{24} \\ a_{31} & a_{32} & a_{33} & a_{34} \\ a_{41} & a_{42} & a_{43} & a_{44} \end{vmatrix}$$

by the first and last rows. There will be $4!/2!2! = 6$ terms. The determinants of order 2 which can be formed from rows 1, 4 are

$$|\mathbf{N}_1| = \begin{vmatrix} a_{11} & a_{12} \\ a_{41} & a_{42} \end{vmatrix}, \quad |\mathbf{N}_2| = \begin{vmatrix} a_{11} & a_{13} \\ a_{41} & a_{43} \end{vmatrix}, \quad |\mathbf{N}_3| = \begin{vmatrix} a_{11} & a_{14} \\ a_{41} & a_{44} \end{vmatrix},$$

$$|\mathbf{N}_4| = \begin{vmatrix} a_{12} & a_{13} \\ a_{42} & a_{43} \end{vmatrix}, \quad |\mathbf{N}_5| = \begin{vmatrix} a_{12} & a_{14} \\ a_{42} & a_{44} \end{vmatrix}, \quad |\mathbf{N}_6| = \begin{vmatrix} a_{13} & a_{14} \\ a_{43} & a_{44} \end{vmatrix}.$$

The corresponding complementary cofactors are

$$|\mathbf{M}_1| = \begin{vmatrix} a_{23} & a_{24} \\ a_{33} & a_{34} \end{vmatrix}, \quad |\mathbf{M}_2| = -\begin{vmatrix} a_{22} & a_{24} \\ a_{32} & a_{34} \end{vmatrix}, \quad |\mathbf{M}_3| = \begin{vmatrix} a_{22} & a_{23} \\ a_{32} & a_{33} \end{vmatrix},$$

$$|\mathbf{M}_4| = \begin{vmatrix} a_{21} & a_{24} \\ a_{31} & a_{34} \end{vmatrix}, \quad |\mathbf{M}_5| = -\begin{vmatrix} a_{21} & a_{23} \\ a_{31} & a_{33} \end{vmatrix}, \quad |\mathbf{M}_6| = \begin{vmatrix} a_{21} & a_{22} \\ a_{31} & a_{32} \end{vmatrix},$$

$$|\mathbf{A}| = \sum_{k=1}^{6} |\mathbf{N}_k| \, |\mathbf{M}_k|.$$

The sign of $|\mathbf{M}_2|$, for example, is found as follows:

$$\sum (i_k + j_k) \quad \text{(for } \mathbf{N}_2\text{)} \quad = 4 + 1 + 1 + 3 = 9;$$

hence the minor must be multiplied by (-1), as shown.

*3–16 Multiplication of determinants. There is a simple multiplication relation for the determinant of the product of square matrices. *If* **A**, **B** *are matrices of order n, then, if* **C** = **AB**,

$$|\mathbf{C}| = |\mathbf{A}| \, |\mathbf{B}|, \tag{3-102}$$

that is, the determinant of the product is the product of the determinants.

To prove (3–102), consider the partitioned matrix of size $2n \times 2n$,

$$\mathbf{D} = \begin{bmatrix} \mathbf{A} & \mathbf{0} \\ -\mathbf{I}_n & \mathbf{B} \end{bmatrix}.$$

Applying Laplace's expansion by the last n rows to the determinant $|\mathbf{D}|$, we obtain

$$|\mathbf{D}| = \begin{vmatrix} \mathbf{A} & \mathbf{0} \\ -\mathbf{I}_n & \mathbf{B} \end{vmatrix} = (-1)^{2n^2 + n(n+1)} |\mathbf{A}| \, |\mathbf{B}| = |\mathbf{A}| \, |\mathbf{B}|, \tag{3-103}$$

since the complementary minor of any submatrix including one of the first n columns will have a column of zeros. The determinant is completely independent of the matrix appearing in the lower left of (3–103). Matrix $-\mathbf{I}_n$ was placed there for the following special reason:

Let us take the columns of **B** to be $\mathbf{b}_j$. Then consider

$$\begin{bmatrix} \mathbf{A} \\ -\mathbf{I}_n \end{bmatrix} \mathbf{b}_1 = \begin{bmatrix} \mathbf{A}\mathbf{b}_1 \\ -\mathbf{b}_1 \end{bmatrix}. \tag{3-104}$$

Equation (3–104) is a linear combination of the first n columns of **D**. We can add (3–104) to column $n+1$ of **D** without changing the value of its determinant. This yields

$$\begin{bmatrix} \mathbf{A}\mathbf{b}_1 \\ 0 \end{bmatrix}$$

as the new column $n+1$. Continuing this process and adding

$$\begin{bmatrix} \mathbf{A} \\ -\mathbf{I}_n \end{bmatrix} \mathbf{b}_j = \begin{bmatrix} \mathbf{A}\mathbf{b}_j \\ -\mathbf{b}_j \end{bmatrix} \tag{3-105}$$

to the $n+j$th column of **D**, we get

$$|\mathbf{D}| = \begin{vmatrix} \mathbf{A} & \mathbf{A}\mathbf{B} \\ -\mathbf{I}_n & 0 \end{vmatrix}. \tag{3-106}$$

We expand by the last n rows, using the Laplace expansion, and obtain

$$|\mathbf{D}| = (-1)^{n^2+n(n+1)}|-\mathbf{I}_n|\,|\mathbf{A}\mathbf{B}| = (-1)^{2n(n+1)}|\mathbf{A}\mathbf{B}| = |\mathbf{A}\mathbf{B}|. \tag{3-107}$$

Thus, from (3–103) we have proved that

$$|\mathbf{A}\mathbf{B}| = |\mathbf{A}|\,|\mathbf{B}|. \tag{3-108}$$

EXAMPLE:

$$\mathbf{A} = \begin{bmatrix} 2 & 3 \\ 1 & 4 \end{bmatrix}, \quad \mathbf{B} = \begin{bmatrix} 1 & 6 \\ 3 & 2 \end{bmatrix}, \quad \mathbf{C} = \mathbf{A}\mathbf{B} = \begin{bmatrix} 11 & 18 \\ 13 & 14 \end{bmatrix};$$

$$|\mathbf{A}| = 5, \quad |\mathbf{B}| = -16, \quad |\mathbf{C}| = -80;$$

$$|\mathbf{C}| = -80 = |\mathbf{A}|\,|\mathbf{B}| = 5(-16).$$

*3–17 Determinant of the product of rectangular matrices. Let **A** be an $m \times n$ matrix and **B** an $n \times m$ matrix with $m < n$. If $\mathbf{C} = \mathbf{A}\mathbf{B}$, $|\mathbf{C}|$ is an mth-order determinant. We shall now show how to express $|\mathbf{C}|$ in terms of the determinants of order m which can be formed from **A** and **B**. Consider the product

$$\begin{bmatrix} \mathbf{I}_m & \mathbf{A} \\ 0 & \mathbf{I}_n \end{bmatrix} \begin{bmatrix} \mathbf{A} & 0 \\ -\mathbf{I}_n & \mathbf{B} \end{bmatrix} = \begin{bmatrix} 0 & \mathbf{A}\mathbf{B} \\ -\mathbf{I}_n & \mathbf{B} \end{bmatrix}. \tag{3-109}$$

Equation (3–108) shows that

$$\begin{vmatrix} \mathbf{I}_m & \mathbf{A} \\ 0 & \mathbf{I}_n \end{vmatrix} \begin{vmatrix} \mathbf{A} & 0 \\ -\mathbf{I}_n & \mathbf{B} \end{vmatrix} = \begin{vmatrix} 0 & \mathbf{AB} \\ -\mathbf{I}_n & \mathbf{B} \end{vmatrix}. \qquad (3\text{–}110)$$

Laplace's expansion indicates immediately that the first determinant has the value unity, and hence

$$\begin{vmatrix} \mathbf{A} & 0 \\ -\mathbf{I}_n & \mathbf{B} \end{vmatrix} = (-1)^{n(m+1)} |\mathbf{AB}|. \qquad (3\text{–}111)$$

It is now necessary to evaluate the determinant of order $n+m$ on the left. We shall use Laplace's expansion by the first m rows. First, note that nonvanishing determinants of order m can be formed only from the columns of $\mathbf{A}$, since the use of any other columns would introduce a column of zeros and hence a vanishing determinant. Thus, there are no more than $n!/m!(n-m)!$ nonvanishing terms in the expansion. The complementary minor to any determinant Δ of order m formed from $\mathbf{A}$ will have $n-m$ columns from $-\mathbf{I}_n$. This complementary minor of order n will be of the form

$$|-\mathbf{e}_{u_1} - \mathbf{e}_{u_2}, \ldots, -\mathbf{e}_{u_{n-m}}, \mathbf{B}|, \qquad (3\text{–}112)$$

where $u_1, u_2, \ldots, u_{n-m}$ refer to the columns of $\mathbf{A}$ not in Δ. We can immediately expand by cofactors, proceeding from the first column to the second, etc., ... to the $(n-m)$th column. Note that, aside from sign, this expansion crosses out rows $u_1, u_2, \ldots, u_{n-m}$ of $\mathbf{B}$ so that, in the end, we obtain a determinant of order m formed from $\mathbf{B}$ *which contains the same rows as the corresponding columns chosen from $\mathbf{A}$ to be in Δ*.

Next, we shall discuss the sign problem. First, the sign of the complementary minor is

$$(-1)^{(1/2)m(m+1)+j_1+j_2+\cdots+j_m},$$

where $j_1, \ldots, j_m$ are the indices of the columns chosen from $\mathbf{A}$ to be in Δ. Then, in the expansion of the complementary minor, the minus signs in the $-\mathbf{e}_{u_i}$ yield $(-1)^{n-m}$. Finally there are the signs coming from the expansion by cofactors. These contribute

$$(-1)^{1+u_1}(-1)^{1+u_2-1}\cdots(-1)^{1+u_{n-m}-(n-m-1)}$$
$$= (-1)^{n-m+u_1+\cdots+u_{n-m}-(1/2)(n-m-1)(n-m)}, \qquad (3\text{–}113)$$

since we expand each time by the first column, and the row index is cut down as a result of preceding expansions. Thus the sign is

$$(-1)^{(1/2)m(m+1)-(1/2)(n-m)(n-m-1)+j_1+\cdots+j_m+u_1+\cdots+u_{n-m}+2(n-m)}. \qquad (3\text{–}114)$$

Note that since the indices of all columns of **A** not in $j_1, \ldots, j_m$ appear in $u_1, \ldots, u_{n-m}$, it must be true that

$$j_1 + \cdots + j_m + u_1 + \cdots + u_{n-m} = 1 + 2 + \cdots + n = \tfrac{1}{2}n(n+1).$$
(3-115)

Cancellation shows that the sign depends only on $(-1)^{n(m+1)}$ and not on the m columns chosen to be in Δ. Hence (3-111) demonstrates that in the expansion of $|\mathbf{AB}|$ a plus sign should be associated with each of the $n!/(n-m)!$ products of two mth-order determinants obtained in the expansion of the left-hand side of (3-111).

We have proved that if **A** is $m \times n$ and **B** is $n \times m$ $(m < n)$, then $|\mathbf{AB}|$ can be represented as the sum of $n!/m!(n-m)!$ terms. Each term is the product of two mth-order determinants formed from **A** and from **B**, respectively. A given mth-order determinant formed from the columns $j_1, \ldots, j_m$ of **A** multiplies the determinant formed from rows $j_1, \ldots, j_m$ of **B**.

EXAMPLE:

$$\mathbf{A} = \begin{bmatrix} 1 & 4 & 5 \\ 2 & 0 & 3 \end{bmatrix}, \quad \mathbf{B} = \begin{bmatrix} 3 & 0 \\ 9 & 2 \\ 1 & 7 \end{bmatrix};$$

$$\mathbf{AB} = \begin{bmatrix} 1 & 4 & 5 \\ 2 & 0 & 3 \end{bmatrix} \begin{bmatrix} 3 & 0 \\ 9 & 2 \\ 1 & 7 \end{bmatrix} = \begin{bmatrix} 44 & 43 \\ 9 & 21 \end{bmatrix}, \quad |\mathbf{AB}| = 537.$$

However, $|\mathbf{AB}|$ can be expressed as the sum of three terms, each of which is the product of a 2×2 determinant from **A** and a 2×2 determinant from **B**. The three pairs of determinants which can be formed from **A** and **B**, respectively, are: From columns 1, 2 of **A** and rows 1, 2 of **B**,

$$|\mathbf{A}_1| = \begin{vmatrix} 1 & 4 \\ 2 & 0 \end{vmatrix} = -8, \quad |\mathbf{B}_1| = \begin{vmatrix} 3 & 0 \\ 9 & 2 \end{vmatrix} = 6;$$

from columns 1, 3 of **A** and rows 1, 3 of **B**,

$$|\mathbf{A}_2| = \begin{vmatrix} 1 & 5 \\ 2 & 3 \end{vmatrix} = -7, \quad |\mathbf{B}_2| = \begin{vmatrix} 3 & 0 \\ 1 & 7 \end{vmatrix} = 21;$$

from columns 2, 3 of **A** and rows 2, 3 of **B**,

$$|\mathbf{A}_3| = \begin{vmatrix} 4 & 5 \\ 0 & 3 \end{vmatrix} = 12, \quad |\mathbf{B}_3| = \begin{vmatrix} 9 & 2 \\ 1 & 7 \end{vmatrix} = 61.$$

Hence,

$$|\mathbf{AB}| = |\mathbf{A}_1|\,|\mathbf{B}_1| + |\mathbf{A}_2|\,|\mathbf{B}_2| + |\mathbf{A}_3|\,|\mathbf{B}_3|$$
$$= -8(6) - 7(21) + 12(61) = -48 - 147 + 732 = 537;$$

the results are indeed identical.

3–18 The matrix inverse. Given any real number $a \neq 0$, there is a number a^{-1} (the inverse or reciprocal of a) such that $a^{-1}a = aa^{-1} = 1$. We shall now investigate whether matrices possess this inverse property and, if so, under what circumstances. We ask: For any given matrix $\mathbf{A}$ does there exist a matrix $\mathbf{B}$ such that

$$\mathbf{AB} = \mathbf{BA} = \mathbf{I}? \tag{3-116}$$

If such a matrix $\mathbf{B}$ does exist, it is called the inverse of $\mathbf{A}$. This inverse is usually written $\mathbf{A}^{-1}$ (it should be noted that $\mathbf{A}^{-1}$ does not mean $1/\mathbf{A}$ or $\mathbf{I}/\mathbf{A}$; we have no rules for dividing matrices). $\mathbf{A}^{-1}$ is merely the symbol given to matrix $\mathbf{B}$ in (3–116).

First of all, we note that $\mathbf{A}$ can have an inverse only if $\mathbf{A}$ is a square matrix, since the products $\mathbf{AB}$ and $\mathbf{BA}$ cannot both represent the same identity matrix unless $\mathbf{A}$ is square. Hence the inverse will also be square.

MATRIX INVERSE: *Given a square matrix* $\mathbf{A}$. *If there exists a square matrix* $\mathbf{A}^{-1}$ *which satisfies the relation*

$$\mathbf{A}^{-1}\mathbf{A} = \mathbf{A}\mathbf{A}^{-1} = \mathbf{I}, \tag{3-117}$$

then $\mathbf{A}^{-1}$ *is called the inverse or reciprocal of* $\mathbf{A}$.

Next we wish to compute the inverse when it exists. As a matter of fact, we have already done this at the end of Section 3–14, where we showed that

$$\mathbf{A}\mathbf{A}^+ = \mathbf{A}^+\mathbf{A} = |\mathbf{A}|\mathbf{I}. \tag{3-118}$$

If we define

$$\mathbf{A}^{-1} = \frac{1}{|\mathbf{A}|}\mathbf{A}^+, \tag{3-119}$$

then this matrix satisfies (3–117) and hence is an inverse of $\mathbf{A}$. The inverse can be formed in this way only if $|\mathbf{A}| \neq 0$.

SINGULAR AND NONSINGULAR MATRICES: *The square matrix* $\mathbf{A}$ *is said to be singular if* $|\mathbf{A}| = 0$, *nonsingular if* $|\mathbf{A}| \neq 0$.

We have shown that every nonsingular matrix has an inverse. It is also true that only nonsingular matrices have inverses. (In other words, if $\mathbf{A}$ is singular, there is no matrix $\mathbf{B}$ such that $\mathbf{AB} = \mathbf{BA} = \mathbf{I}$.) We shall prove this fact in Chapter 4.

If **A** has an inverse, then this inverse is unique. To see this, suppose that **B** is a matrix such that $\mathbf{AB} = \mathbf{BA} = \mathbf{I}$; and that, in addition, there is another matrix **D** such that $\mathbf{AD} = \mathbf{DA} = \mathbf{I}$. Multiplying $\mathbf{AB} = \mathbf{I}$ on the left by **D**, we obtain

$$\mathbf{DAB} = \mathbf{DI} = \mathbf{D}. \tag{3-120}$$

However, by assumption, $\mathbf{DA} = \mathbf{I}$, and expression (3–120) reduces to $\mathbf{B} = \mathbf{D}$; hence the inverse is unique, and it is permissible to speak of $\mathbf{A}^{-1}$ as *the* inverse of **A** [see (3–119)]. Furthermore, if **A** is nonsingular and a matrix **B** is such that $\mathbf{AB} = \mathbf{I}$, then $\mathbf{B} = \mathbf{A}^{-1}$, and consequently, $\mathbf{BA} = \mathbf{I}$ also.* To prove this, we only have to multiply $\mathbf{AB} = \mathbf{I}$ on the left by $\mathbf{A}^{-1}$ and recall that $\mathbf{A}^{-1}\mathbf{A} = \mathbf{I}$. In mathematical terminology, the fact that $\mathbf{BA} = \mathbf{I}$ if $\mathbf{AB} = \mathbf{I}$, and vice versa, indicates that a right inverse for **A** is also a left inverse, and vice versa.

EXAMPLES: (1) Let us compute the general formula for the inverse of a 2×2 nonsingular matrix **A**:

$$\mathbf{A} = \begin{bmatrix} a_{11} & a_{12} \\ a_{21} & a_{22} \end{bmatrix}.$$

The matrix of the cofactors (in this case, the cofactors are determinants of first order) is

$$\begin{bmatrix} a_{22} & -a_{21} \\ -a_{12} & a_{11} \end{bmatrix}.$$

Thus the adjoint (transpose of the above matrix) is

$$\mathbf{A}^{+} = \begin{bmatrix} a_{22} & -a_{12} \\ -a_{21} & a_{11} \end{bmatrix},$$

and

$$\mathbf{A}^{-1} = \frac{1}{|\mathbf{A}|} \begin{bmatrix} a_{22} & -a_{12} \\ -a_{21} & a_{11} \end{bmatrix}.$$

To prove that this is the inverse $\mathbf{A}^{-1}$, we only need to show that $\mathbf{A}^{-1}\mathbf{A} = \mathbf{I}$:

$$\mathbf{A}^{-1}\mathbf{A} = \frac{1}{|\mathbf{A}|} \begin{bmatrix} a_{22} & -a_{12} \\ -a_{21} & a_{11} \end{bmatrix} \begin{bmatrix} a_{11} & a_{12} \\ a_{21} & a_{22} \end{bmatrix}$$

$$= \frac{1}{|\mathbf{A}|} \begin{bmatrix} [a_{22}a_{11} - a_{12}a_{21}] & 0 \\ 0 & [a_{22}a_{11} - a_{12}a_{21}] \end{bmatrix} = \begin{bmatrix} 1 & 0 \\ 0 & 1 \end{bmatrix} = \mathbf{I}.$$

---

* In fact, if $\mathbf{AB} = \mathbf{I}$ for two nth-order matrices **A** and **B**, then both are nonsingular and $\mathbf{B} = \mathbf{A}^{-1}$. It is not necessary to assume that **A** is nonsingular since this follows from $\mathbf{AB} = \mathbf{I}$ (see proof in Chapter 4).

(2) Assume
$$\mathbf{A} = \begin{bmatrix} 1 & 0 \\ 2 & 3 \end{bmatrix}.$$

Then:

$$|\mathbf{A}| = 3, \quad \mathbf{A}^+ = \begin{bmatrix} 3 & 0 \\ -2 & 1 \end{bmatrix}; \quad \text{and} \quad \mathbf{A}^{-1} = \begin{bmatrix} 1 & 0 \\ -2/3 & 1/3 \end{bmatrix}.$$

This is easily proved by computing

$$\mathbf{A}^{-1}\mathbf{A} = \begin{bmatrix} 1 & 0 \\ -2/3 & 1/3 \end{bmatrix}\begin{bmatrix} 1 & 0 \\ 2 & 3 \end{bmatrix} = \begin{bmatrix} 1 & 0 \\ 0 & 1 \end{bmatrix} = \mathbf{I}.$$

(3) If $\mathbf{A} = (a_{11})$, then $|\mathbf{A}| = a_{11}$, $\mathbf{A}^+ = (1)$, and

$$\mathbf{A}^{-1} = \left(\frac{1}{a_{11}}\right), \quad a_{11} \neq 0.$$

When a matrix is a single element, the inverse is also a single element, i.e., the reciprocal of that in $\mathbf{A}$.

Note: It is always possible to determine whether a given matrix is the inverse of some matrix; we simply multiply the two matrices and see whether the identity matrix is obtained. (See footnote page 104.)

3–19 Properties of the inverse. We shall discuss now the inverse of the product of two nonsingular matrices $\mathbf{A}$, $\mathbf{B}$. We shall show that the product of two nth-order nonsingular matrices is nonsingular, and

$$(\mathbf{AB})^{-1} = \mathbf{B}^{-1}\mathbf{A}^{-1}. \tag{3–121}$$

We prove this by noting that

$$\mathbf{B}^{-1}\mathbf{A}^{-1}\mathbf{AB} = \mathbf{B}^{-1}\mathbf{IB} = \mathbf{B}^{-1}\mathbf{B} = \mathbf{I},$$

and

$$\mathbf{ABB}^{-1}\mathbf{A}^{-1} = \mathbf{AIA}^{-1} = \mathbf{AA}^{-1} = \mathbf{I}.$$

Thus $\mathbf{B}^{-1}\mathbf{A}^{-1}$ satisfies (3–117); hence it is the unique inverse. In the same way, it can be shown that the product of any finite number of nonsingular matrices is nonsingular and that the inverse is the product of the inverses in reverse order. In Problem 3–34 the reader is required to prove this statement.

EXAMPLE:

$$\mathbf{A} = \begin{bmatrix} 1 & 0 \\ 2 & 3 \end{bmatrix}, \quad \mathbf{B} = \begin{bmatrix} 2 & 5 \\ 2 & 1 \end{bmatrix}, \quad \mathbf{C} = \mathbf{AB} = \begin{bmatrix} 2 & 5 \\ 10 & 13 \end{bmatrix};$$

$$\mathbf{A}^{-1} = \begin{bmatrix} 1 & 0 \\ -2/3 & 1/3 \end{bmatrix}, \quad \mathbf{B}^{-1} = \begin{bmatrix} -1/8 & 5/8 \\ 1/4 & -1/4 \end{bmatrix},$$

$$\mathbf{C}^{-1} = \begin{bmatrix} -13/24 & 5/24 \\ 10/24 & -2/24 \end{bmatrix};$$

$$\mathbf{C}^{-1} = (\mathbf{AB})^{-1} = \mathbf{B}^{-1}\mathbf{A}^{-1} = \begin{bmatrix} -1/8 & 5/8 \\ 1/4 & -1/4 \end{bmatrix} \begin{bmatrix} 1 & 0 \\ -2/3 & 1/3 \end{bmatrix}$$

$$= \begin{bmatrix} -13/24 & 5/24 \\ 10/24 & -2/24 \end{bmatrix}.$$

Observe the similarity between the formulas for the inverse and the transpose of a product.

If $\mathbf{A}$ is nonsingular, then

$$(\mathbf{A}^{-1})^{-1} = \mathbf{A}. \tag{3-122}$$

This follows immediately from $\mathbf{AA}^{-1} = \mathbf{A}^{-1}\mathbf{A} = \mathbf{I}$ if we consider $\mathbf{A}^{-1}$ to be the given matrix rather than $\mathbf{A}$. Hence, $\mathbf{A}$ is the unique inverse of $\mathbf{A}^{-1}$, and (3-122) holds.

The identity matrix is its own inverse, since $\mathbf{I}_n\mathbf{I}_n = \mathbf{I}_n$ implies that $\mathbf{I}_n^{-1} = \mathbf{I}_n$.

Next, we shall demonstrate that the inverse of the transpose is the transpose of the inverse, that is,

$$(\mathbf{A}')^{-1} = (\mathbf{A}^{-1})'. \tag{3-123}$$

We start with

$$\mathbf{AA}^{-1} = \mathbf{A}^{-1}\mathbf{A} = \mathbf{I}.$$

Taking the transpose and noting $\mathbf{I}' = \mathbf{I}$, we obtain

$$(\mathbf{A}^{-1})'\mathbf{A}' = \mathbf{I} = \mathbf{A}'(\mathbf{A}^{-1})'. \tag{3-124}$$

Thus $(\mathbf{A}^{-1})'$ is the inverse of $\mathbf{A}'$, and (3-123) follows.

EXAMPLE:

$$\mathbf{A} = \begin{bmatrix} 2 & 1 \\ 5 & 3 \end{bmatrix}, \quad |\mathbf{A}| = 1, \quad \mathbf{A}^{-1} = \begin{bmatrix} 3 & -1 \\ -5 & 2 \end{bmatrix};$$

$$(\mathbf{A}^{-1})' = \begin{bmatrix} 3 & -5 \\ -1 & 2 \end{bmatrix}, \quad \mathbf{A}' = \begin{bmatrix} 2 & 5 \\ 1 & 3 \end{bmatrix}, \quad (\mathbf{A}')^{-1} = \begin{bmatrix} 3 & -5 \\ -1 & 2 \end{bmatrix}.$$

Hence, as expected, (3–123) holds in this case.

We have previously seen that

$$\mathbf{AB} = \mathbf{0} \tag{3–125}$$

did not necessarily imply that $\mathbf{A} = \mathbf{0}$ or $\mathbf{B} = \mathbf{0}$. If, however, either $\mathbf{A}$ or $\mathbf{B}$ is nonsingular, then the other is a null matrix. Suppose $\mathbf{A}$ is nonsingular, and (3–125) holds. We premultiply by $\mathbf{A}^{-1}$ and obtain

$$\mathbf{A}^{-1}\mathbf{AB} = \mathbf{0} = \mathbf{IB} = \mathbf{B}; \quad \mathbf{B} = \mathbf{0}.$$

Thus $\mathbf{B}$ is a null matrix. The same proof applies to a nonsingular matrix $\mathbf{B}$. Hence the product of two nonsingular matrices *cannot* be a null matrix.

If we write a nonsingular matrix $\mathbf{B}$ as a row of column vectors,

$$\mathbf{B} = (\mathbf{b}_1, \ldots, \mathbf{b}_n), \tag{3–126}$$

and $\mathbf{B}^{-1}$ as a column of row vectors,

$$\mathbf{B}^{-1} = [\boldsymbol{\beta}^1, \ldots, \boldsymbol{\beta}^n], \tag{3–127}$$

then

$$\mathbf{B}^{-1}\mathbf{B} = \|\boldsymbol{\beta}^i \mathbf{b}_j\| = \mathbf{I}. \tag{3–128}$$

Consequently,

$$\boldsymbol{\beta}^i \mathbf{b}_j = \delta_{ij}, \quad \text{all } i, j. \tag{3–129}$$

Equation (3–129) shows that row i of $\mathbf{B}^{-1}$ is orthogonal to every column of $\mathbf{B}$ except column i.

3–20 Computation of the inverse by partitioning. In this section and in Section 3–21 we shall consider special ways of computing the inverse of a nonsingular matrix. These methods are of theoretical interest; furthermore, variations have been used as numerical procedures for inverting matrices on digital computers.

Suppose we have an $n \times n$ nonsingular matrix **M**. Let **M** be partitioned as follows:*

$$\mathbf{M} = \begin{bmatrix} \alpha & \beta \\ \gamma & \delta \end{bmatrix}, \tag{3-130}$$

where α is an $s \times s$ submatrix, β an $s \times m$ submatrix, γ an $m \times s$ submatrix, and δ an $m \times m$ submatrix ($n = m + s$). $\mathbf{M}^{-1}$ exists and will be partitioned in the same way as **M**, that is,

$$\mathbf{M}^{-1} = \begin{bmatrix} \mathbf{A} & \mathbf{B} \\ \mathbf{C} & \mathbf{D} \end{bmatrix}, \tag{3-131}$$

where **A** is $s \times s$, **B** is $s \times m$, **C** is $m \times s$, and **D** is $m \times m$.

Assume that δ has an inverse and that δ^{-1} is known. Then, since $\mathbf{MM}^{-1} = \mathbf{I}$,

$$\begin{bmatrix} \alpha & \beta \\ \gamma & \delta \end{bmatrix} \begin{bmatrix} \mathbf{A} & \mathbf{B} \\ \mathbf{C} & \mathbf{D} \end{bmatrix} = \begin{bmatrix} \mathbf{I}_s & 0 \\ 0 & \mathbf{I}_m \end{bmatrix}. \tag{3-132}$$

Four equations are obtained for the four unknown submatrices **A, B, D, C**:

$$\alpha \mathbf{A} + \beta \mathbf{C} = \mathbf{I}_s, \tag{3-133}$$

$$\alpha \mathbf{B} + \beta \mathbf{D} = 0, \tag{3-134}$$

$$\gamma \mathbf{A} + \delta \mathbf{C} = 0, \tag{3-135}$$

$$\gamma \mathbf{B} + \delta \mathbf{D} = \mathbf{I}_m. \tag{3-136}$$

From (3-135),

$$\mathbf{C} = -\delta^{-1} \gamma \mathbf{A}. \tag{3-137}$$

Substituting this into (3-133), we obtain

$$\alpha \mathbf{A} - \beta \delta^{-1} \gamma \mathbf{A} = \mathbf{I}_s,$$

or by definition of the inverse,

$$\mathbf{A} = (\alpha - \beta \delta^{-1} \gamma)^{-1}. \tag{3-138}$$

Using (3-136), we arrive at

$$\mathbf{D} = \delta^{-1} - \delta^{-1} \gamma \mathbf{B}. \tag{3-139}$$

---

* In this section we have dropped our convention of using upper-case roman letters for submatrices.

From (3-134) and (3-139),

$$\alpha B + \beta \delta^{-1} - \beta \delta^{-1} \gamma B = 0,$$

$$(\alpha - \beta \delta^{-1} \gamma) B = -\beta \delta^{-1}.$$

Using (3-138), we get
$$B = -A\beta \delta^{-1}. \qquad (3\text{-}140)$$

We have obtained four formulas which can be solved sequentially for **A, B, C, D**. They are

$$A = (\alpha - \beta \delta^{-1} \gamma)^{-1}, \qquad (3\text{-}141)$$

$$B = -A\beta \delta^{-1}, \qquad (3\text{-}142)$$

$$C = -\delta^{-1} \gamma A, \qquad (3\text{-}143)$$

$$D = \delta^{-1} - \delta^{-1} \gamma B. \qquad (3\text{-}144)$$

Since M^{-1} exists, the submatrices **A, B, C, D** exist. Hence if δ^{-1} exists, all the operations can be carried out, and **A, B, C, D** can be computed.

Imagine that we wish to invert a matrix by means of the preceding formulas. If we partition the matrix so that δ is of a size which can be inverted easily, then we shall find it difficult to obtain **A** according to (3-141) if the order of **M** is fairly large. We are able to avoid this difficulty by partitioning in two or more steps. If δ is 1×1 or 2×2, δ^{-1} is easily computed. If **A** is 1×1, then it is given by

$$A = (1/\alpha - \beta \delta^{-1} \gamma). \quad \text{(See Section 3-18, Example 3.)}$$

It is also easy to find **A** if it is a 2×2 matrix. For example, if we wish to obtain the inverse of

$$M = \begin{bmatrix} m_{11} & m_{12} & m_{13} & m_{14} & m_{15} \\ m_{21} & m_{22} & m_{23} & m_{24} & m_{25} \\ m_{31} & m_{32} & m_{33} & m_{34} & m_{35} \\ m_{41} & m_{42} & m_{43} & m_{44} & m_{45} \\ m_{51} & m_{52} & m_{53} & m_{54} & m_{55} \end{bmatrix} = \begin{bmatrix} \alpha & \beta \\ \gamma & \delta \end{bmatrix}, \qquad (3\text{-}145)$$

we might start with the above partition. To find δ^{-1}, we might partition δ as

$$\delta = \begin{bmatrix} m_{33} & m_{34} & m_{35} \\ m_{43} & m_{44} & m_{45} \\ m_{53} & m_{54} & m_{55} \end{bmatrix} = \begin{bmatrix} \alpha' & \beta' \\ \gamma' & \delta' \end{bmatrix}. \qquad (3\text{-}146)$$

Now $(\delta')^{-1}$ is easily found, and hence δ^{-1} can be obtained. Using δ^{-1}, we compute M^{-1}. We have simply applied the partitioning procedure in two steps. To invert a large matrix, the same procedure can be repeated a number of times.

EXAMPLE: Let us compute the inverse of

$$\mathbf{M} = \begin{bmatrix} 2 & 1 & 0 \\ \hline 3 & 0 & 5 \\ 7 & 6 & 4 \end{bmatrix} = \begin{bmatrix} \alpha & \beta \\ \gamma & \delta \end{bmatrix}, \qquad \mathbf{M}^{-1} = \begin{bmatrix} \mathbf{A} & \mathbf{B} \\ \mathbf{C} & \mathbf{D} \end{bmatrix}.$$

We begin by partitioning the matrix as shown. Then

$$\delta = \begin{bmatrix} 0 & 5 \\ 6 & 4 \end{bmatrix}; \qquad \delta^{-1} = \begin{bmatrix} -2/15 & 1/6 \\ 1/5 & 0 \end{bmatrix}.$$

From (3–141):

$$\mathbf{A} = \left[2 - (1, 0) \begin{bmatrix} -2/15 & 1/6 \\ 1/5 & 0 \end{bmatrix} \begin{bmatrix} 3 \\ 7 \end{bmatrix} \right]^{-1} = 30/37.$$

Using (3–142), we obtain

$$\mathbf{B} = -(30/37)(1, 0) \begin{bmatrix} -2/15 & 1/6 \\ 1/5 & 0 \end{bmatrix} = (4/37, -5/37).$$

From (3–143):

$$\mathbf{C} = -\begin{bmatrix} -2/15 & 1/6 \\ 1/5 & 0 \end{bmatrix} \begin{bmatrix} 3 \\ 7 \end{bmatrix} (30/37) = \begin{bmatrix} -23/37 \\ -18/37 \end{bmatrix}.$$

Using (3–144), we arrive at

$$\mathbf{D} = \begin{bmatrix} -2/15 & 1/6 \\ 1/5 & 0 \end{bmatrix} - \begin{bmatrix} -2/15 & 1/6 \\ 1/5 & 0 \end{bmatrix} \begin{bmatrix} 3 \\ 7 \end{bmatrix} (4/37, -5/37)$$

$$= \begin{bmatrix} -8/37 & 10/37 \\ 5/37 & 3/37 \end{bmatrix}.$$

Combining all the results, we see that

$$\mathbf{M}^{-1} = 1/37 \begin{bmatrix} 30 & 4 & -5 \\ -23 & -8 & 10 \\ -18 & 5 & 3 \end{bmatrix}.$$

This result can be checked by showing that $\mathbf{M}^{-1}\mathbf{M} = \mathbf{I}$.

As a special case of the general method for inverting a matrix consider

$$\mathbf{M} = \begin{bmatrix} \mathbf{I} & \mathbf{Q} \\ \mathbf{0} & \mathbf{R} \end{bmatrix}, \quad (3\text{–}147)$$

where $\mathbf{R}^{-1}$ exists and is known, and $\mathbf{I}$ is, of course, the identity submatrix. Using the equations previously determined, we find

$$\mathbf{A} = (\mathbf{I} - \mathbf{Q}\mathbf{R}^{-1}\mathbf{0})^{-1} = \mathbf{I},$$

$$\mathbf{B} = -\mathbf{I}\mathbf{Q}\mathbf{R}^{-1} = -\mathbf{Q}\mathbf{R}^{-1},$$

$$\mathbf{C} = -\mathbf{R}^{-1}\mathbf{0}\mathbf{I} = \mathbf{0},$$

$$\mathbf{D} = \mathbf{R}^{-1} - \mathbf{R}^{-1}\mathbf{0}(-\mathbf{Q}\mathbf{R}^{-1}) = \mathbf{R}^{-1}.$$

Therefore,

$$\mathbf{M}^{-1} = \begin{bmatrix} \mathbf{I} & -\mathbf{Q}\mathbf{R}^{-1} \\ \mathbf{0} & \mathbf{R}^{-1} \end{bmatrix}. \quad (3\text{–}148)$$

3–21 Product form of the inverse. Let us suppose that we have a matrix $\mathbf{B}$ and know $\mathbf{B}^{-1}$. A single column in $\mathbf{B}$ is changed, and we desire to find the inverse of the new matrix. Let us write $\mathbf{B}$ as a row of column vectors,

$$\mathbf{B} = (\mathbf{b}_1, \mathbf{b}_2, \ldots, \mathbf{b}_r, \ldots, \mathbf{b}_n). \quad (3\text{–}149)$$

The column $\mathbf{b}_r$ is removed and replaced by column vector $\mathbf{a}$. We wish to compute the inverse of the new matrix $\mathbf{B}_a$, with $\mathbf{a}$ replacing $\mathbf{b}_r$ in column r:

$$\mathbf{B}_a = (\mathbf{b}_1, \mathbf{b}_2, \ldots, \mathbf{b}_{r-1}, \mathbf{a}, \mathbf{b}_{r+1}, \ldots, \mathbf{b}_n). \quad (3\text{–}150)$$

We shall approach the problem by trying to write $\mathbf{a}$ as a linear combination of the columns of $\mathbf{B}$, that is,

$$\mathbf{a} = \sum_{i=1}^{n} y_i \mathbf{b}_i = \mathbf{B}\mathbf{y},$$

$$\mathbf{y} = [y_1, \ldots, y_n]. \quad (3\text{–}151)$$

Since $\mathbf{B}$ has an inverse, we can premultiply (3–151) by $\mathbf{B}^{-1}$ and obtain

$$\mathbf{y} = \mathbf{B}^{-1}\mathbf{a}. \quad (3\text{–}152)$$

Thus, we can indeed express $\mathbf{a}$ as a linear combination of the columns of $\mathbf{B}$.

If $y_r \neq 0$, then

$$\mathbf{b}_r = -\frac{y_1}{y_r}\mathbf{b}_1 - \frac{y_2}{y_r}\mathbf{b}_2 - \cdots - \frac{y_{r-1}}{y_r}\mathbf{b}_{r-1}$$
$$+ \frac{1}{y_r}\mathbf{a} - \frac{y_{r+1}}{y_r}\mathbf{b}_{r+1} - \cdots - \frac{y_n}{y_r}\mathbf{b}_n. \quad (3\text{-}153)$$

Let us define the column vector

$$\boldsymbol{\eta} = \left[-\frac{y_1}{y_r}, -\frac{y_2}{y_r}, \ldots, -\frac{y_{r-1}}{y_r}, \frac{1}{y_r}, -\frac{y_{r+1}}{y_r}, \ldots, -\frac{y_n}{y_r}\right]. \quad (3\text{-}154)$$

With this definition, (3-153) can be written as the matrix product

$$\mathbf{b}_r = \mathbf{B}_a\boldsymbol{\eta}. \quad (3\text{-}155)$$

On the right, we have $\mathbf{B}_a$ and, on the left, one column of matrix $\mathbf{B}$. The question is: How can we replace $\mathbf{b}_r$ on the left by $\mathbf{B}$? Obviously, we must substitute a matrix for the vector $\boldsymbol{\eta}$; this matrix must have $\boldsymbol{\eta}$ as its rth column, and its remaining columns must be such that, when multiplied by $\mathbf{B}_a$, they yield the proper columns of $\mathbf{B}$. Let the other columns in the matrix replacing $\boldsymbol{\eta}$ be $\boldsymbol{\alpha}_1, \boldsymbol{\alpha}_2, \ldots, \boldsymbol{\alpha}_n$. Then

$$\mathbf{B} = (\mathbf{b}_1, \mathbf{b}_2, \ldots, \mathbf{b}_r, \ldots, \mathbf{b}_n) = \mathbf{B}_a(\boldsymbol{\alpha}_1, \boldsymbol{\alpha}_2, \ldots, \boldsymbol{\eta}, \ldots, \boldsymbol{\alpha}_n)$$
$$= (\mathbf{B}_a\boldsymbol{\alpha}_1, \mathbf{B}_a\boldsymbol{\alpha}_2, \ldots, \mathbf{B}_a\boldsymbol{\eta}, \ldots, \mathbf{B}_a\boldsymbol{\alpha}_n). \quad (3\text{-}156)$$

Consequently,

$$\mathbf{b}_1 = \mathbf{B}_a\boldsymbol{\alpha}_1, \quad \mathbf{b}_2 = \mathbf{B}_a\boldsymbol{\alpha}_2, \ldots, \mathbf{b}_r = \mathbf{B}_a\boldsymbol{\eta}, \ldots, \mathbf{b}_n = \mathbf{B}_a\boldsymbol{\alpha}_n. \quad (3\text{-}157)$$

The first column of $\mathbf{B}_a$ is $\mathbf{b}_1$. Hence if $\boldsymbol{\alpha}_1 = \mathbf{e}_1$, then $\mathbf{B}_a\mathbf{e}_1 = \mathbf{b}_1$. Similarly, the nth column of $\mathbf{B}_a$ is $\mathbf{b}_n$, and $\boldsymbol{\alpha}_n = \mathbf{e}_n$. Thus

$$\boldsymbol{\alpha}_1 = \mathbf{e}_1, \quad \boldsymbol{\alpha}_2 = \mathbf{e}_2, \ldots, \boldsymbol{\alpha}_{r-1} = \mathbf{e}_{r-1}, \quad \boldsymbol{\alpha}_{r+1} = \mathbf{e}_{r+1}, \ldots, \boldsymbol{\alpha}_n = \mathbf{e}_n.$$

Define the matrix

$$\mathbf{E} = (\mathbf{e}_1, \mathbf{e}_2, \ldots, \mathbf{e}_{r-1}, \boldsymbol{\eta}, \mathbf{e}_{r+1}, \ldots, \mathbf{e}_n). \quad (3\text{-}158)$$

Equation (3-156) becomes

$$\mathbf{B} = \mathbf{B}_a\mathbf{E},$$

or

$$\mathbf{I} = \mathbf{B}_a\mathbf{E}\mathbf{B}^{-1}. \quad (3\text{-}159)$$

Consequently,

$$\mathbf{B}_a^{-1} = \mathbf{E}\mathbf{B}^{-1}. \quad (3\text{-}160)$$

We see that $\mathbf{B}_a^{-1}$ will exist if $\mathbf{E}\mathbf{B}^{-1}$ or $\mathbf{E}$ does. Matrix $\mathbf{E}$ will certainly exist provided $y_r \neq 0$.

Consider what happens when $y_r = 0$. It means that the vector **a** is a linear combination of the vectors $\mathbf{b}_1, \mathbf{b}_2, \ldots, \mathbf{b}_{r-1}, \mathbf{b}_{r+1}, \ldots, \mathbf{b}_n$. Hence in $\mathbf{B}_a$ the rth column can be represented as a linear combination of the other columns in $\mathbf{B}_a$. From our study of determinants we know that, in this case, $|\mathbf{B}_a| = 0$, and therefore $\mathbf{B}_a$ has no inverse. Thus when $y_r = 0$, the new matrix has no inverse.

Let us review the procedure for computing the inverse of a matrix which differs only in a single column from a matrix whose inverse we know. A matrix $\mathbf{B}_a$ is formed by replacing column r, $\mathbf{b}_r$, of $\mathbf{B}$ with **a**. We know $\mathbf{B}^{-1}$ and wish to determine $\mathbf{B}_a^{-1}$. First, we compute

$$\mathbf{y} = \mathbf{B}^{-1}\mathbf{a}. \qquad (3\text{-}161)$$

Then we form the column vector

$$\boldsymbol{\eta} = \left[-\frac{y_1}{y_r}, -\frac{y_2}{y_r}, \ldots, -\frac{y_{r-1}}{y_r}, \frac{1}{y_r}, -\frac{y_{r+1}}{y_r}, \ldots, -\frac{y_n}{y_r} \right]. \qquad (3\text{-}162)$$

We replace the rth column of the identity matrix $\mathbf{I}_n$ with $\boldsymbol{\eta}$ and obtain $\mathbf{E}$. Then

$$\mathbf{B}_a^{-1} = \mathbf{E}\mathbf{B}^{-1}. \qquad (3\text{-}163)$$

The procedure outlined above resembles to a considerable extent the technique used for inserting a vector into a basis (Section 2–9). It is exactly that. We can easily see that, when $\mathbf{B}$ has an inverse, the columns of $\mathbf{B}$ form a basis for E^n. We only need to show that the columns are linearly independent. Consider the problem of finding λ_i satisfying

$$\sum \lambda_i \mathbf{b}_i = \mathbf{0}.$$

In matrix form, this is

$$\mathbf{B}\boldsymbol{\lambda} = \mathbf{0};$$

however, since $\mathbf{B}^{-1}$ exists,

$$\mathbf{B}^{-1}\mathbf{B}\boldsymbol{\lambda} = \mathbf{B}^{-1}\mathbf{0} = \mathbf{0} = \boldsymbol{\lambda}. \qquad (3\text{-}164)$$

Thus all λ_i are zero, and the $\mathbf{b}_i$ are linearly independent. A new vector **a** now replaces $\mathbf{b}_r$. If $y_r \neq 0$, then the new set of vectors also forms a basis since $\mathbf{B}_a^{-1}$ exists. The interesting connection between inverses, bases, etc., will be explored more thoroughly in the next chapter.

The method discussed above can be used to invert any nonsingular matrix $\mathbf{B} = (\mathbf{b}_1, \ldots, \mathbf{b}_n)$. We start with the identity matrix $\mathbf{I}_n (\mathbf{I}_n^{-1} = \mathbf{I}_n)$

and, for example, replace the first column of $\mathbf{I}_n$ by column 1 of $\mathbf{B}$. The inverse of the new matrix with $\mathbf{b}_1$ in the first column is

$$\mathbf{B}_1^{-1} = \mathbf{E}_1 \mathbf{I} = \mathbf{E}_1. \tag{3-165}$$

Into column 2 of $\mathbf{B}_1$ we insert column 2, $\mathbf{b}_2$, of $\mathbf{B}$ and obtain $\mathbf{B}_2$. The new inverse is

$$\mathbf{B}_2^{-1} = \mathbf{E}_2 \mathbf{B}_1^{-1} = \mathbf{E}_2 \mathbf{E}_1. \tag{3-166}$$

We continue to insert one column at a time until we get

$$\mathbf{B}^{-1} = \mathbf{E}_n \mathbf{E}_{n-1} \cdots \mathbf{E}_2 \mathbf{E}_1 \tag{3-167}$$

and

$$\mathbf{E}_i = (\mathbf{e}_1, \mathbf{e}_2, \ldots, \boldsymbol{\eta}_i, \ldots, \mathbf{e}_n), \tag{3-168}$$

where $\boldsymbol{\eta}_i$ is in the ith column. Furthermore,

$$\boldsymbol{\eta}_i = \left[-\frac{y_{1i}}{y_{ii}}, -\frac{y_{2i}}{y_{ii}}, \ldots, -\frac{y_{i-1,i}}{y_{ii}}, \frac{1}{y_{ii}}, -\frac{y_{i+1,i}}{y_{ii}}, \ldots, -\frac{y_{ni}}{y_{ii}} \right], \tag{3-169}$$

$$\mathbf{y}_i = \mathbf{B}_{i-1}^{-1} \mathbf{b}_i = \mathbf{E}_{i-1} \mathbf{E}_{i-2} \cdots \mathbf{E}_1 \mathbf{I} \mathbf{b}_i, \quad i \geq 2; \quad \mathbf{y}_1 = \mathbf{b}_1. \tag{3-170}$$

When $\mathbf{B}^{-1}$ is expressed as the product of $\mathbf{E}$ matrices as in (3-167), it is called a product form of the inverse. Form (3-167) is not unique. Different orders of insertion of the $\mathbf{b}_i$ may be used and may be required.

EXAMPLE: Compute the inverse of

$$\mathbf{B} = \begin{bmatrix} 3 & 4 & 3 & 9 \\ 8 & 4 & 8 & 3 \\ 9 & 5 & 1 & 9 \\ 5 & 9 & 5 & 5 \end{bmatrix},$$

and write a product form of the inverse. We begin with the identity matrix and insert the first column of $\mathbf{B}$ into column 1 of $\mathbf{I}$. Then

$$\mathbf{y}_1 = \mathbf{I}\mathbf{b}_1 = \mathbf{b}_1 = [3, 8, 9, 5], \quad \boldsymbol{\eta}_1 = [1/3, -8/3, -9/3, -5/3],$$

$$\mathbf{E}_1 = \begin{bmatrix} 1/3 & 0 & 0 & 0 \\ -8/3 & 1 & 0 & 0 \\ -9/3 & 0 & 1 & 0 \\ -5/3 & 0 & 0 & 1 \end{bmatrix}; \quad \mathbf{B}_1^{-1} = \mathbf{E}_1.$$

Inserting the second column of **B** into column 2 of $\mathbf{B}_1$, we obtain

$$\mathbf{y}_2 = \mathbf{E}_1\mathbf{b}_2 = [4/3,\ -20/3,\ -21/3,\ 7/3],$$

$$\boldsymbol{\eta}_2 = [1/5,\ -3/20,\ -21/20,\ 7/20],$$

$$\mathbf{E}_2 = \begin{bmatrix} 1 & 1/5 & 0 & 0 \\ 0 & -3/20 & 0 & 0 \\ 0 & -21/20 & 1 & 0 \\ 0 & 7/20 & 0 & 1 \end{bmatrix}; \quad \mathbf{B}_2^{-1} = \mathbf{E}_2\mathbf{E}_1 = \begin{bmatrix} -1/5 & 1/5 & 0 & 0 \\ 2/5 & -3/20 & 0 & 0 \\ -1/5 & -21/20 & 1 & 0 \\ -13/5 & 7/20 & 0 & 1 \end{bmatrix}.$$

Inserting the third column of **B** into column 3 of $\mathbf{B}_2$, we get

$$\mathbf{y}_3 = \mathbf{B}_2^{-1}\mathbf{b}_3 = [1,\ 0,\ -8,\ 0], \qquad \boldsymbol{\eta}_3 = [1/8,\ 0,\ -1/8,\ 0],$$

$$\mathbf{E}_3 = \begin{bmatrix} 1 & 0 & 1/8 & 0 \\ 0 & 1 & 0 & 0 \\ 0 & 0 & -1/8 & 0 \\ 0 & 0 & 0 & 1 \end{bmatrix};\ \mathbf{B}_3^{-1} = \mathbf{E}_3\mathbf{B}_2^{-1} = \begin{bmatrix} -9/40 & 11/160 & 1/8 & 0 \\ 2/5 & -3/20 & 0 & 0 \\ 1/40 & 21/160 & -1/8 & 0 \\ -13/5 & 7/20 & 0 & 1 \end{bmatrix}.$$

Finally, we insert the last column of **B** into column 4 of $\mathbf{B}_3$ and arrive at

$$\mathbf{y}_4 = \mathbf{B}_3^{-1}\mathbf{b}_4 = [-111/160,\ 63/20,\ -81/160,\ -347/20],$$

$$\boldsymbol{\eta} = (1/347)[-111/8,\ 63,\ -81/8,\ -20],$$

$$\mathbf{E}_4 = (1/347)\begin{bmatrix} 347 & 0 & 0 & -111/8 \\ 0 & 347 & 0 & 63 \\ 0 & 0 & 347 & -81/8 \\ 0 & 0 & 0 & -20 \end{bmatrix};$$

thus

$$\mathbf{B}^{-1} = \mathbf{B}_4^{-1} = \mathbf{E}_4\mathbf{B}_3^{-1} = (1/347)\begin{bmatrix} -42 & 19 & 347/8 & -111/8 \\ -25 & -30 & 0 & 63 \\ 35 & 42 & -347/8 & -81/8 \\ 52 & -7 & 0 & -20 \end{bmatrix}.$$

A product form of the inverse is

$$\mathbf{B}^{-1} = \mathbf{E}_4\mathbf{E}_3\mathbf{E}_2\mathbf{E}_1.$$

The last few sections have amply demonstrated that the computation of the inverse of a matrix, especially when n is large, is indeed an arduous task—even for large high-speed digital computers. Computing $\mathbf{A}^{-1}$ from $\mathbf{A}^{+}/|\mathbf{A}|$ is usually a rather inefficient procedure since it requires computation of a large number of determinants.

*3–22 Matrix series and the Leontief inverse. Polynomials of real numbers can be generalized to infinite power series, that is, to polynomials containing an infinite number of terms. Such a power series can be written

$$\sum_{n=0}^{\infty} \lambda_n x^n. \tag{3-171}$$

For a fixed x the series is said to converge to a limit S if

$$\lim_{N \to \infty} S_N = S, \tag{3-172}$$

where

$$S_N = \sum_{n=0}^{N} \lambda_n x^n. \tag{3-173}$$

Thus the series converges to a limit S if for any $\epsilon > 0$, however small, there is an N_0 such that

$$|S - S_N| < \epsilon, \quad \text{all } N > N_0. \tag{3-174}$$

When the series (3–171) converges to a limit, then we can write

$$S = \sum_{n=0}^{\infty} \lambda_n x^n, \tag{3-175}$$

and S is called the sum of the series. The limit S will, of course, depend on the value of x. Clearly, the series cannot converge unless

$$\lim_{n \to \infty} \lambda_n x^n = 0, \tag{3-176}$$

since (3–174) cannot hold in any other case. However, the validity of (3–176) does not necessarily imply a convergence of the series. A series which does not converge is said to diverge.

EXAMPLE: Consider the geometric series $\sum_{n=0}^{\infty} x^n$, $|x| < 1$:

$$S_N = 1 + x + \cdots + x^N,$$
$$xS_N = x + x^2 + \cdots + x^{N+1}.$$

Thus

$$S_N = \frac{1-x^{N+1}}{1-x}, \qquad \lim_{N\to\infty} S_N = \frac{1}{1-x}, \qquad |x| < 1.$$

Hence, when $|x| < 1$, the geometric series converges to the sum $S = 1/(1-x)$. If $|x| \geq 1$, then $\lim_{N\to\infty} x^N \neq 0$, that is, the series diverges.

It is possible to develop a theory of matrix power series in analogy with series of real numbers. We can work with powers of matrices only when the matrices are square. Thus, if $\mathbf{A}$ is any square matrix, we can consider the series

$$\sum_{n=0}^{\infty} \lambda_n \mathbf{A}^n, \tag{3-177}$$

where the λ_n are scalars. It is convenient to use the definition

$$\mathbf{A}^0 = \mathbf{I}. \tag{3-178}$$

Any discussion of the convergence of matrix series must be preceded by a definition of the limit of a matrix sequence. This limit is defined in terms of the limits of the sequences of the matrix elements. Thus the sequence $\mathbf{A}_n = \|(a_{ij})_n\|$ will be said to approach a limit $\mathbf{A} = \|a_{ij}\|$ if and only if

$$\lim_{n\to\infty} (a_{ij})_n = a_{ij}, \qquad (\text{each } i, j). \tag{3-179}$$

Note that the limit $\mathbf{A}$ exists if and only if each matrix element approaches a unique limit. Thus $\lim_{n\to\infty} \mathbf{A}_n = \mathbf{A}$ means that the limit of each of the n^2 sequences of real numbers defined by (3-179) must exist and be finite.

Having developed the notion of the limit of a matrix sequence, we can now state that series (3-177) converges and sums to the matrix $\mathbf{S}$ if

$$\lim_{N\to\infty} \mathbf{S}_N = \mathbf{S}, \tag{3-180}$$

where

$$\mathbf{S}_N = \sum_{n=0}^{N} \lambda_n \mathbf{A}^n. \tag{3-181}$$

If the series converges, we can write

$$\mathbf{S} = \sum_{n=0}^{\infty} \lambda_n \mathbf{A}^n. \tag{3-182}$$

Hence, the series converges to $\mathbf{S}$ if

$$\lim_{N\to\infty} \left(\sum_{n=0}^{N} \lambda_n \mathbf{A}^n - \mathbf{S} \right) = \mathbf{0}. \tag{3-183}$$

Again, the series cannot converge unless

$$\lim_{n\to\infty} \lambda_n \mathbf{A}^n = \mathbf{0}, \qquad (3\text{-}184)$$

since otherwise (3-183) cannot hold. However, the validity of (3-184) does not imply the convergence of the series (3-177).

We shall not pursue the subject of matrix series any further. Instead we shall terminate our discussion by showing how the inverse of $\mathbf{I} - \mathbf{A}$ can be written as a power series when $\mathbf{A}$ has certain properties (see below). In economics, a square matrix of the form $\mathbf{I} - \mathbf{A}$ appears in connection with the Leontief input-output model discussed in Chapter 1. In a Leontief system, the matrix $\mathbf{A}$ can be defined so that it has the following properties*:

$$0 \leq a_{ij} < 1 \qquad (\text{all } i, j); \qquad (3\text{-}185)$$

$$\sum_{i=1}^{n} a_{ij} < 1 \qquad (\text{all } j). \qquad (3\text{-}186)$$

Our example of a geometric series showed that $(1 - x)\sum_{i=0}^{\infty} x^i = 1$, $|x| < 1$. This suggests that we can write

$$(\mathbf{I} - \mathbf{A})^{-1} = \sum_{k=0}^{\infty} \mathbf{A}^k = \mathbf{I} + \mathbf{A} + \mathbf{A}^2 + \cdots. \qquad (3\text{-}187)$$

This is not true for any arbitrary $\mathbf{A}$, but it holds when $\mathbf{A}$ satisfies (3-185) and (3-186). Note that

$$\mathbf{B}_k = (\mathbf{I} - \mathbf{A})(\mathbf{I} + \mathbf{A} + \mathbf{A}^2 + \cdots + \mathbf{A}^k) = \mathbf{I} - \mathbf{A}^{k+1}. \qquad (3\text{-}188)$$

Consider the matrix sequence $\mathbf{B}_k$. Then $\mathbf{B}_k$ tends to the limit $\mathbf{I}$ as $k \to \infty$ if

$$\lim_{k\to\infty} (\mathbf{B}_k - \mathbf{I}) = \mathbf{0}. \qquad (3\text{-}189)$$

However, from (3-188):

$$\lim_{k\to\infty} (\mathbf{B}_k - \mathbf{I}) = \lim_{k\to\infty} (-\mathbf{A}^{k+1}). \qquad (3\text{-}190)$$

Thus if

$$\lim_{k\to\infty} (-\mathbf{A}^{k+1}) = \mathbf{0}, \qquad (3\text{-}191)$$

---

* This definition requires that the technological coefficients be measured in monetary rather than physical units.

then
$$(\mathbf{I} - \mathbf{A}) \sum_{k=0}^{\infty} \mathbf{A}^k = \mathbf{I}, \tag{3-192}$$

and, by definition of the inverse, (3–187) is correct.

It remains to be shown that (3–191) holds when the elements of the matrix $\mathbf{A}$ satisfy (3–185) and (3–186). Since (3–186) is true, we can write

$$0 \le \sum_{i=1}^{n} a_{ij} \le r < 1 \qquad \text{(all } j\text{)}. \tag{3-193}$$

Consider any element of $\mathbf{A}^2 = \|a_{ij}^{(2)}\|$, that is,

$$a_{ij}^{(2)} = \sum_{k=1}^{n} a_{ik} a_{kj},$$

and

$$\sum_{i=1}^{n} a_{ij}^{(2)} = \sum_{i=1}^{n} \sum_{k=1}^{n} a_{ik} a_{kj} = \sum_{k=1}^{n} \left(\sum_{i=1}^{n} a_{ik} \right) a_{kj} \le r \sum_{k=1}^{n} a_{kj} \le r^2. \tag{3-194}$$

Hence,
$$\sum_{i=1}^{n} a_{ij}^{(2)} \le r^2 \qquad \text{(all } j\text{)},$$

which implies, since all the elements satisfy (3–185), that

$$a_{ij}^{(2)} \le r^2 \qquad \text{(all } i, j\text{)}. \tag{3-195}$$

Continuing in the same manner, we find that $a_{ij}^{(k)}$, that is, any element of $\mathbf{A}^k$ satisfies

$$a_{ij}^{(k)} \le r^k.$$

Consequently, as $k \to \infty$, each $a_{ij}^{(k)} \to 0$, and (3–191) is valid. We have proved that if (3–185) and (3–193) hold, (3–187) is true also.

Writing $(\mathbf{I} - \mathbf{A})^{-1}$ in the form of a power series (3–187) is of advantage, particularly when $\mathbf{I} - \mathbf{A}$ is of a high order; in such cases, the standard numerical methods of inversion can introduce, through rounding off, rather large errors, implying that $(\mathbf{I} - \mathbf{A})^{-1}$ cannot be determined accurately. This problem of rounding off is considerably reduced if $(\mathbf{I} - \mathbf{A})^{-1}$ is evaluated by selecting a finite number of terms in the power series expansion. The inverse can be computed to any desired degree of accuracy by taking a sufficient number of terms in the expansion. Unfortunately, the series does not always converge rapidly, and a rather large number of terms must be used to obtain a satisfactory degree of accuracy. Berger and Saibel [3]* discuss other power series expansions of $(\mathbf{I} - \mathbf{A})^{-1}$ which converge more rapidly.

---

* Numbers in brackets denote bibliographical references.

References

1. A. C. Aitken, *Determinants and Matrices*. Edinburgh: Oliver and Boyd, 1948.

This little book gives an excellent, concise discussion of matrices and determinants. It is quite readable and approaches the subject from an elementary point of view.

2. R. G. D. Allen, *Mathematical Economics*. London: Macmillan, 1956.
3. W. J. Berger and E. Saibel, "Power Series Inversions of the Leontief Matrix." *Econometrica*, **25**: 1, 1957, pp. 154–165.
4. G. Birkhoff and S. MacLane, *A Survey of Modern Algebra*. New York: Macmillan, 1941.
5. R. Courant and D. Hilbert, *Methods of Mathematical Physics*, Volume I. New York: Interscience, 1953.

Chapter 1 of this famous work gives a brief but powerful discussion of linear algebra. Power series inversions of matrices of the form $\mathbf{I} - \lambda\mathbf{A}$ are also considered.

6. W. L. Ferrar, *Algebra*. Oxford: Oxford University Press, 1941.

This book gives an elementary, clear discussion of matrices, determinants, and quadratic forms.

7. E. A. Guillemin, *The Mathematics of Circuit Analysis*. New York: Wiley, 1949.

Guillemin discusses matrix theory and its application to circuit analysis. A good intuitive feeling for the subject is conveyed to the reader.

8. P. R. Halmos, *Finite-Dimensional Vector Spaces*. Princeton: Van Nostrand, 1958.
9. F. B. Hildebrand, *Methods of Applied Mathematics*. Englewood Cliffs: Prentice-Hall, 1952.

Chapter 1 gives an excellent treatment of matrices and determinants. It is easy to understand and covers a number of useful topics often absent from pure mathematics texts.

10. P. LeCorbeiller, *Matrix Analysis of Electrical Networks*. New York: Wiley, 1950.
11. O. Morgenstern, (ed.), *Economic Activity Analysis*. New York: Wiley, 1954.

Part II presents a rather complete study of the properties of Leontief matrices.

12. S. Perlis, *Theory of Matrices*. Reading, Mass.: Addison-Wesley, 1952.
13. R. M. Thrall and L. Tornheim, *Vector Spaces and Matrices*. New York: Wiley, 1957.

Problems

3-1. When addition is defined, add **A** and **B** in the following cases:

(a) $\mathbf{A} = \begin{bmatrix} 3 & 4 & 5 \\ 2 & 1 & 6 \end{bmatrix}$, $\mathbf{B} = \begin{bmatrix} 9 & 7 & 2 \\ 0 & 1 & 8 \end{bmatrix}$;

(b) $\mathbf{A} = [3, 2]$, $\mathbf{B} = \begin{bmatrix} 4 \\ 6 \end{bmatrix}$;

(c) $\mathbf{A} = \begin{bmatrix} a_{11} & a_{12} \\ a_{21} & a_{22} \end{bmatrix}$, $\mathbf{B} = \begin{bmatrix} b_{11} & b_{12} \\ b_{21} & b_{22} \end{bmatrix}$;

(d) $\mathbf{A} = \begin{bmatrix} 4 \\ 5 \\ 6 \end{bmatrix}$, $\mathbf{B} = \begin{bmatrix} 7 \\ 8 \\ 9 \end{bmatrix}$;

(e) $\mathbf{A} = \begin{bmatrix} 0 & 1 \\ 0 & 6 \end{bmatrix}$, $\mathbf{B} = \begin{bmatrix} 7 & 4 \\ 1 & 0 \end{bmatrix}$;

(f) $\mathbf{A} = \begin{bmatrix} 1 & 0 & 0 \\ 0 & 1 & 0 \\ 0 & 0 & 1 \end{bmatrix}$, $\mathbf{B} = \begin{bmatrix} 0 & 0 & 0 & 0 \\ 0 & 0 & 0 & 0 \\ 0 & 0 & 0 & 0 \end{bmatrix}$.

3-2. Let $\mathbf{A} = \|a_{ij}\|$ be an $m \times n$ matrix and $\mathbf{B} = \|b_{ij}\|$ an $m \times n$ matrix. Show by actually writing out the sums that $\mathbf{A} + \mathbf{B} = \mathbf{B} + \mathbf{A}$.

3-3. Let **A, B, C** be $m \times n$ matrices. Show by actually writing out the sums that $\mathbf{A} + (\mathbf{B} + \mathbf{C}) = (\mathbf{A} + \mathbf{B}) + \mathbf{C}$.

3-4. Consider the addition of sets of linear equations of the form $\mathbf{Ax} = \mathbf{b}$, $\mathbf{Bx} = \mathbf{d}$ and show that our definitions of matrix addition and multiplication by a scalar are consistent with the corresponding operations on sets of simultaneous linear equations. In addition, show how the restrictions on the number of rows and columns are a logical development. Note carefully what addition means in this context. If **x** satisfies both $\mathbf{Ax} = \mathbf{b}$ and $\mathbf{Bx} = \mathbf{d}$, then **x** is a solution to $(\mathbf{A} + \mathbf{B})\mathbf{x} = \mathbf{c} + \mathbf{d}$.

3-5. Find the products **AB** and **BA** (when they are defined):

(a) $\mathbf{A} = \begin{bmatrix} a_{11} & a_{12} \\ a_{21} & a_{22} \end{bmatrix}$, $\mathbf{B} = \begin{bmatrix} b_{11} & b_{12} \\ b_{21} & b_{22} \end{bmatrix}$;

(Cont.)

(b) $\mathbf{A} = \begin{bmatrix} 1 & 2 & 4 & 5 \\ 3 & 1 & 0 & 2 \end{bmatrix}$, $\quad \mathbf{B} = \begin{bmatrix} 1 \\ 4 \\ 8 \\ 9 \end{bmatrix}$;

(c) $\mathbf{A} = \begin{bmatrix} 1 & 3 & 4 \\ 2 & 0 & 7 \\ 5 & 6 & 9 \end{bmatrix}$, $\quad \mathbf{B} = \begin{bmatrix} 10 & 2 & 0 \\ 7 & 1 & 3 \\ 4 & 5 & 6 \end{bmatrix}$;

(d) $\mathbf{A} = \begin{bmatrix} 1 \\ 0 \\ 7 \\ 8 \end{bmatrix}$, $\quad \mathbf{B} = (2, 4, 9, 6, 5, 0)$;

(e) $\mathbf{A} = \begin{bmatrix} 1 \\ 0 \\ 7 \\ 8 \end{bmatrix}$, $\quad \mathbf{B} = (2, 4, 9, 6)$;

(f) $\mathbf{A} = \begin{bmatrix} 3 & 1 \\ 2 & 4 \\ 5 & 6 \end{bmatrix}$, $\quad \mathbf{B} = \begin{bmatrix} 1 \\ 2 \end{bmatrix}$.

3–6. Show by actual computation that $(\mathbf{AB})\mathbf{C} = \mathbf{A}(\mathbf{BC})$ when

$$\mathbf{A} = \begin{bmatrix} 2 & 1 \\ 3 & 4 \end{bmatrix}, \quad \mathbf{B} = \begin{bmatrix} 0 & 3 & 7 \\ 1 & 8 & 9 \end{bmatrix}, \quad \mathbf{C} = \begin{bmatrix} 3 & 7 & 1 \\ 2 & 6 & 1 \\ 1 & 4 & 0 \end{bmatrix}.$$

3–7. Given the diagonal matrices $\mathbf{A} = \|a_i \, \delta_{ij}\|$, $\mathbf{B} = \|b_i \, \delta_{ij}\|$, compute $\mathbf{AB}$ and $\mathbf{BA}$. What is true of $\mathbf{AB}$ and $\mathbf{BA}$?

3–8. Prove that, when the operations are defined, it is always true that $(\mathbf{A} + \mathbf{B})\mathbf{C} = \mathbf{AC} + \mathbf{BC}$.

3–9. Given matrices $\mathbf{A}, \mathbf{B}, \mathbf{C}, \mathbf{D}$, and assuming that all operations are defined, prove from the definition of multiplication that

$$(\mathbf{A} + \mathbf{B})(\mathbf{C} + \mathbf{D}) = \mathbf{A}(\mathbf{C} + \mathbf{D}) + \mathbf{B}(\mathbf{C} + \mathbf{D}) = \mathbf{AC} + \mathbf{AD} + \mathbf{BC} + \mathbf{BD}.$$

Under what conditions are all the operations defined?

3–10. Given matrices **A**, **B**; under what conditions do the following equations hold?

(a) $(\mathbf{A} + \mathbf{B})^2 = \mathbf{A}^2 + 2\mathbf{AB} + \mathbf{B}^2$;

(b) $(\mathbf{A} + \mathbf{B})(\mathbf{A} - \mathbf{B}) = (\mathbf{A} - \mathbf{B})(\mathbf{A} + \mathbf{B}) = \mathbf{A}^2 - \mathbf{B}^2$.

3–11. Given an nth-order diagonal matrix $\mathbf{D} = \|d_i \delta_{ij}\|$ and an $n \times n$ matrix **A**. If $\mathbf{AD} = \mathbf{DA}$, what type of matrix is **A**? Consider a case where all d_i are different, and a case where some d_i may be the same. When two or more d_i are equal, assume that they fall in consecutive rows. Sketch the structure of **A** when some d_i are equal.

3–12. Given two symmetric matrices **A**, **B** of order n. When is the product **AB** also symmetric?

3–13. For any matrix **A** show that $\mathbf{A'A}$ is defined and is a symmetric matrix.

3–14. Given a symmetric matrix **A** and a skew-symmetric matrix **B**, both of order n. Show that **AB** is skew-symmetric if **A** and **B** commute.

3–15. Given two skew-symmetric matrices **A** and **B** of order n. Show that **AB** is skew-symmetric if and only if $\mathbf{AB} = -\mathbf{BA}$. Matrices for which $\mathbf{AB} = -\mathbf{BA}$ are said to anticommute. When is the product of two skew-symmetric matrices symmetric?

3–16. Given the following matrices:

$$\mathbf{A} = \begin{bmatrix} 2 & 1 \\ 3 & 4 \end{bmatrix}, \quad \mathbf{B} = \begin{bmatrix} 1 & 7 & 2 \\ 0 & 6 & 5 \end{bmatrix}.$$

Show by direct computation that $\mathbf{AB} = (\mathbf{Ab}_1, \mathbf{Ab}_2, \mathbf{Ab}_3)$. The $\mathbf{b}_i$ are, of course, the column vectors of **B**.

3–17. Matrices **A** and **B** are partitioned as shown:

$$\mathbf{A} = \begin{bmatrix} 3 & 2 & | & 1 & 4 \\ 4 & 6 & | & 5 & 0 \\ \hline 7 & 1 & | & 0 & 2 \end{bmatrix}, \quad \mathbf{B} = \begin{bmatrix} 1 & 7 \\ 0 & 6 \\ \hline 1 & 2 \\ 5 & 1 \end{bmatrix}.$$

Prove by direct multiplication and by block multiplication that the same result **AB** is obtained either way.

3–18. Write out the expansion of the general fourth-order determinant (use the basic definition). Combine terms so that an expansion by the second row is obtained. Show that the cofactors A_{2j} are simply $(-1)^{j+2}$ times the third-order determinants which were arrived at by crossing out the second row and jth column. Verify thus the expansion by cofactors for this special case.

3–19. Evaluate

$$\begin{vmatrix} 4 & 1 & 2 \\ 7 & 3 & 5 \\ 1 & 6 & 6 \end{vmatrix}.$$

(a) Expand by the first row;
(b) expand by the second column.

3–20. By actual computation show that $|\mathbf{A}| = |\mathbf{A}'|$ when

$$|\mathbf{A}| = \begin{vmatrix} 4 & 1 & 6 \\ 7 & 2 & 9 \\ 3 & 0 & 8 \end{vmatrix}.$$

3–21. Using cofactor expansion by the first row, show that

$$\begin{vmatrix} 1 & 3 & 1 \\ 7 & 0 & 7 \\ 6 & 4 & 6 \end{vmatrix} = 0.$$

3–22. Show by computation that

$$\begin{vmatrix} 5 & 6 & 1 \\ 5 & 2 & 3 \\ 11 & 5 & 0 \end{vmatrix} = \begin{vmatrix} 3 & 6 & 1 \\ 4 & 2 & 3 \\ 8 & 5 & 0 \end{vmatrix} + \begin{vmatrix} 2 & 6 & 1 \\ 1 & 2 & 3 \\ 3 & 5 & 0 \end{vmatrix}.$$

3–23. Show by computation that

$$\begin{vmatrix} 4 & 6 & 0 \\ 2 & 7 & 3 \\ 1 & 5 & 1 \end{vmatrix} = \begin{vmatrix} 4 & 6 & 0 \\ 2+2(3) & 7 & 3 \\ 1+2(1) & 5 & 1 \end{vmatrix}.$$

3–24. Show by computation that

$$\begin{vmatrix} 1 & 3 & 4 \\ 7 & 3 & 1 \\ 2 & 9 & 4 \end{vmatrix} = - \begin{vmatrix} 4 & 3 & 1 \\ 1 & 3 & 7 \\ 4 & 9 & 2 \end{vmatrix}.$$

3–25. Prove by induction that for any positive integer n (when operations are defined),

$$(\mathbf{A}_1 \mathbf{A}_2 \cdots \mathbf{A}_n)' = \mathbf{A}_n' \cdots \mathbf{A}_2' \mathbf{A}_1'.$$

Note: To prove by induction that a relation holds true for all positive integers n, we show first that the relation is true for $n = 1$. Then we demonstrate that, if the relation holds for $n - 1$, it is also valid for n.

3–26. Prove that the only $n \times n$ matrices which commute with all other $n \times n$ matrices are scalar matrices.

3–27. Consider a product of matrices defined as $c_{ij} = a_{ij} b_{ij}$ (suggested in this chapter and then set aside). What are the conditions under which multiplication is defined? Is the multiplication commutative, associative, and distributive?

3-28. Let
$$\mathbf{A} = \begin{bmatrix} 2 & 1 \\ 3 & 4 \end{bmatrix}.$$

Evaluate the polynomial $\lambda_2 \mathbf{A}^2 + \lambda_1 \mathbf{A} + \lambda_0 \mathbf{I}$ when $\lambda_2 = 2, \lambda_1 = 3, \lambda_0 = 5$.

3-29. Show that if $\mathbf{AB} = 0$ and $\mathbf{B} \neq 0$, there is no matrix $\mathbf{C}$ such that $\mathbf{CA} = \mathbf{I}$.

3-30. Expand $(\mathbf{A} + \mathbf{B})^4$ and $(\mathbf{A} - \mathbf{B})^4$. Be careful to note that $\mathbf{A}$ and $\mathbf{B}$ do not commute in general.

3-31. Prove that the determinant of the scalar matrix $\lambda \mathbf{I}_n$ is λ^n and that the determinant of a diagonal matrix is the product of the diagonal elements.

3-32. A triangular matrix is defined as one where all elements below (or equivalently above) the main diagonal are zero. Such a matrix is square, and $a_{ij} = 0$ $(i > j)$. Prove that the determinant of a triangular matrix is the product of the elements on the main diagonal.

3-33. If two nth-order matrices $\mathbf{A}$ and $\mathbf{B}$ differ only in their jth column, prove that
$$2^{1-n} |\mathbf{A} + \mathbf{B}| = |\mathbf{A}| + |\mathbf{B}|.$$

3-34. Prove by induction that for the product of nonsingular matrices,
$$(\mathbf{A}_1 \mathbf{A}_2 \cdots \mathbf{A}_n)^{-1} = \mathbf{A}_n^{-1} \mathbf{A}_{n-1}^{-1} \cdots \mathbf{A}_2^{-1} \mathbf{A}_1^{-1}.$$

3-35. Prove that the inverse of a nonsingular symmetric matrix is symmetric.

3-36. Prove that the inverse of a nonsingular skew-symmetric matrix is skew-symmetric.

3-37. Prove that every skew-symmetric matrix of odd order is singular.

3-38. If $\mathbf{A}^+$ is the adjoint of a symmetric matrix $\mathbf{A}$, prove that $\mathbf{A}^+$ is symmetric, that is, the cofactor of a_{ij} is the same as the cofactor of a_{ji}.

3-39. Show by an example that, in general,
$$(\mathbf{A} + \mathbf{B})^{-1} \neq \mathbf{A}^{-1} + \mathbf{B}^{-1}.$$

3-40. Using $\mathbf{A}^{-1} = (1/|\mathbf{A}|)\mathbf{A}^+$, compute the inverse of the following matrices:

(a) $\mathbf{A} = \begin{bmatrix} 2 & 1 \\ 3 & 4 \end{bmatrix}$; (b) $\mathbf{A} = \begin{bmatrix} 4 & 1 & 2 \\ 0 & 1 & 0 \\ 8 & 4 & 5 \end{bmatrix}$; (c) $\mathbf{A} = \begin{bmatrix} 5 & 1 \\ 6 & 3 \end{bmatrix}$; (d) $\mathbf{A} = (4)$.

3-41. Show that the inverse of the scalar matrix $\mathbf{S} = \lambda \mathbf{I}$ is $\mathbf{S}^{-1} = (1/\lambda)\mathbf{I}$.

3-42. Show that the inverse of the diagonal matrix $\mathbf{D} = \|\lambda_i \delta_{ij}\|$ is $\mathbf{D}^{-1} = \|(1/\lambda_i) \delta_{ij}\|$.

3-43. Show that the inverse of a nonsingular triangular matrix $\mathbf{T}$ is triangular and, by considering $\mathbf{T}^{-1} \mathbf{T} = \mathbf{I}$, obtain a set of equations which can be solved sequentially to yield the elements of $\mathbf{T}^{-1}$. Illustrate this by computing the inverse of

$$\mathbf{T} = \begin{bmatrix} a_{11} & a_{12} & a_{13} \\ 0 & a_{22} & a_{23} \\ 0 & 0 & a_{33} \end{bmatrix}.$$

In particular, observe that if a_{ii} is a diagonal element of $\mathbf{T}$, and t_{ii} is the corresponding diagonal element of $\mathbf{T}^{-1}$, then $t_{ii} = 1/a_{ii}$.

3–44. Show that interchanging rows in a nonsingular matrix $\mathbf{A}$ interchanges the corresponding columns of $\mathbf{A}^{-1}$.

3–45. If $\mathbf{A}$ is a given matrix and $\mathbf{B}$ is a nonsingular matrix (both of nth-order), prove that there is one and only one matrix $\mathbf{X}$ such that $\mathbf{A} = \mathbf{BX}$, and only one matrix $\mathbf{Y}$ such that $\mathbf{A} = \mathbf{YB}$. In fact,

$$\mathbf{X} = \mathbf{B}^{-1}\mathbf{A}; \qquad \mathbf{Y} = \mathbf{AB}^{-1}.$$

3–46. Compute the inverse of the partitioned matrix

$$\hat{\mathbf{B}} = \begin{bmatrix} \mathbf{B} & \mathbf{0} \\ \mathbf{D} & \mathbf{1} \end{bmatrix}$$

when $\mathbf{B}$ is $n \times n$, $\mathbf{D}$ is $1 \times n$, $\mathbf{0}$ is $n \times 1$, $\mathbf{1}$ is 1×1.

3–47. Compute by partitioning the inverse of the following matrix. Also find a product form of the inverse.

$$\mathbf{A} = \begin{bmatrix} 4 & 2 & 1 & 8 \\ 7 & 9 & 4 & 3 \\ 1 & 0 & 5 & 2 \\ 6 & 6 & 1 & 7 \end{bmatrix}.$$

3–48. Obtain a product form of the inverse of

$$\mathbf{A} = \begin{bmatrix} 0 & 1 & 2 \\ 3 & 9 & 7 \\ 2 & 1 & 6 \end{bmatrix}.$$

3–49. If $\mathbf{B}$ is written as a column of row vectors $\mathbf{B} = [\mathbf{b}^1, \ldots, \mathbf{b}^m]$, prove that

$$\mathbf{BC} = [\mathbf{b}^1\mathbf{C}, \ldots, \mathbf{b}^m\mathbf{C}],$$

when the multiplication is defined.

3–50. Show that if $\mathbf{A}$ is the partitioned matrix

$$\mathbf{A} = \begin{bmatrix} \mathbf{A}_{11} & \mathbf{A}_{12} \\ \mathbf{A}_{21} & \mathbf{A}_{22} \end{bmatrix}, \quad \text{then} \quad \mathbf{A}' = \begin{bmatrix} \mathbf{A}'_{11} & \mathbf{A}'_{21} \\ \mathbf{A}'_{12} & \mathbf{A}'_{22} \end{bmatrix}.$$

3–51. Consider that the $m \times n$ matrix $\mathbf{E}_{mn}$ is defined as a matrix all of whose elements have the value of unity. For example, the 2×2 matrix $\mathbf{E}_{22}$ would be

$$\mathbf{E}_{22} = \begin{bmatrix} 1 & 1 \\ 1 & 1 \end{bmatrix}.$$

What is the matrix $\mathbf{E}_{nm}\mathbf{A}$ for any $m \times r$ matrix $\mathbf{A}$? What is the matrix $\mathbf{A}\mathbf{E}_{rn}$?

*3–52. Consider the matrix

$$\mathbf{A} = \begin{bmatrix} 0.1 & 0.2 \\ 0.3 & 0.2 \end{bmatrix}.$$

Evaluate $(\mathbf{I} - \mathbf{A})^{-1}$. Eliminating the remaining terms in the expansion of the inverse, compute $\mathbf{I} + \mathbf{A} + \mathbf{A}^2$.

3–53. Consider all operations defined for vectors and matrices and show in detail that there is a complete equivalence between vectors and matrices of one row or column.

3–54. How does a symmetric matrix simplify the procedure of computing the inverse of a matrix by partitioning (discussed in Section 3–20)? List the simplifications.

3–55. Show that there is a complete correspondence between the scalar matrices $\lambda \mathbf{I}_n$ and the real numbers λ. Consider addition, multiplication, inverses, etc. A correspondence of the type $\lambda \mathbf{I}_n \to \lambda$ is called an *isomorphism*. The systems represent indeed the same thing, except for notation. For example, show that if $\lambda_1 \lambda_2 = \lambda_3$, then $\lambda_1 \mathbf{I}_n (\lambda_2 \mathbf{I}_n) = \lambda_3 \mathbf{I}_n$.

3–56. Consider the 1×1 matrices (λ) containing the single element λ. By examining all the rules for matrix operations, show that there is no difference between a 1×1 matrix (λ) and the real number λ, that is, we can write $(\lambda) = \lambda$.

3–57. Show that if $\mathbf{A}$ is skew-symmetric, $\mathbf{B}'\mathbf{A}\mathbf{B}$ is also skew-symmetric.

3–58. Find the matrix $\mathbf{B}$ whose inverse is

$$\mathbf{B}^{-1} = \begin{bmatrix} 4 & 3 & 6 \\ 1 & 5 & 7 \\ 2 & 9 & 1 \end{bmatrix}.$$

3–59. In Problem 3–58, we replace the first column of $\mathbf{B}$ by the vector $\mathbf{a} = [2, 0, 7]$. Compute the inverse of the new matrix $\mathbf{B}_w$.

3–60. Given the nonsingular matrix $\mathbf{B} = (\mathbf{b}_1, \ldots, \mathbf{b}_n)$. Show that

$$\mathbf{B}^{-1}\mathbf{b}_i = \mathbf{e}_i.$$

3–61. If $\mathbf{B}$ is a nonsingular matrix, column r of $\mathbf{B}$ is $\mathbf{e}_r$, and row r of $\mathbf{B}$ is $\mathbf{e}'_r$, show that column r of $\mathbf{B}^{-1}$ is $\mathbf{e}_r$, and row r of $\mathbf{B}^{-1}$ is $\mathbf{e}'_r$.

---

* Starred problems correspond to starred sections in the text.

*3-62. Suppose that a square matrix **A** can be written in partitioned form as

$$\mathbf{A} = \begin{bmatrix} \mathbf{A}_1 & 0 & \cdots & 0 \\ 0 & \mathbf{A}_2 & \cdots & 0 \\ \vdots & & & \vdots \\ 0 & 0 & \cdots & \mathbf{A}_n \end{bmatrix},$$

where the only nonzero submatrices are square and appear on the main diagonal. Prove that

$$|\mathbf{A}| = |\mathbf{A}_1|\,|\mathbf{A}_2| \cdots |\mathbf{A}_n|.$$

*3-63. Expand by the Laplace expansion method, using the first and third rows

$$\begin{vmatrix} 3 & 4 & 1 & 5 \\ 2 & 8 & 7 & 6 \\ 1 & 0 & 5 & 4 \\ 9 & 9 & 1 & 3 \end{vmatrix}.$$

*3-64. Show that an nth-order determinant can be written

$$\begin{vmatrix} a_{11} & \cdots & a_{1n} \\ \vdots & & \vdots \\ a_{n1} & \cdots & a_{nn} \end{vmatrix} = a_{11}A_{11} + \begin{vmatrix} 0 & a_{12} & \cdots & a_{1n} \\ a_{21} & a_{22} & \cdots & a_{2n} \\ \vdots & & & \vdots \\ a_{n1} & a_{n2} & \cdots & a_{nn} \end{vmatrix}.$$

Each of the submatrices whose determinants yield the cofactors $A_{12}, \ldots, A_{1n}$ contain the elements $a_{21}, \ldots, a_{n1}$ of the first column of **A**. Denote the cofactor of a_{i1} in A_{1j} by $A_{1j;i1}$ (this is a determinant of order $n - 2$). Note that $A_{1j;i1}$ involves the determinant of the submatrix formed from **A** by crossing out rows 1, i and columns 1, j, as does also $A_{11;ij}$ which, in turn, is the cofactor of a_{ij} in A_{11}. Thus $A_{11;ij}$ and $A_{1j;i1}$ can differ only in sign. Show that

$$A_{11;ij} = -A_{1j;i1}.$$

Now prove that $|\mathbf{A}|$ can be expanded in the form

$$|\mathbf{A}| = a_{11}A_{11} - \sum_i \sum_j a_{1j}a_{i1}A_{11;ij} \qquad (i, j \neq 1).$$

This is the famous Cauchy expansion of a determinant.

*3-65. Generalize the Cauchy expansion discussed in Problem 3-64 so that the expansion is made in terms of column h and row k.

*3–66. By means of the Cauchy method expand in terms of the first row and column:

$$\begin{vmatrix} 1 & 2 & 3 & 4 \\ 2 & 0 & 7 & 1 \\ 5 & 9 & 0 & 6 \\ 6 & 1 & 1 & 2 \end{vmatrix}.$$

*3–67. Given:

$$\mathbf{A} = \begin{bmatrix} a_{11} & a_{12} & a_{13} & a_{14} & a_{15} \\ a_{21} & a_{22} & a_{23} & a_{24} & a_{25} \\ a_{31} & a_{32} & a_{33} & a_{34} & a_{35} \end{bmatrix}, \quad \mathbf{B} = \begin{bmatrix} b_{11} & b_{12} & b_{13} \\ b_{21} & b_{22} & b_{23} \\ b_{31} & b_{32} & b_{33} \\ b_{41} & b_{42} & b_{43} \\ b_{51} & b_{52} & b_{53} \end{bmatrix}.$$

Write $|\mathbf{AB}|$ as a sum of products of 3×3 determinants taken from $\mathbf{A}$ and $\mathbf{B}$.

*3–68. Consider $|\mathbf{AA'}|$. Using the theorem on the determinant of the product of rectangular matrices, show what form the expansion of this determinant takes when $\mathbf{A}$ is $m \times n$, $m < n$.

*3–69. A matrix has no numerical value. On occasions, however (for example, when treating infinite series of matrices), it is useful to define a real-valued function of a matrix. This can be done in many ways, e.g.: The modulus of an $m \times n$ matrix $\mathbf{A}$, written $M(\mathbf{A})$, is defined as

$$M(\mathbf{A}) = \max_j \left\{ \sum_{i=1}^{m} |a_{ij}|, \quad j = 1, \ldots, n \right\},$$

where

$|a_{ij}|$ is the absolute value of a_{ij}.

Find $M(\mathbf{A})$ for the following matrices:

$$\mathbf{A} = \begin{bmatrix} 2 & 4 & 3 \\ -7 & 0 & 2 \\ 3 & 1 & 2 \end{bmatrix}, \quad \mathbf{A} = \begin{bmatrix} 6 & 4 & 3 \\ 2 & 1 & 5 \end{bmatrix}, \quad \mathbf{A} = \begin{bmatrix} 0 & 0 \\ 0 & 0 \\ 1 & 0 \end{bmatrix}, \quad \mathbf{A} = \begin{bmatrix} 0 & 0 \\ 0 & 0 \end{bmatrix}.$$

*3–70. Referring to Problem 3–69, prove that $M(\mathbf{A})$ satisfies the following expressions:

(1) $M(\lambda \mathbf{A}) = |\lambda| M(\mathbf{A})$ for any scalar λ;
(2) $M(\mathbf{I}) = 1$;
(3) $M(\mathbf{AB}) \leq M(\mathbf{A}) M(\mathbf{B})$;
(4) $M(\mathbf{A} + \mathbf{B}) \leq M(\mathbf{A}) + M(\mathbf{B})$;
(5) $M(\mathbf{A}) - M(\mathbf{B}) \leq M(\mathbf{A} + \mathbf{B})$;

Hint: For (5) write $\mathbf{A} = (\mathbf{A} + \mathbf{B}) + (-\mathbf{B})$ and apply (4) to this equation.

*3-71. Given an infinite sequence of matrices $\mathbf{A}_k$. Prove that if
$$\lim_{k \to \infty} M(\mathbf{A} - \mathbf{A}_k) = 0,$$
then
$$\lim_{k \to \infty} \mathbf{A}_k = \mathbf{A},$$
and
$$\lim_{k \to \infty} M(\mathbf{A}_k) = M(\mathbf{A}).$$

*3-72. Express $|\mathbf{A}^{-1}|$, $|\mathbf{A}^+|$ in terms of $|\mathbf{A}|$.

*3-73. Consider an $m \times n$ matrix $\mathbf{A}$ such that every column of $\mathbf{A}$ contains either only zero elements or zero elements throughout except one which has the value unity. Show that every minor of $\mathbf{A}$ either has the value 0 or ± 1.

3-74. Given $\mathbf{AB} = \mathbf{AC}$; does it follow that $\mathbf{B} = \mathbf{C}$? Can you provide a counterexample?

3-75. If the matrix $\mathbf{A}$ of Problem 3-62 is nonsingular, compute $\mathbf{A}^{-1}$.

3-76. Derive the cofactor expansion by column j from the result on the expansion by row j, using the fact that $|\mathbf{A}| = |\mathbf{A}'|$.

*3-77. In the general expansion of a determinant of order 4, find all the terms containing $a_{11}a_{22}$ and $a_{12}a_{21}$. Show that, except for sign, the same set of terms is obtained in each case after $a_{11}a_{22}$ or $a_{12}a_{21}$ is factored out. Show that the set of terms from which $a_{12}a_{21}$ is factored out has the opposite sign from that corresponding to $a_{11}a_{22}$. Next, show that the set of terms which multiplies $a_{11}a_{22}$ is the determinant obtained from $\mathbf{A}$ by crossing out rows and columns 1, 2. Do the same thing for terms containing $a_{3u}a_{4v}$ where (u, v) is $(1, 2)$ or $(2, 1)$. Verify that the sign is obtained according to the rules for the Laplace expansion.

Problems Involving Matrices With Complex Elements

3-78. Review the chapter and list the most important results. Show that the results remain true for matrices whose elements are complex numbers.

3-79. Compute $\mathbf{AB}$, $\mathbf{BA}$ for

$$\mathbf{A} = \begin{bmatrix} 2+i & 4+3i & i \\ 7 & 6+9i & 1-i \end{bmatrix}; \quad \mathbf{B} = \begin{bmatrix} 8-4i & 5i \\ 6 & -2i \\ 3+2i & 4+5i \end{bmatrix}.$$

3-80. Compute $|\mathbf{A}|$, $\mathbf{A}^{-1}$ for

$$\mathbf{A} = \begin{bmatrix} 1+i & 2i \\ 4-6i & 2+3i \end{bmatrix}.$$

Show that $\mathbf{AA}^{-1} = \mathbf{A}^{-1}\mathbf{A} = \mathbf{I}$. Note that the definition of the identity matrix does not need to be changed when the theory is generalized to include matrices with complex elements.

3-81. Compute $|\mathbf{A}|$, $\mathbf{A}^{-1}$ for

$$\mathbf{A} = \begin{bmatrix} 3i & 5 & 6+7i \\ 4+2i & 1+i & 3-5i \\ 3 & 2-i & 6+i \end{bmatrix}.$$

3-82. Compute a product form of the inverse for the $\mathbf{A}$ of Problem 3-80.

3-83. Given a matrix $\mathbf{A} = \|a_{ij}\|$ with complex elements, then the matrix $\mathbf{A}^*$ is defined to be $\mathbf{A}^* = \|a_{ij}^*\|$; $\mathbf{A}^*$ is formed from $\mathbf{A}$ by taking the complex conjugate of each element in $\mathbf{A}$. Hence $\mathbf{A}^*$ may be called the conjugate matrix to $\mathbf{A}$. When the elements of a matrix are complex, symmetric matrices are of less interest than so-called Hermitian matrices. A Hermitian matrix is a matrix $\mathbf{A}$ such that $\mathbf{A} = (\mathbf{A}^*)'$. The matrix is equal to its conjugate transpose. Note that if the elements are real, a Hermitian matrix is a symmetric. The matrix $(\mathbf{A}^*)'$ will be denoted by $\overline{\mathbf{A}}$. For any matrix $\mathbf{A}$, $\overline{\mathbf{A}}$ is called the associate matrix of $\mathbf{A}$. Prove that $(\mathbf{ABC})^* = \mathbf{A}^*\mathbf{B}^*\mathbf{C}^*$ and $(\overline{\mathbf{ABC}}) = \overline{\mathbf{C}}\,\overline{\mathbf{B}}\,\overline{\mathbf{A}}$.

3-84. Show that the following is a Hermitian matrix:

$$\mathbf{A} = \begin{bmatrix} 2 & 1+i & 3+4i \\ 1-i & 4 & 2-3i \\ 3-4i & 2+3i & 5 \end{bmatrix}.$$

Prove that the diagonal elements of a Hermitian matrix are always real.

3-85. Compute $\mathbf{A}^2$ for the matrix $\mathbf{A}$ of Problem 3-84. If $\mathbf{A}$ is a Hermitian matrix, what can be said about the elements of $\mathbf{A}^2$?

3-86. Prove that the inverse of a nonsingular Hermitian matrix is Hermitian. Under what conditions is the product of two Hermitian matrices Hermitian?

CHAPTER 4

LINEAR TRANSFORMATIONS, RANK AND ELEMENTARY TRANSFORMATIONS

Be ye transformed by the renewing of your mind.

Rom. XII-2.

4–1 Definition of linear transformations. Transformations of variables in which the new variables are linear combinations of the original ones are often needed in working with linear models. In fact, transformations of this type were used in the preceding chapter to obtain the general definition of matrix multiplication. These linear transformations of variables and matrix theory are closely connected. Furthermore, linear transformations can be given an interesting geometric interpretation which, in turn, contributes materially to our intuitive understanding of many of the properties of matrix operations. We shall see also that a more detailed discussion of linear transformations will allow us to develop some additional significant concepts in matrix algebra.

An especially simple example of a linear transformation of variables is

$$y_1 = a_{11}x_1 + a_{12}x_2$$
$$y_2 = a_{21}x_1 + a_{22}x_2$$
or $\quad \mathbf{y} = \mathbf{Ax}. \quad (4\text{–}1)$

The vector $\mathbf{x} = [x_1, x_2]$ can be viewed as a point in the x_1x_2-plane. The change of variables (4–1) serves to transform each point in the x_1x_2-plane into a point in the y_1y_2-plane. A transformation of this kind is called a *mapping* of the x_1x_2-plane into all or part of the y_1y_2-plane. The vector $\mathbf{x}$ is mapped into the vector $\mathbf{y}$. The point $\mathbf{y}$ is called the image of $\mathbf{x}$. The matrix $\mathbf{A}$ induces the mapping.

There is no reason at all for considering the y_1y_2-plane as necessarily different or distinct from the x_1x_2-plane. They can be assumed to be identical, with the y_1- and y_2-axes being the same as the respective x_1- and x_2-axes. Hence, the x_1x_2- and y_1y_2-planes can be considered to be simply different names for the same thing; according to this interpretation, the transformation (4–1) moves a point in the x_1x_2-plane into another point in the x_1x_2-plane.

Linear transformations of the type (4–1) have some interesting properties. If
$$\mathbf{y}_1 = \mathbf{A}\mathbf{x}_1, \quad \mathbf{y}_2 = \mathbf{A}\mathbf{x}_2,$$
then
$$\mathbf{y}_1 + \mathbf{y}_2 = \mathbf{A}\mathbf{x}_1 + \mathbf{A}\mathbf{x}_2 = \mathbf{A}(\mathbf{x}_1 + \mathbf{x}_2). \tag{4–2}$$

If $\mathbf{y}_1$, $\mathbf{y}_2$ are the images of $\mathbf{x}_1$, $\mathbf{x}_2$, respectively, then $\mathbf{y}_1 + \mathbf{y}_2$ is the image of $\mathbf{x}_1 + \mathbf{x}_2$, that is, the operation of addition is preserved under the transformation. Similarly, if $\mathbf{y}_1$ is the image of $\mathbf{x}_1$, then $\lambda \mathbf{y}_1$ is the image of $\lambda \mathbf{x}_1$ since $\mathbf{A}(\lambda \mathbf{x}_1) = \lambda \mathbf{A}\mathbf{x}_1$. Multiplying $\mathbf{x}$ by a scalar also multiplies the image of $\mathbf{x}$ by the same scalar. The transformation preserves multiplication by a scalar.

If, in (4–1), $\mathbf{A}$, the matrix of the coefficients, is nonsingular, we can write $\mathbf{x} = \mathbf{A}^{-1}\mathbf{y}$. This means that one and only one point in the x_1x_2-plane corresponds to each point in the y_1y_2-plane, just as according to Eq. (4–1), one and only one point in the y_1y_2-plane corresponds to each point in the x_1x_2-plane. Such a transformation is called a one-to-one (1–1) transformation since only one $\mathbf{y}$ corresponds to a given $\mathbf{x}$ and only one $\mathbf{x}$ to a given $\mathbf{y}$. A transformation can be single-valued (that is, only one $\mathbf{y}$ corresponds to a given $\mathbf{x}$) without being 1–1, since there may be two or more points $\mathbf{x}$ which are transformed into the same $\mathbf{y}$. If $\mathbf{A}^{-1}$ exists, then every point $\mathbf{y}$ has a corresponding unique $\mathbf{x}$. Consequently, $\mathbf{A}$ maps the x_1x_2-plane into all of the y_1y_2-plane. Assuming that the x_1x_2-plane and the y_1y_2-plane are identical, we see that a nonsingular linear transformation maps E^2 onto E^2 (all of E^2, not just a part of E^2) in a 1–1 manner.

The ideas developed above carry over to the transformation $\mathbf{y} = \mathbf{A}\mathbf{x}$ when $\mathbf{A}$ is $m \times n$. In this case, each point of E^n is transformed or mapped into a point in E^m. (Of course, if $m \neq n$, then $\mathbf{y}$, $\mathbf{x}$ cannot be considered to be points in the same space.) Again, addition and multiplication by a scalar are preserved under the transformation. Mathematicians prefer to define the concept of a linear transformation abstractly in terms of the properties of preserving addition and multiplication by a scalar:

LINEAR TRANSFORMATION: *A linear transformation T on the space E^n is a correspondence which maps each vector $\mathbf{x}$ of E^n into a vector $T(\mathbf{x})$ of E^m (m can be $>$, $=$, $<$ n) such that for all vectors $\mathbf{x}_1$, $\mathbf{x}_2$ in E^n and all scalars λ_1, λ_2,*

$$T(\lambda_1 \mathbf{x}_1 + \lambda_2 \mathbf{x}_2) = \lambda_1 T(\mathbf{x}_1) + \lambda_2 T(\mathbf{x}_2). \tag{4–3}$$

Equation (4–3) expresses the preservation of both addition and multiplication by a scalar. If we set $\lambda_1 = \lambda_2 = 1$, (4–3) becomes

$$T(\mathbf{x}_1 + \mathbf{x}_2) = T(\mathbf{x}_1) + T(\mathbf{x}_2);$$

i.e., the transformation preserves addition. If we take $\lambda_2 = 0$, (4–3) becomes

$$T(\lambda_1 \mathbf{x}_1) = \lambda_1 T(\mathbf{x}_1);$$

i.e., the transformation preserves multiplication by a scalar, which is often referred to as the *homogeneity property* of linear transformations.

Any matrix transformation is a linear transformation since the rules for matrix operations establish that

$$\mathbf{A}(\lambda_1 \mathbf{x}_1 + \lambda_2 \mathbf{x}_2) = \lambda_1 \mathbf{A}\mathbf{x}_1 + \lambda_2 \mathbf{A}\mathbf{x}_2.$$

In fact, the algebra of matrices is often called the algebra of linear transformations. We shall always identify linear transformations with matrix transformations. The general definition might give the impression that there could exist linear transformations on vector spaces other than matrix transformations. This is not so. In Problem 4–25 you will be required to prove that every linear transformation on a vector space is equivalent to, and can be represented by, a matrix transformation. We shall never have any direct use for the abstract definition (4–3) of a linear transformation. However, (4–3) does indeed provide the general definition of linearity and can be used to define linear differential or difference equations, linear servomechanisms, etc. In general, a physical or economic model can be cast into the form $T(\mathbf{x}) = \mathbf{y}$. This simply means that a set of variables, described by the vector $\mathbf{x}$, is related to another set of variables or known parameters by the transformation T. The model is linear if (4–3) holds. We often refer to T as an *operator* which transforms $\mathbf{x}$ into $\mathbf{y}$. The operator T may involve also other variables, such as time (when $T(\mathbf{x}) = \mathbf{y}$ is a set of differential equations and $\mathbf{x}$ is to be determined as a function of time), etc.

EXAMPLES: (1) The transformation $y = ax$ is linear. To prove this, we only need to show that (4–3) holds. In this case, $T(x) = ax$. Thus,

$$T(\lambda_1 x_1 + \lambda_2 x_2) = a(\lambda_1 x_1 + \lambda_2 x_2) = \lambda_1(ax_1) + \lambda_2(ax_2)$$
$$= \lambda_1 T(x_1) + \lambda_2 T(x_2).$$

(2) The transformation $y = ax^2$ is *not* linear, since

$$T(\lambda_1 x_1 + \lambda_2 x_2) = a(\lambda_1 x_1 + \lambda_2 x_2)^2 = a(\lambda_1 x_1)^2 + a(\lambda_2 x_2)^2 + 2a\lambda_1 \lambda_2 x_1 x_2$$
$$\neq \lambda_1 T(x_1) + \lambda_2 T(x_2) = a\lambda_1 x_1^2 + a\lambda_2 x_2^2.$$

(3) The transformation $y = a_1 x + a_2$, $a_2 \neq 0$, is *not* linear, since

$$T(\lambda_1 x_1 + \lambda_2 x_2) = a_1(\lambda_1 x_1 + \lambda_2 x_2) + a_2 = a_1 \lambda_1 x_1 + a_1 \lambda_2 x_2 + a_2$$
$$\neq \lambda_1 T(x_1) + \lambda_2 T(x_2) = \lambda_1(a_1 x_1 + a_2) + \lambda_2(a_1 x_1 + a_2)$$

for all λ_1, λ_2. The constant a_2 spoils the linearity of the transformation. It should be recalled that no such constants appear in the general matrix transformation $\mathbf{y} = \mathbf{Ax}$. For a transformation to be linear, every constant must multiply a variable.

4–2 Properties of linear transformations.

DOMAIN AND RANGE: *The domain of a transformation is defined to be the set of elements which undergo transformation. The range of a transformation is the set of elements which is formed by the transformation operating on the elements in the domain.*

The range is often called the image of the domain under the transformation. In Eq. (4–1), the domain is the whole x_1x_2-plane, and the range is the set of points in the y_1y_2-plane which is the image of the x_1x_2-plane. When $\mathbf{A}$ is a nonsingular, the range is the whole y_1y_2-plane.

The range of a linear transformation on E^n is a subspace of E^m or, expressed in terms of a matrix transformation: If $\mathbf{A}$ is an $m \times n$ matrix, then the set of points $\mathbf{y} = \mathbf{Ax}$ (for all $\mathbf{x}$ in E^n) is a subspace of E^m. In general, for any linear transformation T, we must demonstrate that if $T(\mathbf{x})$ is in the range, so is $\lambda T(\mathbf{x})$ for any scalar λ. This can be shown, since $\lambda T(\mathbf{x}) = T(\lambda \mathbf{x})$ and $T(\lambda \mathbf{x})$ is the image of $\lambda \mathbf{x}$ and is in the range. Similarly, it must be true that if $T(\mathbf{x}_1)$, $T(\mathbf{x}_2)$ are in the range, the sum $T(\mathbf{x}_1) + T(\mathbf{x}_2)$ is in the range also. Since $T(\mathbf{x}_1) + T(\mathbf{x}_2) = T(\mathbf{x}_1 + \mathbf{x}_2)$ is the image of $\mathbf{x}_1 + \mathbf{x}_2$, it is in the range. It may happen, of course, that the subspace of E^m is E^m itself. A simple example will illustrate the implications of this theorem: Suppose $\mathbf{A}$ is 3×3. Then the set of points $\mathbf{y} = \mathbf{Ax}$ for all $\mathbf{x}$ in E^3 must be either the origin, a line through the origin, a plane through the origin, or all of E^3. In addition, this proof shows that a linear transformation which takes points in E^n into points in E^m also takes a subspace of E^n into a subspace of E^m.

Any $m \times n$ matrix $\mathbf{A}$ can be written as a row of column vectors, $\mathbf{A} = (\mathbf{a}_1, \ldots, \mathbf{a}_n)$. Thus, when a matrix $\mathbf{A}$ maps all of E^n into E^m, we have

$$\mathbf{y} = \mathbf{Ax} = x_1\mathbf{a}_1 + \cdots + x_n\mathbf{a}_n, \tag{4-4}$$

where the x_i can take on all possible values. *The range of the transformation, that is, the subspace generated by the $\mathbf{y}$, is then the subspace of E^m spanned by the columns of $\mathbf{A}$.* We know from Section 2–13 that the dimension of this subspace is the maximum number of linearly independent columns in $\mathbf{A}$. *Thus the dimension of the range is the maximum number of linearly independent columns in $\mathbf{A}$.*

EXAMPLES: (1) The transformation induced on E^4 by

$$\mathbf{A} = \begin{bmatrix} 1 & 5 & 2 & 4 \\ 0 & 0 & 3 & 6 \\ 0 & 0 & 1 & 2 \end{bmatrix}$$

can be written

$$\mathbf{y} = x_1 \begin{bmatrix} 1 \\ 0 \\ 0 \end{bmatrix} + x_2 \begin{bmatrix} 5 \\ 0 \\ 0 \end{bmatrix} + x_3 \begin{bmatrix} 2 \\ 3 \\ 1 \end{bmatrix} + x_4 \begin{bmatrix} 4 \\ 6 \\ 2 \end{bmatrix}$$

$$= (x_1 + 5x_2) \begin{bmatrix} 1 \\ 0 \\ 0 \end{bmatrix} + (x_3 + 2x_4) \begin{bmatrix} 2 \\ 3 \\ 1 \end{bmatrix}.$$

The range of the transformation is the two-dimensional subspace of E^3 spanned by $[1, 0, 0]$ and $[2, 3, 1]$; that is, the range consists of all the vectors $\mathbf{y}$ lying in the plane through the origin and $[1, 0, 0]$, $[2, 3, 1]$ in E^3.

(2) The transformation represented by the matrix

$$\mathbf{A} = \begin{bmatrix} 1 & 0 & 0 \\ 0 & 1 & 0 \end{bmatrix}$$

takes the point $[x_1, x_2, x_3]$ into a point $[x_1, x_2]$ in the x_1x_2-plane. The transformation then projects a point in E^3 on the x_1x_2-plane. The transformation is not 1–1, since every point with the first two components x_1, x_2 goes into $[x_1, x_2]$ regardless of what x_3 happens to be. Thus $\mathbf{A}$ takes E^3 into all of E^2.

(3) The transformation represented by the matrix

$$\mathbf{A} = \begin{bmatrix} 1 & 0 \\ 0 & 1 \\ 0 & 1 \end{bmatrix}$$

takes any point $[x_1, x_2]$ of E^2 into $[x_1, x_2, x_2]$ of E^3 (see Fig. 4–1). In the process, the transformation only rotates the x_1x_2-plane about the x_1-axis through a 45°-angle. The range is a plane and represents a two-dimensional subspace of E^3.

It is important to note that a transformation which takes every point of E^n into a space of higher dimension $E^m (m > n)$ can never have a range of dimension m. We understand intuitively that all of E^m cannot be filled from a space of lower dimension. The dimension of the subspace

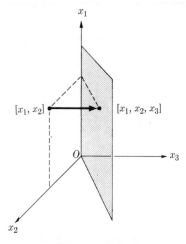

FIGURE 4-1

of E^m, representing the range of the transformation, cannot be greater than n. The dimension of the subspace is the maximum number of linearly independent columns in $\mathbf{A}$, and $\mathbf{A}$ has only n columns.

Let T_1 be a linear transformation which takes E^n into a subspace of E^r, and T_2 a linear transformation which takes E^r into a subspace of E^m. The product $T_3 = T_2 T_1$ of the two linear transformations T_1, T_2 is defined as follows:

$$T_3(\mathbf{x}) = T_2[T_1(\mathbf{x})]; \qquad (4\text{-}5)$$

that is, we obtain $T_3(\mathbf{x})$ by computing $T_1(\mathbf{x}) = \mathbf{y}$ and applying T_2 to $\mathbf{y}$. The product of two linear transformations is also a linear transformation since

$$\begin{aligned}
T_3(\lambda_1 \mathbf{x}_1 + \lambda_2 \mathbf{x}_2) &= T_2[\lambda_1 T_1(\mathbf{x}_1) + \lambda_2 T_1(\mathbf{x}_2)] \\
&= \lambda_1 T_2[T_1(\mathbf{x}_1)] + \lambda_2 T_2[T_1(\mathbf{x}_2)] \qquad (4\text{-}6) \\
&= \lambda_1 T_3(\mathbf{x}_1) + \lambda_2 T_3(\mathbf{x}_2).
\end{aligned}$$

If T_1 takes a point in E^n into a point in E^r, and T_2 takes a point in E^r into a point in E^m, then $T_3 = T_2 T_1$ takes a point in E^n into a point in E^m.

Given any two linear transformations T_1, T_2, with T_1 taking points in E^n into points in E^r, and T_2 taking points in E^s into points in E^m, the product $T_2 T_1$ can be defined if and only if $r = s$. If matrices $\mathbf{A}_1, \mathbf{A}_2$ represent T_1, T_2, respectively, then T_3 is represented by matrix $\mathbf{A}_3 = \mathbf{A}_2 \mathbf{A}_1$. This statement will have to be proved in Problem 4-29. Thus, a product of matrices can be viewed as a sequence of linear transformations. At this point, we wish to note that the product of two linear transformations defines the rules for matrix multiplication. Indeed, in Section 3-3, matrix multiplication has been defined in terms of the product of two linear transformations.

4–3 Rank. The dimension of the range of a linear transformation represented by **A** is the maximum number of linearly independent columns in **A**; as such, it tells us quite a bit about the linear transformation and the matrix **A**. We shall call the number of linearly independent columns in **A** the rank of **A**.

RANK: *The rank (or more precisely the column rank) of an $m \times n$ matrix* **A**, *written $r(\mathbf{A})$, is the maximum number of linearly independent columns in* **A**.

We shall see that, for many purposes, one of the most important aspects of a matrix is its rank. Introduction of this concept enables us to tie up a number of loose ends in matrix theory and to develop some basic ideas in greater detail.

EXAMPLE: The rank of an nth-order identity matrix is n since $\mathbf{I} = (\mathbf{e}_1, \ldots, \mathbf{e}_n)$, and the $\mathbf{e}_i$ are linearly independent.

We are frequently faced with the problem of determining the rank of the matrix $\mathbf{C} = \mathbf{AB}$ from the given ranks of the matrices **A**, **B**. In general, the rank of **C** is not uniquely determined by the ranks of **A**, **B**; however, the following inequality holds:

$$r(\mathbf{AB}) \leq \min\,[r(\mathbf{A}), r(\mathbf{B})]; \qquad (4\text{–}7)$$

that is, *the rank of the product* **AB** *of two matrices cannot be greater than the smaller of the ranks of* **A**, **B**.

The truth of (4–7) becomes evident if we think of **A**, **B** as representing linear transformations. Let us suppose that **A** is $m \times r$ and **B** is $r \times n$. The product **AB** can be viewed as a single linear transformation taking E^n into a subspace of E^m, and also as two linear transformations applied sequentially. First, **B** takes E^n into a subspace of E^r, then **A** takes this subspace of E^r into a subspace of E^m. From a geometrical point of view, it is clear that the dimension of the subspace of E^m cannot be greater than the smaller of the dimensions of: (1) the subspace of E^r obtained by **B** transforming E^n, (2) the subspace of E^m that would result if **A** transformed all of E^r.

Let us now prove Eq. (4–7) analytically. We write

$$\mathbf{z} = \mathbf{Bx}. \qquad (4\text{–}8)$$

Then

$$\mathbf{y} = \mathbf{ABx} = \mathbf{Az} = \sum_{i=1}^{r} z_i \mathbf{a}_i. \qquad (4\text{–}9)$$

Thus all vectors **y** in the subspace of E^m are linear combinations of the columns of **A**. Hence $r(\mathbf{AB})$ cannot be greater than $r(\mathbf{A})$, for otherwise it would be impossible to express every vector **y** as a linear combination of the columns of **A**.

Next, we note that every **z** can be written as a linear combination of $r(\mathbf{B}) = k$ columns from **B**. Let the set of k linearly independent columns be denoted by $\hat{\mathbf{b}}_1, \ldots, \hat{\mathbf{b}}_k$. Hence

$$\mathbf{z} = \sum_{i=1}^{k} \alpha_i \hat{\mathbf{b}}_i \qquad (4\text{--}10)$$

for every **z** in the subspace of E^r. Using Eq. (4–10), we can write every **y** in the subspace of E^m as

$$\mathbf{y} = \sum_{i=1}^{k} \alpha_i \mathbf{A}\hat{\mathbf{b}}_i, \qquad (4\text{--}11)$$

that is, every **y** is a linear combination of k vectors. Although we assumed that the $\hat{\mathbf{b}}_i$ were linearly independent, this is not necessarily true for the $\mathbf{A}\hat{\mathbf{b}}_i$. Hence $r(\mathbf{AB}) \leq r(\mathbf{B})$, and Eq. (4–7) is proved.

The problems at the end of the chapter will show that in some cases the strict equality and in others the strict inequality sign will hold in Eq. (4–7). However:

If a matrix of rank k is multiplied in either order by a nonsingular matrix, the rank of the product is k. This can be proved easily from Eq. (4–7). Let $r(\mathbf{AB}) = R$; assume that matrix **A** is nonsingular and that $r(\mathbf{B}) = k$. From (4–7)

$$R \leq k. \qquad (4\text{--}12)$$

But

$$\mathbf{B} = \mathbf{A}^{-1}(\mathbf{AB}). \qquad (4\text{--}13)$$

Applying (4–7) to (4–13), we obtain

$$k \leq R. \qquad (4\text{--}14)$$

Comparison of (4–12) and (4–14) yields

$$k = R. \qquad (4\text{--}15)$$

If **B** is nonsingular and $r(\mathbf{A}) = k$, the proof can be established in exactly the same way.

From the preceding result we immediately see that *all nth-order nonsingular matrices have the same rank.* Let **A**, **B** be two nth-order nonsingular matrices, and $r(\mathbf{A}) = k_1$, $r(\mathbf{B}) = k_2$. From (4–15)

$$r(\mathbf{AB}) = k_1, \qquad r(\mathbf{AB}) = k_2;$$

hence

$$k_1 = k_2. \qquad (4\text{--}16)$$

4–4 Rank and determinants. An interesting relation exists between the rank of a matrix **A** and the order of its nonvanishing minors, which is of considerable theoretical importance and, in addition, provides a means for computing the rank of a matrix.

*The rank of an $m \times n$ matrix **A** is k if and only if every minor in **A** of order $k + 1$ vanishes, while there is at least one minor of order k which does not vanish.* To prove the necessity, assume that $r(\mathbf{A}) = k$. Then any $k + 1$ columns of **A** are linearly dependent, and any column in **A** can be expressed as a linear combination of some set of k columns. Hence, the columns of any submatrix of order $k + 1$ can be expressed as linear combinations of k columns from **A**. Each of the determinants obtained by expanding the determinant of this submatrix vanishes because its associated matrix has two identical columns, and hence the determinant of the submatrix vanishes. This holds true for every submatrix of order $k + 1$, that is, all minors of order $k + 1$ vanish.

Next, we must show that there is at least one minor of order k which does not vanish. Assume that the opposite holds, that is, that all determinants of order k vanish. Select k columns from **A** which are linearly independent. Without loss of generality, we can consider them to be the first k columns. Assume that all determinants of order k in the first k columns and, in particular, the determinant of the submatrix formed from rows $1, \ldots, k$, vanish. Thus, expanding in cofactors by row k, we find

$$\sum_i a_{ki} A_{ki} = 0. \tag{4-17}$$

The same cofactors are obtained if we form a $k \times k$ submatrix from the first $k - 1$ rows and any other row j, $j = k + 1, \ldots, m$, and expand by row j. This implies that

$$\sum_i a_{ji} A_{ki} = 0, \quad j = k+1, \ldots, m. \tag{4-18}$$

Furthermore,

$$\sum_i a_{ji} A_{ki} = 0, \quad j = 1, \ldots, k-1, \tag{4-19}$$

since we are expanding by one row and are using the cofactors of another. Combining (4–17), (4–18), and (4–19), we obtain

$$\sum_i A_{ki} \mathbf{a}_i = \mathbf{0}. \tag{4-20}$$

This implies that columns $\mathbf{a}_1, \ldots, \mathbf{a}_k$ are linearly dependent if the A_{ki} are not all zero. In fact, if there were any minor of order $k - 1$ in the first k columns, which did not vanish, we could rearrange the rows and columns so that at least one A_{ki} in (4–20) would be different from zero. If all determinants of order $k - 1$ formed from the first k columns of **A** were zero, we could repeat the same procedure with $k-1$ columns and show that the $k-1$ columns are linearly dependent if the determinants of order $k-2$ do not vanish.

Conceivably, we may in this way arrive at a point where all determinants of order 2 vanish. Now we have reduced the problem to showing that any set of two columns is linearly dependent. This linear dependence would follow even if all determinants of order 1 vanished, since then the columns would be composed of zeros only. Thus, using the above procedure, we have contradicted our assumption that the k columns were linearly independent. Hence, there must be at least one nonvanishing determinant of order k in any set of k linearly independent columns. Thus we have proved that if $r(\mathbf{A}) = k$, then all determinants of order $k+1$ in $\mathbf{A}$ vanish, and that there is at least one determinant of order k which does not vanish. To prove the sufficiency, let us assume that all determinants of order $k+1$ vanish and that there is one determinant of order k which does not vanish. The proof of the necessity showed that any $k+1$ columns whose determinants of order $k+1$ all vanish are linearly dependent. Hence $r(\mathbf{A}) \leq k$. Now let us consider the columns associated with any determinant of order k which does not vanish. These columns cannot be linearly dependent or, as shown in the necessity, all determinants of order k formed from them would vanish. Hence $r(\mathbf{A}) = k$.

Note: If all determinants of order $k+1$ in $\mathbf{A}$ vanish, all determinants of order $k+r$ also vanish $(r \geq 1)$. Cofactor expansion immediately demonstrates the correctness of this statement.

The above result gives us a means for computing the rank of any matrix: We look for the largest nonvanishing determinant in $\mathbf{A}$; the order of this determinant is the rank of $\mathbf{A}$. To find nonvanishing determinants, it is not necessary to consider only adjacent rows or columns. Any rows and columns can be chosen to form the determinant.

We have now a way of testing whether any given set of m vectors is linearly independent. A matrix is formed with the vectors as columns. The rank of this matrix is found by the method just described (or by any other method). If the rank is m, then the vectors are linearly independent. If the rank is $k < m$, the vectors are not linearly independent; in addition, we have determined the maximum number of linearly independent vectors in the set.

EXAMPLES:

(1) $\mathbf{A} = \begin{bmatrix} 1 & 3 \\ 4 & 2 \end{bmatrix}$ has rank 2 since $\begin{vmatrix} 1 & 3 \\ 4 & 2 \end{vmatrix} \neq 0$.

(2) $\mathbf{A} = \begin{bmatrix} 2 & 3 \\ 1 & 1.5 \end{bmatrix}$ has rank 1 since $\begin{vmatrix} 2 & 3 \\ 1 & 1.5 \end{vmatrix} = 0$, but there are determinants of order 1, let us say $|2|$, which do not vanish.

(3) $\mathbf{A} = \begin{bmatrix} 0 & 0 & 0 \\ 0 & 0 & 0 \end{bmatrix}$ has rank 0 because all elements of $\mathbf{A}$ vanish.

(4) $\mathbf{A} = \begin{bmatrix} 2 & 0 & 7 \\ 3 & 3 & 6 \\ 2 & 2 & 4 \end{bmatrix}$ has rank 2 since $|\mathbf{A}| = 0$, but there are some determinants of order 2 which do not vanish.

The rank of a matrix has been defined as the number of linearly independent columns of $\mathbf{A}$. Strictly speaking, this is a definition of the *column rank*, since we may equally well speak of a *row rank* of $\mathbf{A}$ and define it as the maximum number of linearly independent rows in $\mathbf{A}$. The reader may feel that we should have been more meticulous, that is, used the term "column rank" whenever we referred to the "rank" of a matrix. We shall now prove that *the row rank of $\mathbf{A}$ is equal to the column rank and hence the maximum number of linearly independent columns is equal to the maximum number of linearly independent rows. The rank of a matrix is thus a unique number which can be found by computing the maximum number of linearly independent rows or columns.* For this reason, it was not necessary to distinguish between the two types of rank.

To prove this equality, let us suppose that the column rank of $\mathbf{A}$ is k and the row rank of $\mathbf{A}$ is k'. But the row rank of $\mathbf{A}$ is equal to the column rank of $\mathbf{A}'$. Since, by assumption, the column rank of $\mathbf{A}'$ is k', all minors of order $k' + 1$ in $\mathbf{A}'$, and hence in $\mathbf{A}$, vanish. Thus $k' \geq k$. However, there is at least one minor of order k' from $\mathbf{A}'$, and hence from $\mathbf{A}$, which does not vanish. Thus $k \geq k'$ and, therefore, $k' = k$. We have proved that the row rank of $\mathbf{A}$ is equal to its column rank.

Let us summarize the results of our study of the concept of rank: If $r(\mathbf{A}) = k$, then at least one set of k columns and rows of $\mathbf{A}$ is linearly independent (there may be, of course, a number of sets of k rows and columns in $\mathbf{A}$ which contain linearly independent vectors), and no $k + 1$ rows or columns are linearly independent. Furthermore, there is at least one determinant of order k in $\mathbf{A}$ which does not vanish. All determinants of order $k + 1$ do vanish.

A particularly interesting case is presented by an nth-order matrix $\mathbf{A}$. If $\mathbf{A}$ is nonsingular, $|\mathbf{A}| \neq 0$. Hence, the columns (or rows) of $\mathbf{A}$ are linearly independent and form a basis for E^n, and $r(\mathbf{A}) = n$. Conversely, if we have n linearly independent vectors from E^n, we can form a matrix $\mathbf{A}$ with the vectors as its columns. This matrix $\mathbf{A}$ will have rank n, and thus $|\mathbf{A}| \neq 0$; $\mathbf{A}$ is nonsingular and has an inverse.

The results of this section and of Section 4–3 can be used to clear up one question left unanswered in Chapter 3. There we stated, but did not prove, the fact that only nonsingular matrices have inverses; that is, if **A** is an nth-order matrix and there exists an nth-order matrix **B** such that $\mathbf{AB} = \mathbf{I}_n$, then **A** is nonsingular. In the preceding section, we have shown that $r(\mathbf{I}_n) = n$, so that if $\mathbf{AB} = \mathbf{I}_n$, then $r(\mathbf{A}) = n$, and $r(\mathbf{B}) = n$. In this section, we have seen that if $r(\mathbf{A}) = n$, $|\mathbf{A}| \neq 0$, and hence **A** is nonsingular. Thus we have proved that only nonsingular matrices have inverses. We have also demonstrated that if the product of any two nth-order matrices yields the identity matrix, both matrices are nonsingular. It then follows from this result and the proofs furnished in Chapter 3 that if two nth-order matrices **A**, **B** satisfy $\mathbf{AB} = \mathbf{I}_n$, then (1) $\mathbf{BA} = \mathbf{I}_n$; (2) $\mathbf{A}^{-1} = \mathbf{B}$, $\mathbf{B}^{-1} = \mathbf{A}$; (3) $|\mathbf{A}| \neq 0$, $|\mathbf{B}| \neq 0$.

It is now clear how to express any n-component vector **b** as a linear combination of n linearly independent vectors $\mathbf{a}_1, \ldots, \mathbf{a}_n$ which form a basis for E^n. We wish to find the y_i such that

$$\mathbf{b} = \sum_{i=1}^{n} y_i \mathbf{a}_i. \tag{4-21}$$

We define
$$\mathbf{A} = (\mathbf{a}_1, \ldots, \mathbf{a}_n), \qquad \mathbf{y} = [y_1, \ldots, y_n]. \tag{4-22}$$

Then
$$\mathbf{b} = \mathbf{A}\mathbf{y},$$

or
$$\mathbf{y} = \mathbf{A}^{-1}\mathbf{b}. \tag{4-23}$$

The vector **y** is found by premultiplying **b** by $\mathbf{A}^{-1}$.

EXAMPLE: Write **b** as a linear combination of $\mathbf{a}_1$, $\mathbf{a}_2$ when

$$\mathbf{b} = [3, 2], \qquad \mathbf{a}_1 = [4, 1], \qquad \mathbf{a}_2 = [2, 5];$$

$$\mathbf{A} = (\mathbf{a}_1, \mathbf{a}_2) = \begin{bmatrix} 4 & 2 \\ 1 & 5 \end{bmatrix}, \qquad \mathbf{A}^{-1} = \frac{1}{18}\begin{bmatrix} 5 & -2 \\ -1 & 4 \end{bmatrix},$$

$$\mathbf{y} = \mathbf{A}^{-1}\mathbf{b} = \frac{1}{18}\begin{bmatrix} 5 & -2 \\ -1 & 4 \end{bmatrix}\begin{bmatrix} 3 \\ 2 \end{bmatrix} = \frac{1}{18}\begin{bmatrix} 11 \\ 5 \end{bmatrix}; \quad \text{hence } y_1 = \frac{11}{18}, \; y_2 = \frac{5}{18}.$$

This result is easily verified:

$$\begin{bmatrix} 3 \\ 2 \end{bmatrix} = \left(\frac{11}{18}\right)\begin{bmatrix} 4 \\ 1 \end{bmatrix} + \left(\frac{5}{18}\right)\begin{bmatrix} 2 \\ 5 \end{bmatrix}.$$

The method outlined above for expressing a vector in terms of a set of basis vectors is especially convenient if $\mathbf{A}^{-1}$ is available.

In the following two sections, we shall develop a more efficient procedure for computing the rank of a matrix and, in addition, gain some interesting theoretical information.

4–5 Elementary transformations.

There are simple operations which can be performed on the rows and columns of a matrix without changing its rank. By performing such operations, it is possible to convert a matrix into one whose rank can be read off by simply looking at the matrix. These operations then provide another way to compute the rank of a matrix; as a matter of fact, they yield a rather efficient numerical procedure. In addition, they lead to some interesting theoretical results.

Three types of operations on the rows of a matrix (called elementary row operations) are of importance; they are:

(1) Interchange of two rows;
(2) Multiplication of a row by any scalar $\lambda \neq 0$;
(3) Addition to the ith row of λ times the jth row (λ any scalar, and $j \neq i$).

These elementary row operations have one especially interesting property: They can be performed on the matrix **A** by premultiplication of **A** by a matrix **E**; that is, if **B** is obtained from **A** by some elementary row operation on **A**, then there exists a matrix **E** such that $\mathbf{B} = \mathbf{EA}$.

EXAMPLES: (1) Exchange the first and third rows in **A**,

$$\mathbf{A} = \begin{bmatrix} 2 & 1 \\ 3 & 4 \\ 5 & 6 \end{bmatrix};$$

consider

$$\mathbf{E} = \begin{bmatrix} 0 & 0 & 1 \\ 0 & 1 & 0 \\ 1 & 0 & 0 \end{bmatrix}; \quad \mathbf{EA} = \begin{bmatrix} 0 & 0 & 1 \\ 0 & 1 & 0 \\ 1 & 0 & 0 \end{bmatrix} \begin{bmatrix} 2 & 1 \\ 3 & 4 \\ 5 & 6 \end{bmatrix} = \begin{bmatrix} 5 & 6 \\ 3 & 4 \\ 2 & 1 \end{bmatrix}.$$

Premultiplication of **A** by **E** interchanges the first and third rows in **A**.

(2) Multiply the second row of **A** by 5. If

$$\mathbf{E} = \begin{bmatrix} 1 & 0 & 0 \\ 0 & 5 & 0 \\ 0 & 0 & 1 \end{bmatrix}; \quad \mathbf{EA} = \begin{bmatrix} 2 & 1 \\ 15 & 20 \\ 5 & 6 \end{bmatrix}.$$

Premultiplication of **A** by **E** multiplies the second row of **A** by 5.

(3) Add twice the second row to the third row. When

$$\mathbf{E} = \begin{bmatrix} 1 & 0 & 0 \\ 0 & 1 & 0 \\ 0 & 2 & 1 \end{bmatrix}; \quad \mathbf{EA} = \begin{bmatrix} 2 & 1 \\ 3 & 4 \\ 11 & 14 \end{bmatrix}.$$

Premultiplication of **A** by **E** adds twice the second row of **A** to the third row. Careful study of the **E** matrices will reveal that they can be obtained from the identity matrix by performing on it the appropriate elementary operation.

Our simple examples have shown that the elementary operations can be performed on **A** through premultiplication by a matrix **E**; next, we shall demonstrate the general applicability of this procedure. First, if **E** is to perform an elementary operation on **A**, then **EA** must have the same number of rows and columns as **A**; hence **E** must be square. Let us suppose that we wish to interchange rows i and j of the $m \times n$ matrix **A**. If $\mathbf{e}'_i$ is the transpose of $\mathbf{e}_i$, then

$$\mathbf{e}'_i \mathbf{A} = \mathbf{a}^i \quad (\mathbf{a}^i \text{ is the } i\text{th row of } \mathbf{A}). \tag{4-24}$$

If, in the identity matrix $\mathbf{I}_m$, rows i and j are interchanged, then $\mathbf{e}'_i$ appears as row j and $\mathbf{e}'_j$ as row i. For the other rows, $\mathbf{e}'_k$ appears as row k. Assuming **E** to be the identity matrix with rows i and j interchanged, we see that **EA** does indeed merely interchange rows i and j in **A**. In general, we see that the matrix **E** which, when postmultiplied by **A**, interchanges the ith and jth rows in **A**, is the matrix obtained by interchanging the ith and jth row in **I**. The matrix which induces an interchange of rows i and j in **A** will be denoted by $\mathbf{E}_{ij}$.

If we wish to multiply the ith row of **A** by $\lambda \neq 0$, it is immediately obvious that the operative matrix **E** is the identity matrix whose ith row is multiplied by λ; that is, $\mathbf{e}'_i$ is replaced by $\lambda \mathbf{e}'_i$. The matrix which, when postmultiplied by **A**, multiplies the ith row of **A** by λ will be denoted by $\mathbf{E}_i(\lambda)$. Again $\mathbf{E}_i(\lambda)$ is found by performing the appropriate elementary operation on the identity matrix.

Finally, consider the addition of λ times row j of **A** to row i ($j \neq i$). Observe that

$$(\mathbf{e}'_i + \lambda \mathbf{e}'_j)\mathbf{A} = \mathbf{a}^i + \lambda \mathbf{a}^j. \tag{4-25}$$

The matrix **E** which performs the required operation is simply the identity matrix with row i replaced by $\mathbf{e}'_i + \lambda \mathbf{e}'_j$. Once again, **E** is found by applying the proper elementary transformation to the identity matrix. The matrix **E** which, when postmultiplied by **A**, adds λ times row j of **A** to row i will be denoted by $\mathbf{E}_i(\lambda|j)$.

The preceding paragraphs have shown that any elementary row operation can be performed on a matrix simply through multiplication by a matrix **E** which, in turn, is obtained by applying the appropriate operation to the identity matrix. The matrices $\mathbf{E}_{ij}$, $\mathbf{E}_i(\lambda)$, $\mathbf{E}(\lambda|j)$ are called elementary matrices.

The elementary matrices $\mathbf{E}_{ij}$, $\mathbf{E}_i(\lambda)$, $\mathbf{E}_i(\lambda|j)$ are nonsingular and inverses may be easily obtained as follows: If we interchange rows i and j in $\mathbf{E}_{ij}$, an identity matrix results. Hence

$$\mathbf{E}_{ij}\mathbf{E}_{ij} = \mathbf{I} \quad \text{or} \quad \mathbf{E}_{ij}^{-1} = \mathbf{E}_{ij}. \tag{4-26}$$

When the ith row of $\mathbf{E}_i(\lambda)$ is multiplied by $1/\lambda$, the identity matrix is obtained:

$$\mathbf{E}_i\left(\frac{1}{\lambda}\right)\mathbf{E}_i(\lambda) = \mathbf{I} \quad \text{or} \quad \mathbf{E}_i^{-1}(\lambda) = \mathbf{E}_i\left(\frac{1}{\lambda}\right). \tag{4-27}$$

If λ times row j is subtracted from the ith row of $\mathbf{E}_i(\lambda|j)$, the identity matrix results. Hence,

$$\mathbf{E}_i(-\lambda|j)\mathbf{E}_i(\lambda|j) = \mathbf{I} \quad \text{or} \quad \mathbf{E}_i^{-1}(\lambda|j) = \mathbf{E}_i(-\lambda|j). \tag{4-28}$$

The inverse of an elementary matrix is an elementary matrix.

EXAMPLES: Suppose that:

(1) $\mathbf{E}_{23} = \begin{bmatrix} 1 & 0 & 0 \\ 0 & 0 & 1 \\ 0 & 1 & 0 \end{bmatrix}$;

according to (4-26),

$$\mathbf{E}_{23}^{-1} = \mathbf{E}_{23},$$

since

$$\begin{bmatrix} 1 & 0 & 0 \\ 0 & 0 & 1 \\ 0 & 1 & 0 \end{bmatrix} \begin{bmatrix} 1 & 0 & 0 \\ 0 & 0 & 1 \\ 0 & 1 & 0 \end{bmatrix} = \begin{bmatrix} 1 & 0 & 0 \\ 0 & 1 & 0 \\ 0 & 0 & 1 \end{bmatrix}.$$

(2) $\mathbf{E}_1(\lambda) = \begin{bmatrix} \lambda & 0 & 0 \\ 0 & 1 & 0 \\ 0 & 0 & 1 \end{bmatrix}$;

according to (4-27),

$$\mathbf{E}_1^{-1}(\lambda) = \begin{bmatrix} \lambda^{-1} & 0 & 0 \\ 0 & 1 & 0 \\ 0 & 0 & 1 \end{bmatrix},$$

since

$$\begin{bmatrix} \lambda^{-1} & 0 & 0 \\ 0 & 1 & 0 \\ 0 & 0 & 1 \end{bmatrix} \begin{bmatrix} \lambda & 0 & 0 \\ 0 & 1 & 0 \\ 0 & 0 & 1 \end{bmatrix} = \begin{bmatrix} 1 & 0 & 0 \\ 0 & 1 & 0 \\ 0 & 0 & 1 \end{bmatrix}.$$

(3) $\mathbf{E}_1(4|2) = \begin{bmatrix} 1 & 4 & 0 \\ 0 & 1 & 0 \\ 0 & 0 & 1 \end{bmatrix}$;

according to (4–28),

$$\mathbf{E}_1^{-1}(4|2) = \begin{bmatrix} 1 & -4 & 0 \\ 0 & 1 & 0 \\ 0 & 0 & 1 \end{bmatrix},$$

$$\begin{bmatrix} 1 & -4 & 0 \\ 0 & 1 & 0 \\ 0 & 0 & 1 \end{bmatrix} \begin{bmatrix} 1 & 4 & 0 \\ 0 & 1 & 0 \\ 0 & 0 & 1 \end{bmatrix} = \begin{bmatrix} 1 & 0 & 0 \\ 0 & 1 & 0 \\ 0 & 0 & 1 \end{bmatrix};$$

the rule gives the proper $\mathbf{E}_1^{-1}(4|2)$.

Elementary column operations on a matrix can be defined in the same way as row operations, that is, they are: (1) interchange of two columns; (2) multiplication of column i by a scalar $\lambda \neq 0$; (3) addition of λ times column j to column i ($j \neq i$). These operations can be performed on any matrix $\mathbf{A}$ through postmultiplying $\mathbf{A}$ by a matrix $\mathbf{F}$. The matrix $\mathbf{F}$ is found by performing the required elementary operations on the columns of the identity matrix. $\mathbf{F}_{ij}$ will denote the elementary matrix which, when premultiplied by $\mathbf{A}$, interchanges columns i, j of $\mathbf{A}$. The symbol $\mathbf{F}_i(\lambda)$ will be used to denote the elementary matrix which, when premultiplied by $\mathbf{A}$, multiplies the ith column of $\mathbf{A}$ by λ. The elementary matrix which, when premultiplied by $\mathbf{A}$, adds λ times column j to column i of $\mathbf{A}$, will be denoted by $\mathbf{F}_i(\lambda|j)$. The elementary row matrices are nonsingular; so are the elementary column matrices $\mathbf{F}_{ij}$, $\mathbf{F}_i(\lambda)$, $\mathbf{F}_i(\lambda|j)$.

EXAMPLE: Find the matrix $\mathbf{F}$ which, when premultiplied by $\mathbf{A}$, interchanges columns 1 and 3 and multiplies column 2 by 2:

$$\mathbf{A} = \begin{bmatrix} 3 & 2 & 1 \\ 1 & 5 & 4 \end{bmatrix}.$$

Matrix $\mathbf{F}_{13}$ which interchanges columns 1, 3 is obtained by interchanging columns 1 and 3 in the identity matrix:

$$\mathbf{F}_{13} = \begin{bmatrix} 0 & 0 & 1 \\ 0 & 1 & 0 \\ 1 & 0 & 0 \end{bmatrix}.$$

Matrix $\mathbf{F}_2(2)$ which multiplies column 2 of $\mathbf{A}$ by 2 is found by multiplying column 2 of $\mathbf{I}$ by 2:

$$\mathbf{F}_2(2) = \begin{bmatrix} 1 & 0 & 0 \\ 0 & 2 & 0 \\ 0 & 0 & 1 \end{bmatrix}.$$

Hence the required matrix $\mathbf{F}$ is $\mathbf{F}_{13}\mathbf{F}_2(2)$, or

$$\mathbf{F} = \mathbf{F}_{13}\mathbf{F}_2(2) = \begin{bmatrix} 0 & 0 & 1 \\ 0 & 1 & 0 \\ 1 & 0 & 0 \end{bmatrix} \begin{bmatrix} 1 & 0 & 0 \\ 0 & 2 & 0 \\ 0 & 0 & 1 \end{bmatrix} = \begin{bmatrix} 0 & 0 & 1 \\ 0 & 2 & 0 \\ 1 & 0 & 0 \end{bmatrix};$$

$$\mathbf{AF} = \begin{bmatrix} 3 & 2 & 1 \\ 1 & 5 & 4 \end{bmatrix} \begin{bmatrix} 0 & 0 & 1 \\ 0 & 2 & 0 \\ 1 & 0 & 0 \end{bmatrix} = \begin{bmatrix} 1 & 4 & 3 \\ 4 & 10 & 1 \end{bmatrix}.$$

The matrix $\mathbf{F}$ does indeed perform the desired elementary column operations; it is obtained by applying the elementary column operations to the identity matrix.

4–6 Echelon matrices and rank. By means of a series of elementary row operations, any $m \times n$ matrix $\mathbf{A}$ can be reduced to a so-called echelon matrix which has the following structure:

(1) The first k rows, $k \geq 0$, are nonzero (that is, one or more elements in the row are not zero) and all the elements of the remaining $m - k$ rows are zero.

(2) In the ith row, $i = 1, \ldots, k$ (if $k \geq 1$), the first nonzero element (reading from left to right) equals unity. The symbol c_i will denote the column in which the element unity occurs.

(3) Then the arrangement of the rows is such that $c_1 < c_2 < \cdots < c_k$.

$$\mathbf{H} = \begin{bmatrix} 0 & 1 & h_{13} & h_{14} & h_{15} & h_{16} \\ 0 & 0 & 0 & 1 & h_{25} & h_{26} \\ 0 & 0 & 0 & 0 & 1 & h_{36} \\ 0 & 0 & 0 & 0 & 0 & 0 \end{bmatrix} \quad \text{is a typical example of an echelon matrix.} \quad (4\text{–}29)$$

We shall furnish constructive proof that any matrix can be converted into an echelon matrix by elementary row operations. The term "constructive," as applied here, means that the proof will actually describe in detail how this reduction is effected. Starting with matrix $\mathbf{A}$, we move to the first column, let us say j, which has at least one element different from zero. If a nonzero element does not occur in the first row, we interchange the first row and any other row where a nonzero element appears. This element, whose value will be denoted by α, can be converted to unity by performing the elementary operation of dividing row 1 by α. If there are other nonzero elements in column j, they can be reduced to zero by subtracting β times the first row from the row where the nonzero element occurs (β represents the value of the nonzero element). Columns 1 to j and row 1 are now taken care of.

Starting from column j, we move to the next column, h, where at least one nonzero element appears below the first row. If the element in row 2 is zero, we interchange row 2 and any row (with an index >2) having a nonzero element. The nonzero element in the new row 2, column h, is converted to unity by dividing row 2 by the value of this element. Any other nonzero elements in column h (with row index >2) can be converted to zero by subtracting from that row a constant times row 2. Continuing in this way, we ultimately obtain an echelon matrix.

In the actual process of reducing any matrix **A** to an echelon matrix, it is not necessary to find the elementary matrices and carry out the matrix multiplication. The elementary operations can be performed directly. It is very important, however, to know that the reduction can be carried out by premultiplying **A** by a matrix **E** which is the product of elementary matrices. Furthermore, **E** is nonsingular. We can write, therefore,

$$\mathbf{H} = \mathbf{EA}, \tag{4-30}$$

where **H** is an echelon matrix. Since **E** is nonsingular, $r(\mathbf{H}) = r(\mathbf{A})$ (see Section 4–3). This result is interesting because the rank of **H** can now be read off at a glance. The rank of **H** is simply the number of nonzero rows in **H**.

To prove that $r(\mathbf{H})$ is k, the number of nonzero rows in **H**, observe first that $r(\mathbf{H})$ cannot be greater than k. Now it is only necessary to establish the linear independence of the k nonzero rows. Let $\mathbf{h}^1, \ldots, \mathbf{h}^k$ denote the nonzero rows of **H**. We try to determine λ_i such that

$$\sum \lambda_i \mathbf{h}^i = \mathbf{0}. \tag{4-31}$$

Row $\mathbf{h}^1$ has an element unity in column c_1, while all other rows $\mathbf{h}^i$ have zeros in this column. Thus, $\lambda_1 = 0$. In the same way, $\lambda_2 = \lambda_3 = \cdots = \lambda_k = 0$. Therefore, the first k rows are linearly independent and $r(\mathbf{H}) = k$.

The reduction of a matrix to an echelon matrix is a fairly efficient numerical procedure for computing the rank of a matrix. It is usually much more efficient than the process of finding the largest nonvanishing determinant in **A**, especially if **A** is large. However, the task of determining the rank of a large matrix is almost always difficult, irrespective of the technique applied to the problem.

150   LINEAR TRANSFORMATIONS   [CHAP. 4

EXAMPLE: By converting **A** to an echelon matrix determine the rank of

$$\mathbf{A} = \begin{bmatrix} 0 & 0 & 1 & 2 & 8 & 9 \\ 0 & 0 & 4 & 6 & 5 & 3 \\ 0 & 2 & 3 & 1 & 4 & 7 \\ 0 & 3 & 0 & 9 & 3 & 7 \\ 0 & 0 & 5 & 7 & 3 & 1 \end{bmatrix}.$$

Column 2 is the first one to have a nonzero element. Either row 3 or row 4 can be interchanged with row 1. Let us interchange rows 1 and 3. Now a 2 is the first nonzero element in the new row 1. This 2 is reduced to a 1 by dividing the new row 1 by 2. To convert the 3 (row 4, column 2) to a 0, 3 times the final version of row 1 is subtracted from row 4. After all these operations are performed, the following matrix is obtained:

$$\begin{bmatrix} 0 & 1 & 1.5 & 0.5 & 2 & 3.5 \\ 0 & 0 & 4 & 6 & 5 & 3 \\ 0 & 0 & 1 & 2 & 8 & 9 \\ 0 & 0 & -4.5 & 7.5 & -3 & -3.5 \\ 0 & 0 & 5 & 7 & 3 & 1 \end{bmatrix}.$$

The third column consists entirely of nonzero elements. We divide row 2 by 4 and subtract this new row 2 from row 3; then we subtract -4.5 times row 2 from row 4, and 5 times row 2 from row 5, and arrive at

$$\begin{bmatrix} 0 & 1 & 1.5 & 0.5 & 2 & 3.5 \\ 0 & 0 & 1 & 1.5 & 1.25 & 0.75 \\ 0 & 0 & 0 & 0.5 & 6.75 & 8.25 \\ 0 & 0 & 0 & 14.25 & 2.625 & -0.125 \\ 0 & 0 & 0 & -0.5 & -3.25 & -2.75 \end{bmatrix}.$$

Moving next to the fourth column and third row, we obtain, after dividing row 3 by 0.5 and making the appropriate subtractions from rows 4 and 5,

$$\begin{bmatrix} 0 & 1 & 1.5 & 0.5 & 2 & 3.5 \\ 0 & 0 & 1 & 1.5 & 1.25 & 0.75 \\ 0 & 0 & 0 & 1 & 13.50 & 16.50 \\ 0 & 0 & 0 & 0 & -189.75 & -235.25 \\ 0 & 0 & 0 & 0 & 3.50 & 5.50 \end{bmatrix}.$$

Continuing in this manner, we carry out the last two steps and obtain

$$\begin{bmatrix} 0 & 1 & 1.5 & 0.5 & 2 & 3.5 \\ 0 & 0 & 1 & 1.5 & 1.25 & 0.75 \\ 0 & 0 & 0 & 1 & 13.50 & 16.50 \\ 0 & 0 & 0 & 0 & 1 & 1.238 \\ 0 & 0 & 0 & 0 & 0 & 1 \end{bmatrix}.$$

Consequently, the rank of this matrix, and hence of the original matrix, is 5. The preceding computations will have given the reader some awareness of the difficulty of determining the rank of a matrix. He may also feel that it would have been easier to find the largest nonvanishing determinant. In our example, the amount of work required is about the same, whatever our choice of procedure. For larger matrices, however, the systematic reduction of a matrix to its echelon form is vastly preferable.

To determine the rank of a matrix, it is only necessary to reduce it to an echelon matrix. However, by applying elementary column operations to the echelon matrix, we can carry the reduction even further. If the leading unity element appears in column j of row 1, then all the elements of row 1 in the columns following j can be reduced to zero by subtracting from any given column r with index $r > j$ the appropriate multiple of column j. Since the only nonzero element of column j appears in row 1, this operation does not affect any nonzero elements in column r with row index > 1. If the leading unity element of row 2 occurs in column q, $q > j$, the same procedure can be used to reduce all nonzero elements of row 2 in the columns following q to zero. Note that, because of the transformations on row 1, the only nonzero element of column q occurs in row 2. The same reduction is then carried out for the remaining rows. The resulting matrix is such that each of the first k rows has only one nonzero element, and this element is unity. The remaining $m - k$ rows are composed entirely of zeros. Furthermore, no column can contain more than a single nonzero entry. Precisely k columns will be unit vectors and the remaining $n - k$ columns will be composed entirely of zeros. Finally, we can interchange the columns so that the unity element appears in row i and column i, $i = 1, \ldots, k$. By a series of elementary column operations (characterized by the matrix $\mathbf{F}$) we have reduced the echelon matrix $\mathbf{H}$ to the unique form

$$\mathbf{HF} = \begin{bmatrix} \mathbf{I}_k & \mathbf{0} \\ \mathbf{0} & \mathbf{0} \end{bmatrix}. \tag{4-32}$$

But $\mathbf{H} = \mathbf{EA}$; thus, by a series of elementary row and column operations characterized by the nonsingular matrices **E**, **F**, any matrix **A** can be converted to the form

$$\mathbf{EAF} = \begin{bmatrix} \mathbf{I}_k & \mathbf{0} \\ \mathbf{0} & \mathbf{0} \end{bmatrix}, \tag{4-33}$$

if $r(\mathbf{A}) = k$. Depending on the size of **A**, some or all of the **0** submatrices in the right-hand side of (4-33) may not appear. If **A** is an nth-order nonsingular matrix, $\mathbf{EAF} = \mathbf{I}_n$.

EQUIVALENCE TRANSFORMATION: *A transformation of the type*

$$\mathbf{B} = \mathbf{EAF}, \tag{4-34}$$

where **E**, **F** *are nonsingular matrices, is called an equivalence transformation on* **A**.

Matrix **B** is said to be equivalent to **A**.

REFERENCES

The references dealing with matrices (see Chapter 3) apply to this chapter also.

1. R. A. FRAZER, W. J. DUNCAN, and A. R. COLLAR, *Elementary Matrices and Some Applications to Dynamics and Differential Equations.* Cambridge: Cambridge University Press, 1938.

Problems

4-1. Show whether the following transformations are linear:

(a) $y = 2x + \dfrac{4}{x}$, (b) $y = \ln x$, (c) $y = e^x$

(d) $y = \sin x$, (e) $\ln y = a \ln x$, (f) $y = 3x_1 + 2x_2$.

4-2. Illustrate geometrically the effect of the linear transformation

$$\mathbf{A} = \begin{bmatrix} 3 & 1 \\ 2 & 5 \end{bmatrix}$$

on the square whose vertices are [0, 0], [1, 0], [1, 1], [0, 1].

4-3. Interpret geometrically the meaning of the linear transformation on E^2,

$$y = 2x_1 + 3x_2.$$

Hint: Consider a line normal to the lines which are given by

$$y = 2x_1 + 3x_2.$$

4-4. Interpret geometrically the transformation produced on E^2 by

$$\mathbf{A} = \begin{bmatrix} 1 & 0 \\ 0 & 1 \\ 1 & 0 \end{bmatrix}.$$

What is the rank of the transformation? What is the dimension of the range?

4-5. Find the rank of the following matrices:

(a) $\mathbf{A} = (1)$, (b) $\mathbf{A} = \begin{bmatrix} 2 & 4 \\ 1 & 2 \end{bmatrix}$, (c) $\mathbf{A} = \begin{bmatrix} 3 & 2 \\ 1 & 1 \end{bmatrix}$, (d) $\mathbf{A} = \begin{bmatrix} 0 \\ 0 \end{bmatrix}$,

(e) $\mathbf{A} = \begin{bmatrix} 2 \\ 1 \end{bmatrix}$, (f) $\mathbf{A} = \begin{bmatrix} 3 & 2 & 1 \\ 4 & 6 & 2 \\ 0 & 0 & 0 \end{bmatrix}$, (g) $\mathbf{A} = \begin{bmatrix} 3 & 2 & 1 \\ 3 & 2 & 1 \\ 0 & 0 & 0 \end{bmatrix}$.

4-6. Find the rank of $\mathbf{A}$, $\mathbf{B}$, $\mathbf{C}$ where $\mathbf{C} = \mathbf{AB}$.

(a) $\mathbf{A} = \begin{bmatrix} 2 & 1 \\ 4 & 2 \end{bmatrix}$, $\mathbf{B} = \begin{bmatrix} 3 & 1 \\ 2 & 1 \end{bmatrix}$; (b) $\mathbf{A} = \begin{bmatrix} 2 & 1 \\ 4 & 2 \end{bmatrix}$, $\mathbf{B} = \begin{bmatrix} 4 & 2 \\ 3 & 1 \end{bmatrix}$;

(c) $\mathbf{A} = \begin{bmatrix} 3 & 1 \\ 2 & 1 \end{bmatrix}$, $\mathbf{B} = \begin{bmatrix} 4 & 2 \\ 3 & 1 \end{bmatrix}$; (d) $\mathbf{A} = \begin{bmatrix} 1 & 4 \\ 0 & 0 \end{bmatrix}$, $\mathbf{B} = \begin{bmatrix} 4 & 0 \\ -1 & 0 \end{bmatrix}$.

4-7. Find the rank of the following matrix by reducing it to an echelon matrix:

$$\mathbf{A} = \begin{bmatrix} 2 & 1 & 4 & 0 & 6 & 7 & 10 \\ 3 & 5 & 9 & 1 & 0 & 4 & 3 \\ 7 & 7 & 3 & 2 & 8 & 1 & 1 \\ -9 & -8 & 7 & -3 & -10 & 9 & 11 \\ 4 & 2 & -20 & 2 & 4 & -20 & -24 \end{bmatrix}.$$

4-8. Given the matrix

$$\mathbf{A} = \begin{bmatrix} 4 & 5 & 6 & -2 \\ -2 & 7 & 0 & 8 \\ 3 & 1 & 3 & 6 \\ 3 & 2 & 1 & 5 \end{bmatrix}.$$

(a) Find the matrix which, when postmultiplied by $\mathbf{A}$, interchanges rows 1 and 3, and 2 and 4. Carry out the multiplication to show that the required interchange is actually accomplished.

(b) Find the matrix which adds 6 times the third row to the first, multiplies the second row by 10, and then interchanges the second and third rows. Carry out the multiplication to show that the required transformations are obtained.

4-9. Given the matrix of Problem 4-8.

(a) Find the matrix which, when premultiplied by $\mathbf{A}$, interchanges columns 1 and 3 and multiplies columns 2 and 4 by 8. Carry out the multiplication.

(b) Find the matrix which adds 2 times the third column plus 4 times the second column to the first column. Carry out the multiplication to demonstrate that the correct result has been obtained.

4-10. Find the matrices $\mathbf{E}$, $\mathbf{F}$ such that $\mathbf{I}_3 = \mathbf{EAF}$:

$$\mathbf{A} = \begin{bmatrix} 4 & 1 & -2 \\ 2 & 5 & 1 \\ 3 & 2 & 0 \end{bmatrix}.$$

4-11. Show that any matrix $\mathbf{A}$ with $r(\mathbf{A}) = r$ can be written

$$\mathbf{A} = \mathbf{R}_1 \begin{bmatrix} \mathbf{I}_r & \mathbf{0} \\ \mathbf{0} & \mathbf{0} \end{bmatrix} \mathbf{R}_2.$$

What are $\mathbf{R}_1$ and $\mathbf{R}_2$?

4-12. If $\mathbf{A}$, $\mathbf{B}$ are $m \times n$ matrices with the same rank, show that there exist nonsingular matrices $\mathbf{R}_1$, $\mathbf{R}_2$ such that

$$\mathbf{B} = \mathbf{R}_1 \mathbf{A} \mathbf{R}_2.$$

What are $\mathbf{R}_1$, $\mathbf{R}_2$?

4-13. Which of the matrices representing elementary row transformations commute and which do not?

4-14. Evaluate the determinant of each of the matrices representing elementary row and column operations. Note that the values of the determinants are independent of the order of the matrices.

4-15. If $\mathbf{E}$, $\mathbf{F}$ are elementary matrices and $\mathbf{A}$ is a square matrix, prove that

$$|\mathbf{EA}| = |\mathbf{E}| \, |\mathbf{A}|, \qquad |\mathbf{AF}| = |\mathbf{A}| \, |\mathbf{F}|.$$

4-16. Prove that the determinant of the product of two or more elementary matrices is the product of the determinants.

4-17. Prove that if $\mathbf{A}$ is a nonsingular matrix, it can be written as the product of elementary matrices. Hint: see Problem 4-11.

4-18. Prove that if $\mathbf{A}$, $\mathbf{B}$ are nth-order matrices,

$$|\mathbf{AB}| = |\mathbf{A}| \, |\mathbf{B}|.$$

Hint: First, assume that $\mathbf{A}$, $\mathbf{B}$ are nonsingular and use the results of Problems 4-16 and 4-17. What is true if either $\mathbf{A}$, or $\mathbf{B}$, or both happen to be singular?

4-19. We have seen that, by a sequence of elementary row operations, any matrix $\mathbf{A}$ could be reduced to an echelon matrix. Show that, by additional elementary row transformations, a nonsingular matrix $\mathbf{A}$ can be reduced to an identity matrix; that is, prove that if $\mathbf{A}$ is an nth-order nonsingular matrix, then there exists a nonsingular matrix $\mathbf{E}$ such that

$$\mathbf{I}_n = \mathbf{EA}.$$

4-20. Using the result of Problem 4-19, find a matrix $\mathbf{E}$ such that $\mathbf{I}_3 = \mathbf{EA}$ when

$$\mathbf{A} = \begin{bmatrix} 2 & 7 & 9 \\ 3 & 0 & 1 \\ 4 & 6 & 5 \end{bmatrix}.$$

What is $\mathbf{E}$?

4-21. In Chapter 3, several methods were presented for evaluating determinants. None of the expansions, however, yielded a very efficient numerical procedure. The use of elementary transformations provides the key to a reasonably efficient technique. Note that if some row or column of a determinant has only one element which differs from zero, we can immediately expand by that row or column and reduce the determinant to one whose order is one less than that of the original. Let us choose any nonzero element in the determinant, for example, a_{11}. Divide the first row or column by a_{11}. Reduce the a_{1j} ($j \neq 1$)

to zero by adding a suitable multiple of column 1. Thus

$$\begin{vmatrix} a_{11} & a_{12} & \cdots & a_{1n} \\ a_{21} & a_{22} & \cdots & a_{2n} \\ \vdots & \vdots & & \vdots \\ a_{n1} & a_{n2} & \cdots & a_{nn} \end{vmatrix} = a_{11} \begin{vmatrix} 1 & \dfrac{a_{12}}{a_{11}} & \cdots & \dfrac{a_{1n}}{a_{11}} \\ a_{21} & a_{22} & \cdots & a_{2n} \\ \vdots & & & \vdots \\ a_{n1} & a_{n2} & \cdots & a_{nn} \end{vmatrix} = a_{11} \begin{vmatrix} 1 & 0 & \cdots & 0 \\ a_{21} & a'_{22} & \cdots & a'_{2n} \\ \vdots & \vdots & & \vdots \\ a_{n1} & a'_{n2} & \cdots & a'_{nn} \end{vmatrix}$$

$$= a_{11} \begin{vmatrix} a'_{22} & \cdots & a'_{2n} \\ \vdots & & \vdots \\ a'_{n2} & \cdots & a'_{nn} \end{vmatrix} ; \quad a'_{ij} = a_{ij} - \frac{a_{1j}}{a_{11}} a_{i1}.$$

The same procedure is repeated with $\begin{vmatrix} a'_{22} & \cdots & a'_{2n} \\ \vdots & & \vdots \\ a'_{n2} & \cdots & a'_{nn} \end{vmatrix}$.

This technique is called pivotal condensation. The element reduced to unity at any stage is the pivot. In this case, the element a_{11} was the initial pivot. It is often advisable to choose as pivot the largest element in absolute value Why? Apply the pivotal condensation method to the evaluation of the determinant

$$\begin{vmatrix} 4 & 5 & -1 & 3 & -2 \\ 8 & 9 & 0 & 7 & 6 \\ -4 & 3 & 2 & 8 & -7 \\ 5 & 1 & 1 & -6 & 2 \\ 3 & 10 & 1 & -1 & 9 \end{vmatrix}.$$

4-22. If we wish to reduce **A** to an echelon matrix and, at the same time, find matrix **E** which carries out the reduction, reduce columns 1 through n of matrix $(\mathbf{A}, \mathbf{I})$ to an echelon matrix. After reduction, **E** is found in the columns originally occupied by **I** since

$$\mathbf{E}(\mathbf{A}, \mathbf{I}) = (\mathbf{H}, \mathbf{E}).$$

Reduce

$$\begin{bmatrix} 1 & 2 & 3 \\ 4 & 5 & 6 \\ 7 & 8 & 9 \end{bmatrix}$$

to an echelon matrix and, at the same time, find the matrix **E** which, when postmultiplied by **A**, yields the echelon matrix.

4-23. The results of Problem 4-22 can be used to develop a numerical procedure for finding the inverse of a nonsingular matrix. Problem 4-19 shows that there exists a nonsingular matrix $\mathbf{E}$ such that $\mathbf{I} = \mathbf{EA}$, and hence $\mathbf{E} = \mathbf{A}^{-1}$. Thus, if we form $(\mathbf{A}, \mathbf{I})$ and reduce the first n columns to an identity matrix by elementary row operations, the last n columns will contain $\mathbf{A}^{-1}$. Using this method, compute the inverse of

$$\mathbf{A} = \begin{bmatrix} 4 & 1 & -2 \\ 2 & 6 & 1 \\ 3 & 2 & 0 \end{bmatrix}.$$

4-24. Express the vector $\mathbf{b} = [2, 7]$ in terms of the following bases:

(a) $\mathbf{a}_1 = [3, 4]$, $\quad \mathbf{a}_2 = [2, 5]$, $\quad$ (b) $\mathbf{a}_1 = [1, 1]$, $\quad \mathbf{a}_2 = [0, 1]$,

(c) $\mathbf{a}_1 = [7, 2]$, $\quad \mathbf{a}_2 = [3, 1]$, $\quad$ (d) $\mathbf{a}_1 = [2, 1]$, $\quad \mathbf{a}_2 = [3, 5]$.

4-25. Prove that every linear transformation of E^n into a subspace of E^m is equivalent to, and can be represented by, a matrix transformation. Hint: The linear transformation T is completely characterized by the way it transfers the unit vectors, since

$$T(\mathbf{x}) = \sum_{j=1}^{n} x_j T(\mathbf{e}_j).$$

However, $T(\mathbf{e}_j)$ is an element of E^m and can be written

$$T(\mathbf{e}_j) = \sum_{i=1}^{m} a_{ij} \boldsymbol{\epsilon}_i,$$

where the $\boldsymbol{\epsilon}_i$ are the unit vectors for E^m. Does the matrix $\mathbf{A} = \|a_{ij}\|$ characterize the linear transformation? Show that if $\mathbf{y} = T(\mathbf{x})$, then $\mathbf{y} = \mathbf{Ax}$. Using the preceding results, prove that, for fixed bases in E^n, E^m, the matrix which characterizes the linear transformation is unique.

4-26. Consider the linear transformation of E^3 into E^3 described by

$$T(\mathbf{e}_1) = 8\boldsymbol{\epsilon}_1 + 6\boldsymbol{\epsilon}_2 + \boldsymbol{\epsilon}_3, \quad T(\mathbf{e}_2) = 2\boldsymbol{\epsilon}_1 + 5\boldsymbol{\epsilon}_3, \quad T(\mathbf{e}_3) = \boldsymbol{\epsilon}_1 + 2\boldsymbol{\epsilon}_2,$$

where the $\boldsymbol{\epsilon}_i$ are also unit vectors. What is the matrix $\mathbf{A}$ which represents this transformation in such a way that $\mathbf{y} = \mathbf{Ax}$? If, instead of using the unit vectors $\mathbf{e}_1, \mathbf{e}_2, \mathbf{e}_3$, we use the vectors $\mathbf{v}_1 = [6, 1, 1]$, $\mathbf{v}_2 = [3, 7, 5]$, $\mathbf{v}_3 = [0, 1, 6]$ for a basis, what is the matrix which characterizes the linear transformation?

4-27. A linear transformation takes the vector $\mathbf{a}_1 = [3, 4]$ into $\mathbf{b}_1 = [-5, 6]$ and the vector $\mathbf{a}_2 = [-1, 2]$ into $\mathbf{b}_2 = [4, 1]$. What matrix represents this linear transformation when the unit vectors are used as a basis?

4-28. A linear transformation T takes $[1, 1]$ into $[0, 1, 2]$ and $[-1, 1]$ into $[2, 1, 0]$. What matrix $\mathbf{A}$ represents T relative to the basis $[1, 1]$, $[-1, 1]$ in E^2 and $\boldsymbol{\epsilon}_1, \boldsymbol{\epsilon}_2, \boldsymbol{\epsilon}_3$, the unit vectors, in E^3?

4-29. Let T_1 be a linear transformation which takes E^n into a subspace of E^r and let **B** characterize this transformation for fixed bases in E^n and E^r. Furthermore, assume that T_2 is a linear transformation which takes E^r into a subspace of E^m and that **A** characterizes this transformation for fixed bases in E^r and E^m. Under what conditions does **AB** characterize the product T_2T_1? Carry out the details to show that if appropriate assumptions are made, **AB** does represent T_2T_1.

4-30. Consider the elementary operation of multiplying a row or column of a matrix by a scalar λ; why do we require that $\lambda \neq 0$?

4-31. Show that if columns i and j are interchanged in a nonsingular matrix **B**, the inverse of the new matrix can be found by interchanging rows i and j in $\mathbf{B}^{-1}$. Hint: The new matrix can be written **BF**.

4-32. Let **A** be a matrix obtained from a nonsingular matrix **B** by a given series of elementary operations. Discuss how $\mathbf{A}^{-1}$ can be found from $\mathbf{B}^{-1}$. Consider individually each type of elementary operation and the sequence in which they are performed.

4-33. A set of basis vectors $\mathbf{b}_1, \ldots, \mathbf{b}_n$ for E^n is said to be triangular if, for $\mathbf{b}_j = [b_{1j}, \ldots, b_{nj}]$, $b_{ij} = 0$ $(i > j)$. This means that the matrix $\mathbf{B} = (\mathbf{b}_1 \ldots, \mathbf{b}_n)$ is triangular (see Problem 3-32). Show how a triangular basis may be obtained for E^n from any set of basis vectors.

4-34. Given a triangular basis for E^n, $\mathbf{b}_1, \ldots, \mathbf{b}_n$; show that if any other vector **x** is expressed as a linear combination of the $\mathbf{b}_j$, $\mathbf{x} = \sum \lambda_j \mathbf{b}_j$, the λ_j can be computed sequentially and the inverse of $\mathbf{B} = (\mathbf{b}_1, \ldots, \mathbf{b}_n)$ need not be found. Hint: Only $\mathbf{b}_n$ has an nth component different from zero. Thus $\lambda_n = x_n/b_{nn}$. What is λ_{n-1}, etc?

4-35. An nth-order matrix **A** is called decomposable if by interchanging some rows and the corresponding columns it is possible to obtain a null matrix in the lower left-hand corner so that **A** can be written ($\mathbf{A}_{11}$, $\mathbf{A}_{22}$ square)

$$\mathbf{A} = \begin{bmatrix} \mathbf{A}_{11} & \mathbf{A}_{12} \\ 0 & \mathbf{A}_{22} \end{bmatrix}.$$

By analogy, **A** is called indecomposable if it is not possible to obtain the required single zero element in the lower left position. Note that if rows i, j are interchanged, then columns i, j are interchanged also. It may turn out that $\mathbf{A}_{11}$, $\mathbf{A}_{22}$ in the foregoing expression are also decomposable. Any decomposable matrix can then be reduced to the triangular form

$$\mathbf{A} = \begin{bmatrix} \mathbf{A}_1 & \mathbf{B}_{12} & \cdots & \mathbf{B}_{1k} \\ 0 & \mathbf{A}_2 & \cdots & \mathbf{B}_{2k} \\ \vdots & & & \vdots \\ 0 & 0 & \cdots & \mathbf{A}_k \end{bmatrix},$$

where $\mathbf{A}_1, \ldots, \mathbf{A}_k$ are indecomposable. Show that the following matrix is decomposable and write it in triangular form:

$$\begin{bmatrix} 1 & 0 & 0 & 0 & 0 \\ 4 & 4 & 6 & 8 & 1 \\ 9 & 0 & 1 & 2 & 0 \\ 3 & 0 & 0 & 4 & 0 \\ 1 & 3 & 5 & 7 & 0 \end{bmatrix}.$$

4–36. Prove that the square decomposable matrix **A** can be decomposed by the equivalence transformation **EAF** with $\mathbf{F} = \mathbf{E}'$, **E** being the product of the elementary matrices $\mathbf{E}_{ij}$.

4–37. Show that reducing a decomposable matrix to the triangular form discussed in Problem 4–35 does not change the value of the determinant of **A**.

4–38. Is the sequence of elementary row and column operations, i.e., the matrices **E**, **F** which reduce **A** to the form of Eq. (4–33), unique? Can you provide an example where they are not unique?

4–39. Devise a matrix **B** such that if the same row and column operations which reduce **A** to the form of Eq. (4–33) are performed on **B**, then **E** and **F** can be found in **B** also.

4–40. Express the vector $\mathbf{b} = [2, 7]$ in terms of the following triangular bases:

(a) $\mathbf{a}_1 = [3, 4]$, $\quad \mathbf{a}_2 = [2, 0]$; $\qquad$ (b) $\mathbf{a}_1 = [1, 1]$, $\quad \mathbf{a}_2 = [0, 1]$;

(c) $\mathbf{a}_1 = [7, 2]$, $\quad \mathbf{a}_2 = [3, 0]$; $\qquad$ (d) $\mathbf{a}_1 = [2, 1]$, $\quad \mathbf{a}_2 = [3, 0]$.

4–41. Consider the matrices

$$\mathbf{A} = \begin{bmatrix} 3 & 1 & 2 \\ 2 & 1 & 1 \end{bmatrix}; \quad \mathbf{B} = \begin{bmatrix} 2 & 0 \\ -3 & 2 \\ -1 & -1 \end{bmatrix}.$$

Show that $\mathbf{AB} = \mathbf{I}_2$. Why do we not want to call **B** the inverse of **A**? Interpret this geometrically.

4–42. Consider the process of reducing a matrix **A** to row echelon form. If $r(\mathbf{A}) = k$, this can be considered to be a k-stage process. At stage s, $s = 1, \ldots, k$, the first nonzero element in row s is converted to unity. If this element lies in column r, all elements in column r with row index $> s$ are reduced to zero. Denote the elements in the matrix at the beginning of stage s by u_{ij}. Show that the elements $\hat{u}_{ij}$ of the matrix at the end of stage s are given by

$$\hat{u}_{ij} = u_{ij}, \quad i < s; \quad \hat{u}_{sj} = \frac{u_{sj}}{u_{sr}}, \quad \text{all } j; \quad \hat{u}_{ij} = u_{ij} - \frac{u_{sj}}{u_{sr}} u_{ir}, \quad i > s, \quad \text{all } j.$$

The matrix at the beginning of stage s will be the same as the matrix at the end of stage $s - 1$ unless it is necessary to interchange rows so that the first nonzero element of row s will have as low a column index as possible.

4-43. It is desirable to have some sort of automatic check on the numerical work necessary in reducing a matrix to row echelon form or in finding the inverse of a matrix by means of the technique discussed in Problem 4-23. A very simple check, called a sum check, can be used. This check requires only a slight amount of additional effort. Suppose that we wish to reduce **A** to row echelon form. Let t_i be the sum of the elements in the ith row of **A**, that is, $t_i = \sum_j a_{ij}$, or if $\mathbf{t} = [t_1, \ldots, t_m]$, then $\mathbf{t} = \sum_{j=1}^n \mathbf{a}_j$. Consider the matrix $\mathbf{B} = (\mathbf{A}, \mathbf{t})$ which has the same number of rows as **A**, but one additional column **t**. We then reduce **B** to row echelon form. The check is as follows: At each stage, the element in column $n + 1$ of row s should be the sum of the elements in the first n columns of row s. Thus after each stage, we sum the first n elements in row s; the result should be the number which appears as element $n + 1$ of row s. This is done for every row s. Prove that this is true. Hint: See Problem 4-42.

4-44. By means of the sum check show that the numerical computations in the example of Section 4-6 are correct.

4-45. Discuss in detail the way the sum check is used in inverting a matrix by the technique of Problem 4-23. Illustrate this by adding a sum check column when computing the inverse of **A** in that problem.

Problems Involving Complex Numbers

4-46. List the important results of this chapter and show that they all hold if the elements of the matrices are complex numbers. Demonstrate that even geometrical interpretations can be given a meaning in $V_n(c)$, that is, in the n-dimensional vector space of all n-tuples with complex components.

4-47. Find the rank of the following matrix **A** by reducing it to row echelon form:

$$\mathbf{A} = \begin{bmatrix} 2+i & 3-4i & 5 & 6+2i \\ 7 & 3-i & 2i & 9-3i \\ -6i & 8 & 7+6i & 4-8i \\ 5-3i & 2i & 6+i & 1-i \end{bmatrix}.$$

4-48. Invert the following matrix **A**, using the technique suggested in Problem 4-23:

$$\mathbf{A} = \begin{bmatrix} 4+3i & 2 & 6-i \\ -5+i & 9-2i & 3 \\ 4i & -6 & 8+i \end{bmatrix}.$$

4-49. Find nonsingular matrices **E**, **F** such that $\mathbf{B} = \mathbf{EAF}$, where

$$\mathbf{A} = \begin{bmatrix} i & 5 \\ 6-7i & 2+3i \end{bmatrix}, \quad \mathbf{B} = \begin{bmatrix} 3 & 6i \\ 1-i & 4+2i \end{bmatrix}.$$

4-50. Find the matrix **E** which, when postmultiplied by **A**, interchanges rows 1, 3, multiplies row 2 by $7 - i$, and subtracts $-2 + 3i$ times row 4 from row 5. Show by actual multiplication that **E** does indeed perform these operations.

$$\mathbf{A} = \begin{bmatrix} 4 - i & 2 \\ 3i & -4 - 2i \\ 3 + 2i & -5i \\ 5 - 6i & 7 + 2i \\ 7i & 8 - i \end{bmatrix}.$$

4-51. Let **A**, **B**, **C** be matrices with real or complex elements such that $\mathbf{C} = \mathbf{AB}$. Assume that **A** is $m \times r$ and **B** is $r \times n$. If $r(\mathbf{C}) = m$, show that $r(\mathbf{A}) = m$. What restriction does this place on r? What can be said about the rank of **B**? What restriction does this place on n? If $m = n$, show that $r(\mathbf{A}) = r(\mathbf{B}) = m$. What restriction is thereby placed on r?

CHAPTER 5

SIMULTANEOUS LINEAR EQUATIONS

"Wavering between the profit and the loss
In this brief transit where the dreams cross"

T. S. Eliot.

5–1 Introduction. Sets of simultaneous linear equations appear in most linear models. Frequently, the number of equations will be equal to the number of variables (as in the Leontief economic and the statistical regression models of Chapter 1); in such cases, as is to be expected, we are usually able to solve for unique values of the variables. If there are more variables than equations (the constraints of linear programming problems may provide such an example), we expect, in general, to obtain an infinite number of solutions. Sometimes, we have more equations than variables. Thus for the d-c circuit (Chapter 1), it is possible to write down many more equations than there are variables. However, not all of these equations are independent since some of them can be obtained from the others. Under such circumstances, it is desirable to find enough independent equations to be able to solve for all the variables.

This chapter deals with the theory of simultaneous linear equations. We shall be concerned with deriving criteria for the existence and uniqueness of solutions and with the properties of solutions. We shall begin by discussing a fairly efficient numerical technique for solving simultaneous equations.

5–2 Gaussian elimination. In the real world, it is normally expected that n equations (linear or not) relating n variables can be solved to yield a set of numerical values for the n variables. Frequently, physical intuition leads us to assume also that the solution will be unique. Let us suppose that we have n linear equations relating n variables. They can be written:

$$a_{11}x_1 + \cdots + a_{1n}x_n = b_1,$$
$$\vdots \qquad\qquad\qquad\qquad (5\text{--}1)$$
$$a_{n1}x_1 + \cdots + a_{nn}x_n = b_n,$$

or

$$\mathbf{Ax} = \mathbf{b} \qquad\qquad (5\text{--}2)$$

when Eq. (5–1) is written in matrix form. We wish to solve this system of equations, that is, find the values $x_1, \ldots, x_n$ which will satisfy the equations.

5-2] GAUSSIAN ELIMINATION

A procedure which immediately suggests itself is that of successive elimination of the variables. Without any loss in generality, we can assume $a_{11} \neq 0$ since the equations can be rearranged and the variables renamed to conform with our premise. Then we solve explicitly for x_1 and obtain

$$x_1 = -\frac{a_{12}}{a_{11}} x_2 - \cdots - \frac{a_{1n}}{a_{11}} x_n + \frac{b_1}{a_{11}}. \tag{5-3}$$

Dividing the first equation by a_{11} to reduce the coefficient of x_1 to unity, and using Eq. (5-3) to eliminate x_1 in the remaining $n-1$ equations, we have

$$x_1 + \frac{a_{12}}{a_{11}} x_2 + \cdots + \frac{a_{1n}}{a_{11}} x_n = \frac{b_1}{a_{11}},$$

$$\left[a_{22} - a_{21}\left(\frac{a_{12}}{a_{11}}\right)\right] x_2 + \cdots + \left[a_{2n} - a_{21}\left(\frac{a_{1n}}{a_{11}}\right)\right] x_n = b_2 - a_{21} \frac{b_1}{a_{11}},$$

$$\vdots$$

$$\left[a_{n2} - a_{n1}\left(\frac{a_{12}}{a_{11}}\right)\right] x_2 + \cdots + \left[a_{nn} - a_{n1}\left(\frac{a_{1n}}{a_{11}}\right)\right] x_n = b_n - a_{n1} \frac{b_1}{a_{11}},$$

$$\tag{5-4}$$

or

$$x_1 + a'_{12} x_2 + \cdots + a'_{1n} x_n = b'_1,$$

$$a'_{22} x_2 + \cdots + a'_{2n} x_n = b'_2,$$

$$\vdots \tag{5-5}$$

$$a'_{n2} x_2 + \cdots + a'_{nn} x_n = b'_n.$$

If at least one of the a'_{ij} $(i, j = 2, \ldots, n)$ differs from zero, we can assume, without loss of generality, that $a'_{22} \neq 0$. The reduction process is continued by dividing the second equation of (5-5) by a'_{22} and by using this equation to eliminate x_2 in equations $3, \ldots, n$. Then, as expected, x_3 is eliminated from equations $4, \ldots, n$ until, finally, we obtain the system

$$x_1 + h_{12} x_2 + \cdots + h_{1n} x_n = g_1,$$

$$x_2 + \cdots + h_{2n} x_n = g_2,$$

$$\cdots \tag{5-6}$$

$$x_{n-1} + h_{n-1\,n} x_n = g_{n-1},$$

$$x_n = g_n.$$

We obtain immediately $x_n = g_n$. This value of x_n is then substituted into the preceding $n-1$ equations. Thus

$$x_{n-1} = g_{n-1} - h_{n-1} g_n.$$

This result is also substituted into the remaining $n - 2$ equations, etc. By continuing this process of back substitution, we obtain all the x_i values. This procedure is called gaussian reduction or gaussian elimination.

We can now make an interesting observation: Since the final set of equations can be written in the form of

$$\begin{bmatrix} 1 & h_{12} & \cdots & h_{1n} \\ 0 & 1 & \cdots & h_{2n} \\ \vdots & & & \vdots \\ 0 & 0 & \cdots & 1 \end{bmatrix} \begin{bmatrix} x_1 \\ x_2 \\ \vdots \\ x_n \end{bmatrix} = \begin{bmatrix} g_1 \\ g_2 \\ \vdots \\ g_n \end{bmatrix}, \tag{5-7}$$

or as

$$\mathbf{Hx} = \mathbf{g}, \tag{5-8}$$

it becomes clear that $\mathbf{H}$ represents the echelon matrix that would be obtained from $\mathbf{A}$, following the rules of Section 4–6.* We only have to recall the result of the operations performed on the elements of $\mathbf{A}$ in order to see that the gaussian reduction scheme does indeed convert $\mathbf{A}$ into an echelon matrix $\mathbf{H}$. If matrix $\mathbf{E}$ represents the combination of elementary row operations which takes $\mathbf{A}$ into $\mathbf{H}$, then

$$\mathbf{H} = \mathbf{EA}, \quad \mathbf{g} = \mathbf{Eb}. \tag{5-9}$$

This observation supplies us with some useful information. We know that (5–1) will reduce to the form of (5–6) if and only if the rank of $\mathbf{A}$ is n. If rank $\mathbf{A}$ is less than n, the method fails for one of two reasons: Either we shall not be able to determine the values of all the variables or there will be an apparent inconsistency evidenced by the fact that all h_{ij} vanish in some row, while g_i does not vanish. At present, we are not sure what this failure implies. This will become clear later. However, we know that the method will work if $r(\mathbf{A}) = n$.

Instead of eliminating x_k only in equations $k + 1, \ldots, n$, we could equally well eliminate x_k in equations $1, \ldots, k - 1$ also, so that x_k would appear only in the kth equation. Now, back substitution is not needed. This modification of gaussian elimination is called the Gauss-Jordan method. Both reduction schemes are iterative procedures, and we would think of them first in attempting to solve a set of linear equations. Interestingly enough, they are fairly efficient numerical procedures, and modifications of them are among the methods used for solving systems of linear equations either by hand or on high-speed computers. The

---

* At this point, it should be obvious that the definition of elementary operations on matrices follows logically from the manipulations of simultaneous linear equations.

following example will illustrate both the Gauss and the Gauss-Jordan methods in a simple case.

EXAMPLE: Solve the following set of linear equations:

$$2x_1 + x_2 + 4x_3 = 16,$$
$$3x_1 + 2x_2 + x_3 = 10,$$
$$x_1 + 3x_2 + 3x_3 = 16.$$

(a) *Gauss reduction*: We use the first equation to solve for x_1 and substitute this into the second and third equations. This yields

$$x_1 + \tfrac{1}{2}x_2 + 2x_3 = 8,$$
$$\tfrac{1}{2}x_2 - 5x_3 = -14,$$
$$\tfrac{5}{2}x_2 + x_3 = 8.$$

Using the second equation from this new set, we eliminate x_2 in the third equation. The first equation remains unchanged. Thus

$$x_1 + \tfrac{1}{2}x_2 + 2x_3 = 8,$$
$$x_2 - 10x_3 = -28,$$
$$26x_3 = 78.$$

From the third equation $x_3 = 3$. Substituting this into the first two equations, we find

$$x_1 + \tfrac{1}{2}x_2 = 2,$$
$$x_2 = 2.$$

Hence

$$x_1 = 1, \quad x_2 = 2, \quad x_3 = 3.$$

(b) *Gauss-Jordan reduction*: The first step is the same as under (a):

$$x_1 + \tfrac{1}{2}x_2 + 2x_3 = 8,$$
$$\tfrac{1}{2}x_2 - 5x_3 = -14,$$
$$\tfrac{5}{2}x_2 + x_3 = 8.$$

Using the second equation, we solve for x_2. The result is now substituted into both the first and third equations. This gives

$$x_1 + 7x_3 = 22,$$
$$x_2 - 10x_3 = -28,$$
$$26x_3 = 78.$$

On obtaining $x_3 = 3$, we immediately find $x_1 = 1$, $x_2 = 2$.

5–3 Cramer's rule. Let us again consider the system of n simultaneous linear equations in n unknowns (5–1). Repeating the matrix form, we have

$$\mathbf{A}\mathbf{x} = \mathbf{b}, \tag{5-10}$$

where

$$\mathbf{A} = \|a_{ij}\|; \quad \mathbf{x} = [x_1, \ldots, x_n], \quad \mathbf{b} = [b_1, \ldots, b_n].$$

Now assume $r(\mathbf{A}) = n$ so that $\mathbf{A}$ is nonsingular. This means that $|\mathbf{A}| \neq 0$ and $\mathbf{A}^{-1}$ exists. Premultiplying (5–10) by $\mathbf{A}^{-1}$, we arrive at

$$\mathbf{A}^{-1}\mathbf{A}\mathbf{x} = \mathbf{I}\mathbf{x} = \mathbf{x} = \mathbf{A}^{-1}\mathbf{b}. \tag{5-11}$$

Hence, when $\mathbf{A}$ is nonsingular, we obtain a unique solution $\mathbf{x} = \mathbf{A}^{-1}\mathbf{b}$ to the set of equations (5–1). The solution is unique since the inverse is unique. Thus, in Eq. (5–11), we have arrived at an explicit solution to the set of equations through the use of the inverse matrix. We are, however, no closer to a numerical solution than we were at the outset unless we happen to know $\mathbf{A}^{-1}$. Nevertheless, because of its explicit character the solution of (5–11) is of great use in theoretical work. The value of the originally unknown vector is expressed as $\mathbf{x} = \mathbf{A}^{-1}\mathbf{b}$, and $\mathbf{A}^{-1}\mathbf{b}$ can be computed from the known quantities $\mathbf{A}$, $\mathbf{b}$.

Equation (5–11) can be cast into a more interesting form. We recall from Section 3–18 that the inverse is given by

$$\mathbf{A}^{-1} = |\mathbf{A}|^{-1}\mathbf{A}^+; \tag{5-12}$$

$$\mathbf{A}^+ = \begin{bmatrix} A_{11} & \cdots & A_{n1} \\ \vdots & & \vdots \\ A_{1n} & \cdots & A_{nn} \end{bmatrix}, \tag{5-13}$$

where A_{ij} is the cofactor of element a_{ij} in $\mathbf{A}$. Thus we can write Eq. (5–11) in component form as

$$x_i = |\mathbf{A}|^{-1} \sum_{j=1}^{n} A_{ji}b_j = |\mathbf{A}|^{-1} \sum_{j=1}^{n} b_j A_{ji}. \tag{5-14}$$

However, we remember that

$$\sum_{j=1}^{n} a_{ji} A_{ji}$$

is the expansion of $|\mathbf{A}|$ by column i. On comparison, we see that

$$\sum_{j=1}^{n} b_j A_{ji}$$

is the expansion of the determinant formed from $\mathbf{A}$ by removing the ith column and replacing it with column **b**. This yields what is called Cramer's rule: *To obtain the value of x_i, we divide by $|A|$ the determinant of the matrix formed from **A** by replacing the ith column with **b**.* Hence

$$x_1 = \frac{1}{|A|} \begin{vmatrix} b_1 & a_{12} & \cdots & a_{1n} \\ b_2 & a_{22} & \cdots & a_{2n} \\ \vdots & \vdots & & \vdots \\ b_n & a_{n2} & \cdots & a_{nn} \end{vmatrix}, \ldots, x_n = \frac{1}{|A|} \begin{vmatrix} a_{11} & \cdots & a_{1n-1} & b_1 \\ a_{21} & \cdots & a_{2n-1} & b_2 \\ \vdots & & \vdots & \vdots \\ a_{n1} & \cdots & a_{nn-1} & b_n \end{vmatrix}. \quad (5\text{-}15)$$

This, of course, is the method learned in elementary algebra for solving equations by determinants. Cramer's rule is singularly inefficient for solving a set of equations numerically. The evaluation of the $n+1$ determinants involves too much work, especially, when n is fairly large. The gaussian reduction method is considerably more efficient. However, Cramer's rule is very useful in theoretical studies, as is $\mathbf{x} = \mathbf{A}^{-1}\mathbf{b}$, because it allows an explicit expression for the solution.

EXAMPLE: Solve
$$3x_1 + 2x_2 = 7,$$
$$4x_1 + x_2 = 1;$$

$$|A| = \begin{vmatrix} 3 & 2 \\ 4 & 1 \end{vmatrix} = -5 \neq 0.$$

Thus a unique solution exists and is given by

$$x_1 = -\frac{1}{5} \begin{vmatrix} 7 & 2 \\ 1 & 1 \end{vmatrix} = -\frac{1}{5}(7-2) = -1,$$

$$x_2 = -\frac{1}{5} \begin{vmatrix} 3 & 7 \\ 4 & 1 \end{vmatrix} = -\frac{1}{5}(3-28) = 5.$$

This solution may be easily verified by substituting the above values into the original equations.

5–4 Rules of rank. In the preceding two sections, we have examined ways of solving a set of n simultaneous linear equations in n unknowns. We now wish to investigate the conditions which determine whether solutions do or do not exist. We have seen that a set of n equations in n unknowns has a unique solution if $r(\mathbf{A}) = n$. We have also noted that difficulties arose if $r(\mathbf{A}) < n$ (although this case was not considered in any detail). To be completely general, let us consider a set of m simultaneous linear equations in n unknowns. No restriction will be made as to whether $m > n$, $m = n$, $m < n$. This set can be written

$$a_{11}x_1 + \cdots + a_{1n}x_n = b_1,$$
$$a_{21}x_1 + \cdots + a_{2n}x_n = b_2,$$
$$\vdots$$
$$a_{m1}x_1 + \cdots + a_{mn}x_n = b_m,$$
(5–16)

or

$$\mathbf{Ax} = \mathbf{b},$$

where $\mathbf{A}$ is an $m \times n$ matrix.

Next we shall define a new $m \times (n + 1)$ matrix $\mathbf{A}_b$ which contains $\mathbf{A}$ in the first n columns and $\mathbf{b}$ in column $n + 1$, that is,

$$\mathbf{A}_b = (\mathbf{A}, \mathbf{b}) = \begin{bmatrix} a_{11} \cdots a_{1n} & b_1 \\ a_{21} \cdots a_{2n} & b_2 \\ \vdots & \vdots \\ a_{m1} \cdots a_{mn} & b_m \end{bmatrix}.$$
(5–17)

Matrix $\mathbf{A}_b$ is a very important quantity; its rank as well as the rank of $\mathbf{A}$ determine whether the set of equations (5–16) has a solution. $\mathbf{A}_b$ is called the augmented matrix of the system.

Since every determinant in $\mathbf{A}$ also occurs in $\mathbf{A}_b$, the rank of $\mathbf{A}$ cannot exceed that of $\mathbf{A}_b$. Hence two possibilities exist: (a) $r(\mathbf{A}) < r(\mathbf{A}_b)$, (b) $r(\mathbf{A}) = r(\mathbf{A}_b)$. It should be noted that $r(\mathbf{A}_b)$ cannot be greater than $r(\mathbf{A}) + 1$. The following paragraph will demonstrate that the cases (a) and (b) play a crucial role in determining whether Eq. (5–16) has a solution.

If (a) holds, that is, if $r(\mathbf{A}) < r(\mathbf{A}_b)$, then there do not exist any x_j satisfying (5–16). The largest nonvanishing determinant in $\mathbf{A}_b$ must contain the column $\mathbf{b}$, since $r(\mathbf{A}) < r(\mathbf{A}_b)$. Hence $\mathbf{b}$ is linearly independent of the columns of $\mathbf{A}$, and thus there are no x_j such that

$$\sum_{j=1}^{n'} x_j \mathbf{a}_j = \mathbf{b},$$

where the $\mathbf{a}_j$ represent the columns of $\mathbf{A}$; that is, there do not exist any x_j satisfying (5–16). Hence, there is no solution, and the equations are *inconsistent*.

However, if (b) holds, that is, $r(\mathbf{A}) = r(\mathbf{A}_b) = k$, then there is always at least one solution. Since $r(\mathbf{A}) = k$ and $r(\mathbf{A}_b) = k$, every column of $\mathbf{A}_b$ can be expressed as a linear combination of k linearly independent columns of $\mathbf{A}$. (Without loss of generality, we can assume that they are the first k columns of $\mathbf{A}$.) Since $\mathbf{b}$ is a column of $\mathbf{A}_b$, there must exist numbers x_j such that

$$\sum_{j=1}^{k} x_j \mathbf{a}_j = \mathbf{b};$$

hence, there is at least one solution to the system of equations (5–16). Thus, we have proved that: *If $r(\mathbf{A}) < r(\mathbf{A}_b)$, the equations (5–16) are inconsistent and there is no solution. Conversely, if $r(\mathbf{A}) = r(\mathbf{A}_b)$, there is always at least one solution to the set* (5–16).

It is important to note that the existence of a solution does not depend on $r(\mathbf{A})$ = minimum (m, n). Conceivably, we could have 100 equations and 1000 variables with $r(\mathbf{A}) = r(\mathbf{A}_b) = 1$; from the preceding discussion we know that there would be at least one solution.

EXAMPLES:

(1) Is there a solution to

$$3x_1 + 4x_2 = 7,$$
$$2.25x_1 + 3x_2 = 5.25?$$

$$\mathbf{A} = \begin{bmatrix} 3 & 4 \\ 2.25 & 3 \end{bmatrix}, \quad r(\mathbf{A}) = 1, \quad \mathbf{A}_b = \begin{bmatrix} 3 & 4 & 7 \\ 2.25 & 3 & 5.25 \end{bmatrix}, \quad r(\mathbf{A}_b) = 1.$$

Thus, the set of equations has a solution. In fact, there are an infinite number of solutions. These are given by

$$x_1 = \frac{7}{3} - \frac{4}{3} x_2$$

for any x_2, since the second equation is just $\frac{3}{4}$ times the first. Note that geometrically the two equations represent the same straight line.

(2) Does a solution exist for the set

$$3x_1 + 4x_2 = 7,$$
$$2.25x_1 + 3x_2 = 1?$$

$$r(\mathbf{A}) = 1, \quad \mathbf{A}_b = \begin{bmatrix} 3 & 4 & 7 \\ 2.25 & 3 & 1 \end{bmatrix}, \quad r(\mathbf{A}_b) = 2.$$

No solution exists. If the first equation is multiplied by $\frac{3}{4}$, the left-hand side becomes the left-hand side of the second equation. However, the right-hand side of the first equation does not become the right-hand side of the second equation. The equations are clearly inconsistent. Illustrate this graphically (two parallel lines which do not intersect).

(3) Is there a solution to
$$3x_1 + 2x_2 + x_3 = 7,$$
$$x_1 + 0.5x_2 - x_3 = 4,$$
$$x_1 + 0.75x_2 + x_3 = 5?$$

$$\mathbf{A} = \begin{bmatrix} 3 & 2 & 1 \\ 1 & 0.5 & -1 \\ 1 & 0.75 & 1 \end{bmatrix}, \quad \mathbf{A}_b = \begin{bmatrix} 3 & 2 & 1 & 7 \\ 1 & 0.5 & -1 & 4 \\ 1 & 0.75 & 1 & 5 \end{bmatrix}, \quad |\mathbf{A}| = 0, \quad r(\mathbf{A}) = 2.$$

However, $r(\mathbf{A}_b) = 3$. There is no solution. If the second equation is added to twice the third, we obtain

$$3x_1 + 2x_2 + x_3 = 14;$$

this is inconsistent with the first equation. What is the geometrical interpretation?

5–5 Further properties. We have noted in one of the preceding examples that if a system of linear equations has a solution, this solution need not be unique. If the system $\mathbf{Ax} = \mathbf{b}$ has two distinct solutions $\mathbf{x}_1$ and $\mathbf{x}_2$, then $\lambda \mathbf{x}_1 + (1 - \lambda)\mathbf{x}_2$ is also a solution for any number λ. To prove this, assume that

$$\mathbf{Ax}_1 = \mathbf{b}, \quad \mathbf{Ax}_2 = \mathbf{b}. \tag{5–18}$$

Then
$$\lambda \mathbf{Ax}_1 = \mathbf{A}(\lambda \mathbf{x}_1) = \lambda \mathbf{b},$$
$$(1 - \lambda)\mathbf{Ax}_2 = \mathbf{A}(1 - \lambda)\mathbf{x}_2 = (1 - \lambda)\mathbf{b}. \tag{5–19}$$

Adding the two equations, we obtain

$$\mathbf{A}[\lambda \mathbf{x}_1 + (1 - \lambda)\mathbf{x}_2] = \lambda \mathbf{b} + (1 - \lambda)\mathbf{b} = \mathbf{b}. \tag{5–20}$$

Hence, $\lambda \mathbf{x}_1 + (1 - \lambda)\mathbf{x}_2$ is a solution if $\mathbf{x}_1, \mathbf{x}_2$ are solutions. This result illustrates immediately that *if a system* $\mathbf{Ax} = \mathbf{b}$ *has two distinct solutions, then there exists an infinite number of solutions*. This follows since in (5–20) λ can take on any value.

Let us suppose that we are given the system $\mathbf{Ax} = \mathbf{b}$, with $\mathbf{A}$ an $m \times n$ matrix and

$$r(\mathbf{A}) = r(\mathbf{A}_b) = k < m. \tag{5–21}$$

The rows of $\mathbf{A}_b$ will be denoted by $(\mathbf{a}^i, b_i)$. If we choose k rows of $\mathbf{A}$ which are linearly independent, then the same k rows of $\mathbf{A}_b$ are also linearly independent. Let us assume that these are the first k rows of $\mathbf{A}_b$. Then, according to Eq. (5–21), any other row in $\mathbf{A}_b$ is a linear combination of the first k rows, that is,

$$(\mathbf{a}^r, b_r) = \sum_{i=1}^{k} \lambda_{ir}(\mathbf{a}^i, b_i), \qquad r = k+1, \ldots, m; \qquad (5\text{–}22)$$

or, separating out b_r, we obtain

$$\mathbf{a}^r = \sum_{i=1}^{k} \lambda_{ir}\mathbf{a}^i, \qquad (5\text{–}23)$$

$$b_r = \sum_{i=1}^{k} \lambda_{ir} b_i. \qquad (5\text{–}24)$$

If $\mathbf{x}$ satisfies the first k equations of $\mathbf{A}\mathbf{x} = \mathbf{b}$, that is, $\mathbf{a}^i\mathbf{x} = b_i$, $i = 1, \ldots, k$, then from Eqs. (5–23) and (5–24),

$$\mathbf{a}^r\mathbf{x} = \sum_{i=1}^{k} \lambda_{ir}\mathbf{a}^i\mathbf{x} = \sum_{i=1}^{k} \lambda_{ir}b_i = b_r, \qquad r = k+1, \ldots, m. \quad (5\text{–}25)$$

Thus: Any $\mathbf{x}$ which satisfies k equations $\mathbf{a}^i\mathbf{x} = b_i$ for which the corresponding rows $\mathbf{a}^i$ in $\mathbf{A}$ are linearly independent satisfies all m equations. In other words, all but k equations can be ignored when seeking the solutions to $\mathbf{A}\mathbf{x} = \mathbf{b}$.

Let us imagine that k equations have been selected for which the corresponding rows of $\mathbf{A}$ are linearly independent. This assumption implies that there must be at least k variables, that is, $n \geq k$. If $n = k$, then the matrix of the coefficients of this set of k equations must be nonsingular, and, according to Cramer's rule, there is a unique solution. If $n > k$, the k equations can be written

$$\mathbf{A}_1 \mathbf{x}_\alpha + \mathbf{R}\mathbf{x}_\beta = \mathbf{b}^*, \qquad (5\text{–}26)$$

where $\mathbf{A}_1$ is a $k \times k$ nonsingular matrix, and $\mathbf{R}$ is a $k \times (n-k)$ matrix. Furthermore, $\mathbf{x}_\alpha = [x_1, \ldots, x_k]$, $\mathbf{x}_\beta = [x_{k+1}, \ldots, x_n]$, and $\mathbf{b}^*$ contains the k components of $\mathbf{b}$ corresponding to the k equations selected. We have named the variables so that the first k variables have associated with them a nonsingular matrix. Then

$$\mathbf{x}_\alpha = \mathbf{A}_1^{-1}\mathbf{b}^* - \mathbf{A}_1^{-1}\mathbf{R}\mathbf{x}_\beta. \qquad (5\text{–}27)$$

Hence, given *any* $\mathbf{x}_\beta$, we can solve uniquely for $\mathbf{x}_\alpha$ in terms of $\mathbf{x}_\beta$. Therefore, arbitrary values can be assigned to the $n - k$ variables in $\mathbf{x}_\beta$; values for the remaining k variables in $\mathbf{x}$ can be found by Eq. (5–27), so that $\mathbf{x} = [\mathbf{x}_\alpha, \mathbf{x}_\beta]$ is a solution to $\mathbf{A}\mathbf{x} = \mathbf{b}$. All solutions to the set of equations can be generated by assigning all possible values to the set of variables in $\mathbf{x}_\beta$.

Summing up the discussion of the preceding paragraph, *we see that, in a system of m simultaneous linear equations in n unknowns, if $r(\mathbf{A}) = r(\mathbf{A}_b) = k < m$, then any* **x** *which satisfies k of the equations for which the corresponding rows of* **A** *are linearly independent satisfies all equations of the set. Furthermore, if $k < n$, $n - k$ of the variables can be assigned arbitrary values, and the remaining k variables can be solved for provided the columns of* **A** *associated with the k variables are linearly independent.* An explicit form of the general solution is given by Eq. (5-27). The reader should now be able to see why the Gauss reduction procedure discussed in Section 5-2 might not lead to a unique solution of a set of n equations in n unknowns. If $r(\mathbf{A}) < r(\mathbf{A}_b)$, there is no solution; then a nonzero g_i appears in the reduction, while all h_{ij} in that row vanish. When $r(\mathbf{A}) = r(\mathbf{A}_b) = k < n$, arbitrary values can be assigned to $n - k$ of the variables. This can be clearly seen in the Gauss reduction since one or more rows of h_{ij} will be composed entirely of zeros, and the corresponding $g_i = 0$. Hence, back substitution will not eliminate all variables. It is important to note that both the Gauss and the Gauss-Jordan methods are equally useful in solving numerically sets of m simultaneous linear equations in n unknowns. Problem 5-24 illustrates this point.

The preceding discussion shows: (1) Whenever $r(\mathbf{A}) = r(\mathbf{A}_b) < n$ (the number of variables), an infinite number of solutions will satisfy the equations since in this case some variables, with arbitrary values assigned to them, can always be transferred to the right-hand side. (2) There will be a unique solution if and only if $r(\mathbf{A}) = r(\mathbf{A}_b) = n$. Thus for a system of simultaneous linear equations, there is either a *unique* solution, an *infinite* number of solutions, or *no* solution at all.

If $r(\mathbf{A}) = r(\mathbf{A}_b) = k < m$, $m - k$ of the equations are linear combinations of the remaining k equations. These $m - k$ equations are called *redundant* since they do not place any additional constraints on the variables; they could be dropped from the set without any effect on the solutions. When formulating a system of equations, we try to avoid redundant equations. However, a large system involving many variables and equations may make it extremely difficult to determine whether any new equation is linearly independent of the others.

EXAMPLE: Find a solution to the system

$$2x_1 + 7x_3 = 4,$$
$$3x_1 + 3x_2 + 6x_3 = 3,$$
$$2x_1 + 2x_2 + 4x_3 = 2.$$

First, let us check whether $r(\mathbf{A}) = r(\mathbf{A}_b)$ to make sure that a solution does exist. We note that $r(\mathbf{A}) = r(\mathbf{A}_b) = 2$. Thus there is a solution. However, since $r(\mathbf{A}) \neq 3$, the solution will not be unique; there exists an infinite number of solutions to the system. The determinant of order 2 in the upper left-hand corner of $\mathbf{A}$ does not vanish. Hence, we shall use the first two equations to solve for x_1 and x_2 by Cramer's rule, setting x_3 to any arbitrary value. Let us suppose that $x_3 = 2$. Then we must solve

$$\begin{bmatrix} 2 & 0 \\ 3 & 3 \end{bmatrix} \begin{bmatrix} x_1 \\ x_2 \end{bmatrix} = \begin{bmatrix} 4 - 7x_3 \\ 3 - 6x_3 \end{bmatrix} = \begin{bmatrix} -10 \\ -9 \end{bmatrix};$$

hence

$$x_1 = \frac{1}{6} \begin{vmatrix} -10 & 0 \\ -9 & 3 \end{vmatrix} = -5, \quad x_2 = \frac{1}{6} \begin{vmatrix} 2 & -10 \\ 3 & -9 \end{vmatrix} = 2.$$

Thus one solution to the system has the values: $x_1 = -5$, $x_2 = 2$, $x_3 = 2$. We can check this result by substituting these values into all three equations. We see that:

Equation 1: $-10 + 14 = 4$.
Equation 2: $-15 + 6 + 12 = 3$.
Equation 3: $-10 + 4 + 8 = 2$.

The second or third equation can be considered redundant since the second equation is 3/2 of the third.

5–6 Homogeneous linear equations. We shall now examine the special case of $\mathbf{b} = \mathbf{0}$, that is, the right-hand side of Eq. (5–16) vanishes. A system of linear equations of this type is called homogeneous. It can be written:

$$\begin{aligned} a_{11}x_1 + \cdots + a_{1n}x_n &= 0, \\ &\vdots \\ a_{m1}x_1 + \cdots + a_{mn}x_n &= 0, \end{aligned} \quad (5\text{–}28)$$

or, in matrix form,

$$\mathbf{Ax} = \mathbf{0}. \quad (5\text{–}29)$$

We see immediately that a set of homogeneous linear equations always has a solution since, with $\mathbf{b} = \mathbf{0}$, it must be true that

$$r(\mathbf{A}) = r(\mathbf{A}_b).$$

We note also that $\mathbf{x} = \mathbf{0}$ *is always a solution* (called a *trivial* solution). It is of interest to determine when solutions other than $\mathbf{x} = \mathbf{0}$ exist. From Section 5–5, we know that if $r(\mathbf{A}) = k < n$, arbitrary values can be assigned to $n - k$ of the variables, and hence a nontrivial solution always exists. Thus we can prove the following important theorem: *A* necessary and sufficient condition for a system of homogeneous linear equations $\mathbf{Ax} = \mathbf{0}$ in n variables to have a solution other than $\mathbf{x} = \mathbf{0}$ is that $r(\mathbf{A}) < n$. If $r(\mathbf{A}) < n$, then it follows from the above argument that there is a solution which is not trivial. If $r(\mathbf{A}) = n$, then from Eq. (5–27) $\mathbf{x}_\alpha = \mathbf{x}$, and

$$\mathbf{x} = \mathbf{A}_1^{-1}\mathbf{0} = \mathbf{0}.$$

Thus there is only one solution and it is trivial. This result has two useful corollaries: (1) *If there are fewer equations than unknowns, the system (5–28) always has a nontrivial solution.* (2) *If the number of equations is equal to the number of unknowns, a necessary and sufficient condition for a nontrivial solution is that the determinant of the coefficients vanish, that is,* $|\mathbf{A}| = 0$. We are familiar with this fact from elementary algebra. If we have a set of n homogeneous linear equations in n unknowns, there is no solution other than $\mathbf{x} = \mathbf{0}$ if $|\mathbf{A}| \neq 0$. This result can, of course, also be obtained directly from (5–11), for

$$\mathbf{x} = \mathbf{A}^{-1}\mathbf{0} = \mathbf{0}.$$

However, if $|\mathbf{A}| = 0$, then there exists a solution different from $\mathbf{x} = \mathbf{0}$.

Let us consider any solution $\mathbf{x} \neq \mathbf{0}$ to $\mathbf{Ax} = \mathbf{0}$. Since for any scalar λ,

$$\lambda \mathbf{Ax} = \mathbf{A}\lambda\mathbf{x} = \lambda\mathbf{0} = \mathbf{0}, \qquad (5\text{–}30)$$

it follows that if $\mathbf{x}$ is a solution to the set of equations, so is $\lambda\mathbf{x}$. Hence, the appearance of one nontrivial solution automatically implies the existence of an infinite number of nontrivial solutions. If $n = 2$ and $\mathbf{x}$ is a nontrivial solution, then, geometrically speaking, the fact that $\lambda\mathbf{x}$ is a solution means that any point on the line through $\mathbf{x}$ and the origin is also a solution.

The arguments in the preceding paragraph can be pursued further. In general, we are considering a homogeneous system of m equations in n unknowns. The vector $\mathbf{x}$ satisfying $\mathbf{Ax} = \mathbf{0}$ is a point in E^n. We have already shown that $\mathbf{x} = \mathbf{0}$ is a solution to $\mathbf{Ax} = \mathbf{0}$, and that if $\mathbf{x}$ is a solution, so is $\lambda\mathbf{x}$. Furthermore, if $\mathbf{x}_1$ and $\mathbf{x}_2$ are distinct solutions, then $\mathbf{x}_3 = \mathbf{x}_1 + \mathbf{x}_2$ is also a solution. This follows since

$$\mathbf{Ax}_1 = \mathbf{0}, \qquad \mathbf{Ax}_2 = \mathbf{0};$$

adding the two expressions, we obtain

$$\mathbf{Ax}_1 + \mathbf{Ax}_2 = \mathbf{A}(\mathbf{x}_1 + \mathbf{x}_2) = \mathbf{Ax}_3 = \mathbf{0}.$$

Thus we have proved (see definition of a subspace) that the set of all solutions to $\mathbf{Ax} = \mathbf{0}$ forms a subspace of E^n. We shall now show that q, the dimension of this subspace, is $n - k$ where n, as usual, denotes the number of columns in $\mathbf{A}$ and k is the rank of $\mathbf{A}$. To prove this, let $\mathbf{x}_1, \ldots, \mathbf{x}_q$ be q vectors which span the q-dimensional subspace. According to Section 2–10, this set of vectors can be extended to form a basis for E^n. Let $\mathbf{y}_1, \ldots, \mathbf{y}_{n-q}$ be the $n - q$ additional vectors which, along with $\mathbf{x}_1, \ldots, \mathbf{x}_q$, form a basis for E^n. Then any vector $\mathbf{z}$ in E^n can be written as a linear combination of the basis vectors, that is,

$$\mathbf{z} = \sum_{i=1}^{n-q} \sigma_i \mathbf{y}_i + \sum_{j=1}^{q} \beta_j \mathbf{x}_j.$$

However,

$$\mathbf{A}\mathbf{z} = \sum_{i=1}^{n-q} \sigma_i \mathbf{A}\mathbf{y}_i = \sum_{j=1}^{n} z_j \mathbf{a}_j, \qquad (5\text{-}31)$$

since by definition

$$\mathbf{A}\mathbf{x}_j = \mathbf{0}.$$

Equation (5–31) indicates that the $n - q$ vectors, $\mathbf{A}\mathbf{y}_i$, span the subspace of E^m which is generated by the columns of $\mathbf{A}$ (see Section 4–2). Let us show that the vectors $\mathbf{A}\mathbf{y}_i$ are linearly independent. We shall assume that there exist λ_i not all zero such that

$$\sum \lambda_i \mathbf{A}\mathbf{y}_i = \mathbf{0} = \mathbf{A}[\sum \lambda_i \mathbf{y}_i] = \mathbf{0}.$$

This expression implies that $\sum \lambda_i \mathbf{y}_i$ is an element of the subspace spanned by $\mathbf{x}_1, \ldots, \mathbf{x}_q$, that is,

$$\sum_{i=1}^{n-q} \lambda_i \mathbf{y}_i = \sum_{j=1}^{q} \gamma_j \mathbf{x}_j, \quad \text{or} \quad \sum_{i=1}^{n-q} \lambda_i \mathbf{y}_i - \sum_{j=1}^{q} \gamma_j \mathbf{x}_j = \mathbf{0}.$$

This result contradicts the original assumption that the $\mathbf{y}_i$ and $\mathbf{x}_i$ form a basis for E^n. Hence, the vectors $\mathbf{A}\mathbf{y}_i$ are linearly independent and form a basis for the subspace of E^m generated by the columns of $\mathbf{A}$. Since there are $n - q$ vectors, the rank of $\mathbf{A}$ is $n - q = k$. Thus we have proved that

$$q = n - k. \qquad (5\text{-}32)$$

Consequently, the dimension of the subspace of E^n generated by the solutions to $\mathbf{A}\mathbf{x} = \mathbf{0}$ is $n - k$. We could have guessed this intuitively since arbitrary values can be assigned to the $n - k$ of the variables.

We shall now examine equation $\mathbf{A}\mathbf{x} = \mathbf{0}$ in a slightly different way. Let us note that $\mathbf{A}$ maps E^n into a subspace of E^m. The set of vectors $\mathbf{x}$ in E^n which are taken into the origin of E^m is the set of solutions to $\mathbf{A}\mathbf{x} = \mathbf{0}$; we have shown that this set of vectors is a subspace of E^n.

This collection of vectors is sometimes called the null space of **A**. *The dimension of this subspace of* E^n *which depends only on* **A** *is called the nullity of* **A**. We have proved that *the nullity of* **A** *plus the rank of* **A** *is equal to the number of columns in* **A**, *that is*, $q + k = n$. The study of homogeneous linear equations reveals thus another property of matrices and linear transformations.

Let us again view **Ax** = **0** as a set of homogeneous equations. *A set of basis vectors spanning the subspace of* E^n *generated by the solutions to* **Ax** = **0** *is called a fundamental system for* **Ax** = **0**. Since the dimension of the subspace is $n - k$, we have $n - k$ vectors in a fundamental system; and since the number of bases is infinite, an infinite number of different fundamental systems can be generated.

EXAMPLE: Illustrate geometrically the subspace formed by the solutions to

$$3x_1 + 4x_2 = 0,$$

$$2.25x_1 + 3x_2 = 0.$$

Since $|\mathbf{A}| = 0$ and $r(\mathbf{A}) = 1$, there are nontrivial solutions. The second equation is $\frac{3}{4}$ times the first, and hence the general solution is

$$x_2 = -\tfrac{3}{4}x_1.$$

The value of x_1 can be chosen arbitrarily. The dimension of the subspace formed by the solutions is thus 1; it is a line through the origin with slope $-\frac{3}{4}$ (see Fig. 5-1). A single vector forms a basis for this subspace. If $x_1 = 4$, then $x_2 = -3$; hence $\bar{\mathbf{x}} = [4, -3]$ is a basis vector, and $\bar{\mathbf{x}}$ is a fundamental system for the set of equations.

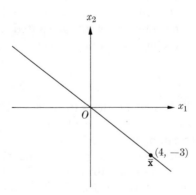

FIGURE 5-1

5–7 Geometric interpretation.

We have already seen that the solutions to a system of m homogeneous linear equations in n unknowns generate a subspace of E^n. This is true even if the only solution is $\mathbf{x} = \mathbf{0}$; the single vector $\mathbf{0}$ yields a zero-dimensional subspace of E^n. If $n = 3$, then the subspace spanned by the solutions to $\mathbf{Ax} = \mathbf{0}$ is either a plane through the origin, a line through the origin, or the origin itself. (In the completely degenerate case, when $\mathbf{A} = \mathbf{0}$, the subspace is all of E^3.)

EXAMPLES: (1) The solutions to
$$2x_1 + 3x_2 + 4x_3 = 0$$
lie on a plane through the origin.

(2) The solutions to
$$2x_1 + 3x_2 + 4x_3 = 0,$$
$$x_1 + 2x_2 + 2x_3 = 0$$
lie on a line through the origin (the intersection of two planes which pass through the origin).

(3) The solution of
$$x_1 + 2x_3 = 0,$$
$$x_1 + x_2 = 0,$$
$$x_1 - 2x_2 = 0$$
is unique and trivial, that is, $\mathbf{x} = \mathbf{0}$ is the only solution (the intersection of three planes which pass through the origin).

Let us consider the geometric interpretation of the solutions to $\mathbf{Ax} = \mathbf{b} \neq \mathbf{0}$. In this case, $\mathbf{x} = \mathbf{0}$ is not a solution; hence we are sure that the set of solutions does not form a subspace of E^n. However, if we are given one solution $\mathbf{x}_1$ to $\mathbf{Ax} = \mathbf{b}$, then any other solution $\mathbf{x}_2$ can be written

$$\mathbf{x}_2 = \mathbf{x}_1 + \mathbf{x}_2 - \mathbf{x}_1 = \mathbf{x}_1 + \mathbf{y}, \qquad \mathbf{y} = \mathbf{x}_2 - \mathbf{x}_1, \qquad (5\text{-}33)$$

and

$$\mathbf{Ay} = \mathbf{Ax}_2 - \mathbf{Ax}_1 = \mathbf{b} - \mathbf{b} = \mathbf{0}. \qquad (5\text{-}34)$$

Equations (5–33) and (5–34) show that if we know one solution $\mathbf{x}_1$ to $\mathbf{Ax} = \mathbf{b}$, any other solution $\mathbf{x}_2$ can be written $\mathbf{x}_2 = \mathbf{x}_1 + \mathbf{y}$, where $\mathbf{y}$ is a solution to the homogeneous set of equations $\mathbf{Ay} = \mathbf{0}$. In other words, all solutions to $\mathbf{Ax} = \mathbf{b}$ can be generated by knowing a *single* solution to $\mathbf{Ax} = \mathbf{b}$ and *all* solutions to the homogeneous system $\mathbf{Ay} = \mathbf{0}$.

The preceding results illustrate that the solutions to $\mathbf{Ax} = \mathbf{b}$ will generate a space having the same dimension as the subspace spanned by the solutions of $\mathbf{Ax} = \mathbf{0}$. The space spanned by the solutions of $\mathbf{Ax} = \mathbf{b}$ is not a subspace since $\mathbf{0}$ is not a solution. The solutions are of the form

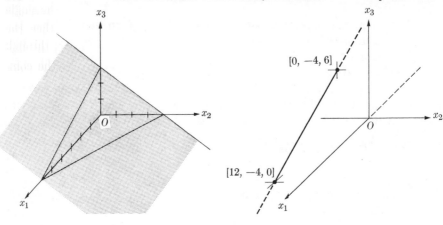

FIGURE 5-2   FIGURE 5-3

$\mathbf{x} = \mathbf{x}_1 + \mathbf{y}$, and $\mathbf{x}_1$ translates the space away from the origin.* If $n = 3$, then the solutions to $\mathbf{Ax} = \mathbf{b}$ lie on a plane, on a line, or, when the solution is unique, it is represented by a point, that is, a space of zero dimension. In general, the dimension of the space generated by the solutions of $\mathbf{Ax} = \mathbf{b}$ is the nullity of $\mathbf{A}$.

EXAMPLES: (1) The solutions to

$$2x_1 + 3x_2 + 4x_3 = 12$$

lie on the plane shown in Fig. 5-2.

(2) The solutions to

$$2x_1 + 3x_2 + 4x_3 = 12,$$
$$x_1 + 2x_2 + 2x_3 = 4$$

lie on the line shown in Fig. 5-3 (the intersection of two planes).

5-8 Basic solutions. We now wish to study the solutions to a set of m equations $\mathbf{Ax} = \mathbf{b}$ in $n > m$ unknowns, which have as many of the variables equal to zero as possible. From Section 5-5, we know that if $r(\mathbf{A}) = k$ and we select any k linearly independent columns from $\mathbf{A}$, we can assign arbitrary values to the $n - k$ variables not associated with these columns. The remaining k variables will be uniquely determined in terms of the $n - k$ variables. Thus for such a system, we can set $n - k$

---

* The set of solutions to $\mathbf{Ax} = \mathbf{b}$ is often referred to as an *affine* subspace of E^n. It has the same dimension as the subspace generated by the solutions to $\mathbf{Ax} = \mathbf{0}$, the only difference being that the affine subspace is translated away from the origin.

variables to zero; the other k variables will, in general, be different from zero since the equations must be satisfied (in certain cases, however, one or more of these k variables will be zero). To be specific, we shall assume that for our set of equations,

$$r(\mathbf{A}) = r(\mathbf{A}_b) = m. \tag{5-35}$$

This implies that none of the equations is redundant (if there were redundant equations in the original set, we assume that they have been dropped).

Then the columns of matrix $\mathbf{A}$ can be named so that $\mathbf{A}$ can be written

$$\mathbf{A} = (\mathbf{B}, \mathbf{R}), \tag{5-36}$$

where $\mathbf{B}$ is an $m \times m$ nonsingular matrix [this follows since $r(\mathbf{A}) = m$] and $\mathbf{R}$ is an $m \times (n - m)$ matrix. The vector $\mathbf{x}$ can be partitioned as follows:

$$\mathbf{x} = [\mathbf{x}_B, \mathbf{x}_R], \quad \mathbf{x}_B = [x_1, x_2, \ldots, x_m], \quad \mathbf{x}_R = [x_{m+1}, \ldots, x_n]. \tag{5-37}$$

Thus

$$\mathbf{Ax} = \mathbf{Bx}_B + \mathbf{Rx}_R = \mathbf{b}. \tag{5-38}$$

All the solutions to this set of equations can be generated by assigning arbitrary values to $\mathbf{x}_R$. Let us now set $\mathbf{x}_R = \mathbf{0}$. Since $\mathbf{B}$ has an inverse, we obtain

$$\mathbf{x}_B = \mathbf{B}^{-1}\mathbf{b}. \tag{5-39}$$

This type of solution to the system of equations is called a basic solution.

BASIC SOLUTION: *Given a system of m simultaneous linear equations in n unknowns, $\mathbf{Ax} = \mathbf{b}$ ($m < n$) and $r(\mathbf{A}) = m$: If any $m \times m$ nonsingular matrix is chosen from $\mathbf{A}$, and if all the $n - m$ variables not associated with the columns of this matrix are set equal to zero, the solution to the resulting system of equations is called a basic solution.*

A basic solution has no more than m nonzero variables. It can be written $\mathbf{x} = [\mathbf{x}_B, \mathbf{0}]$, with $\mathbf{x}_B$ given by Eq. (5-39). *The m variables which can be different from zero are called basic variables.* Hence, in a basic solution $n - m$ variables are set equal to zero, and the remaining m variables are uniquely determined since, by assumption, the matrix of their coefficients is nonsingular.

The term "basic solution" refers to the fact that the columns of $\mathbf{B}$ form a basis for E^m. Basic solutions are of great importance in linear programming.

How many basic solutions are possible in a system of m equations and n unknowns? This question is analogous to asking how many combinations of n variables there are when taken m at a time (the order of the variables in any basic solution is, of course, irrelevant). This number N is the standard formula for combinations, that is,

$$N = \frac{n!}{m!(n-m)!}; \qquad (5\text{-}40)$$

it represents the maximum number of possible basic solutions. However, the columns of $\mathbf{A}$ associated with the m basic variables must be linearly independent so that an inverse exists. Since any m columns from $\mathbf{A}$ will not necessarily be linearly independent, we shall not always obtain the maximum number of possible basic solutions.

It is of interest to know whether any x_i in the vector $\mathbf{x}_B$ is zero. If this is the case, more than $n - m$ variables will be zero in the solution. When this happens, we say that the basic solution is *degenerate*.

DEGENERACY: *A basic solution to* $\mathbf{Ax} = \mathbf{b}$ *is degenerate if one or more of the basic variables vanishes.*

A necessary and sufficient condition for the existence and nondegeneracy of all possible basic solutions of $\mathbf{Ax} = \mathbf{b}$ *is the linear independence of every set of m columns from the augmented matrix* $\mathbf{A}_b = (\mathbf{A}, \mathbf{b})$. To prove the necessity let us suppose that all basic solutions exist and that none is degenerate. Then for any set of m columns, say $\mathbf{a}_1, \ldots, \mathbf{a}_m$ of $\mathbf{A}$,

$$\sum_{i=1}^{m} x_i \mathbf{a}_i = \mathbf{b}, \qquad (5\text{-}41)$$

and no $x_i = 0$. Since we assumed the existence of all basic solutions, every set of m columns from $\mathbf{A}$ must be linearly independent. In Section 2–9 we made the point that any vector in a basis can be replaced by a given vector $\mathbf{b}$ if the coefficient of the vector to be replaced does not vanish in the expression of $\mathbf{b}$ as a linear combination of the basis vectors. According to Eq. (5–41), $\mathbf{b}$ can replace any $\mathbf{a}_i$ in the basis. Hence $\mathbf{b}$ and any $m - 1$ columns from $\mathbf{A}$ are linearly independent. The necessity is proved.

To prove the sufficiency let us suppose that any m columns from $\mathbf{A}_b$ are linearly independent. This immediately tells us that all basic solutions exist. When $\mathbf{b}$ is expressed as a linear combination of $\mathbf{a}_1, \ldots, \mathbf{a}_m$, we arrive at Eq. (5–41). However, since $\mathbf{a}_2, \ldots, \mathbf{a}_m, \mathbf{b}$ are linearly independent, the coefficient x_1 of $\mathbf{a}_1$ cannot vanish, because $\mathbf{b}$ can replace $\mathbf{a}_1$, and a basis is maintained. Similarly, since $\mathbf{a}_1, \mathbf{a}_3, \ldots, \mathbf{a}_m, \mathbf{b}$ are linearly independent, the coefficient x_2 of $\mathbf{a}_2$ cannot vanish. Thus we see that none of the x_i can vanish for any basic solution. Hence all basic solutions exist and are nondegenerate.

The preceding theorem has a corollary: *A necessary and sufficient condition for any given basic solution $x_B = B^{-1}b$ to be nondegenerate is the linear independence of b and every $m - 1$ columns from B*. If a solution is nondegenerate, b can replace any column of B and still maintain a basis since $x_i \neq 0$. Hence any $m - 1$ columns of B and b are linearly independent. Conversely, if b and any $m - 1$ columns of B are linearly independent, b can replace any column of B and still maintain a basis. Hence no x_i can vanish.

Since the condition for the nondegeneracy of a basic solution is quite stringent, we may expect to find cases where the condition is violated and degeneracy occurs. This is quite true. The possibility of degeneracy complicates somewhat the theory of linear programming.

EXAMPLES: (1) When $m = 2$, a basic solution has all but two x_i equal to zero. Degeneracy will occur if the b vector lies along the same line as any column a_i from A. Not all possible basic solutions will exist if two columns from A are collinear. Let us consider Fig. 5–4. The system

$$Ax = (a_1, a_2, a_3, a_5)x = b_1$$

will be degenerate for any basic solution including a_3 since b_1 is collinear with a_3, that is, vector b_1 can be expressed in terms of a_3 alone; hence the x_i corresponding to the other vector in the basis will vanish. However, all basic solutions exist. For the system

$$Ax = (a_1, a_2, a_3, a_5)x = b_2,$$

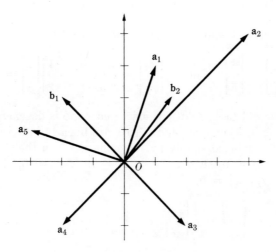

FIGURE 5–4

all basic solutions exist and are nondegenerate. The system

$$\mathbf{Ax} = (\mathbf{a}_1, \mathbf{a}_2, \mathbf{a}_3, \mathbf{a}_4)\mathbf{x} = \mathbf{b}_2$$

does not have a basic solution involving $\mathbf{a}_2, \mathbf{a}_4$ since they lie on the same line and do not form a basis. All existing basic solutions, however, are nondegenerate. In this case, one part of our theorem is violated, namely the part concerning the existence of all basic solutions.

(2) Find the basic solutions to

$$x_1 + 2x_2 + x_3 = 4,$$
$$2x_1 + x_2 + 5x_3 = 5.$$

The possible number of basic solutions is

$$\frac{3!}{2!1!} = 3.$$

First, we set $x_3 = 0$ and solve for x_1 and x_2. This yields

$$\begin{bmatrix} 1 & 2 \\ 2 & 1 \end{bmatrix} \begin{bmatrix} x_1 \\ x_2 \end{bmatrix} = \begin{bmatrix} 4 \\ 5 \end{bmatrix} \quad \text{or} \quad \begin{bmatrix} x_1 \\ x_2 \end{bmatrix} = -\frac{1}{3} \begin{bmatrix} 1 & -2 \\ -2 & 1 \end{bmatrix} \begin{bmatrix} 4 \\ 5 \end{bmatrix} = \begin{bmatrix} 2 \\ 1 \end{bmatrix}.$$

Then we set $x_2 = 0$ and solve for x_1 and x_3:

$$\begin{bmatrix} 1 & 1 \\ 2 & 5 \end{bmatrix} \begin{bmatrix} x_1 \\ x_3 \end{bmatrix} = \begin{bmatrix} 4 \\ 5 \end{bmatrix} \quad \text{or} \quad \begin{bmatrix} x_1 \\ x_3 \end{bmatrix} = \frac{1}{3} \begin{bmatrix} 5 & -1 \\ -2 & 1 \end{bmatrix} \begin{bmatrix} 4 \\ 5 \end{bmatrix} = \begin{bmatrix} 5 \\ -1 \end{bmatrix}.$$

Finally we set $x_1 = 0$ and solve for x_2 and x_3, that is:

$$\begin{bmatrix} 2 & 1 \\ 1 & 5 \end{bmatrix} \begin{bmatrix} x_2 \\ x_3 \end{bmatrix} = \begin{bmatrix} 4 \\ 5 \end{bmatrix} \quad \text{or} \quad \begin{bmatrix} x_2 \\ x_3 \end{bmatrix} = \frac{1}{9} \begin{bmatrix} 5 & -1 \\ -1 & 2 \end{bmatrix} \begin{bmatrix} 4 \\ 5 \end{bmatrix} = \begin{bmatrix} 5/3 \\ 2/3 \end{bmatrix}.$$

In this example, all basic solutions exist and none is degenerate. Hence any two vectors in the augmented matrix are linearly independent. The situation can be changed simply by replacing the 5 in the **b** vector with an 8. Thus if $x_3 = 0$, then $x_1 = 4, x_2 = 0$; if $x_2 = 0$, then $x_1 = 4, x_3 = 0$. Finally, if $x_1 = 0$,

$$\begin{bmatrix} x_2 \\ x_3 \end{bmatrix} = \frac{1}{9} \begin{bmatrix} 5 & -1 \\ -1 & 2 \end{bmatrix} \begin{bmatrix} 4 \\ 8 \end{bmatrix} = \begin{bmatrix} 4/3 \\ 4/3 \end{bmatrix}.$$

In this case all three basic solutions exist, but two are degenerate (illustrate this graphically).

References

The following texts discuss, at least in part, systems of linear equations. None, however, deals with basic solutions.

1. A. G. Aitken, *Determinants and Matrices*. Edinburgh: Oliver and Boyd, 1948.
2. R. G. D. Allen, *Mathematical Economics*. London: Macmillan and Co., Ltd., 1956.
3. G. Birkhoff, and S. MacLane, *A Survey of Modern Algebra*. New York: Macmillan, 1941.
4. M. Bôcher, *Introduction to Higher Algebra*. New York: Macmillan, 1907.
5. P. S. Dwyer, *Linear Computations*. New York: John Wiley and Sons, 1951.
6. W. L. Ferrar, *Algebra*. Oxford: Oxford University Press, 1941.
7. F. B. Hildebrand, *Introduction to Numerical Analysis*. New York: McGraw-Hill, 1956.
8. F. B. Hildebrand, *Methods of Applied Mathematics*. New York: Prentice-Hall, 1952.
9. S. Perlis, *Theory of Matrices*. Reading, Mass.: Addison-Wesley, 1952.
10. R. M. Thrall, and L. Tornheim, *Vector Spaces and Matrices*. New York: John Wiley and Sons, 1957.

Problems

Solve Problems 5-1 through 5-3 by (a) gaussian reduction, (b) Gauss-Jordan reduction, (c) Cramer's rule.

5-1. $3x_1 + 2x_2 + 4x_3 = 7,$
$2x_1 + x_2 + x_3 = 4,$
$x_1 + 3x_2 + 5x_3 = 2.$

5-2. $x_1 + 2x_2 + 3x_3 + 4x_4 = 5,$
$2x_1 + x_2 + 4x_3 + x_4 = 2,$
$3x_1 + 4x_2 + x_3 + 5x_4 = 6,$
$2x_1 + 3x_2 + 5x_3 + 2x_4 = 3.$

5-3. $4x_1 + 2x_2 + 5x_3 + 7x_4 + x_5 = 8,$
$x_1 + 4x_2 + x_3 + x_4 + 5x_5 = 4,$
$2x_1 + 3x_2 + 4x_3 + 5x_4 + 6x_5 = 3,$
$3x_1 + 9x_2 + 7x_3 + x_4 + 8x_5 = 16,$
$7x_1 + x_2 + x_3 + 6x_4 + x_5 = 9.$

5-4. Discuss in detail what happens if the gaussian elimination method is used for solving a system of n equations in n unknowns where (a) there is no solution, (b) the solution is not unique.

5-5. Derive Cramer's rule for solving the system

$$a_{11}x_1 + \cdots + a_{1n}x_n = b_1,$$
$$\vdots$$
$$a_{n1}x_1 + \cdots + a_{nn}x_n = b_n,$$

without use of matrix theory. Hint: Multiply the ith equation by A_{i1}, the cofactor of a_{i1}, and add the equations.

5–6. Using the fact that $\mathbf{A}_b$ can be transformed to an echelon matrix, prove that there is at least one solution to $\mathbf{Ax} = \mathbf{b}$ if $r(\mathbf{A}) = r(\mathbf{A}_b)$, and no solution if $r(\mathbf{A}) < r(\mathbf{A}_b)$.

5–7. Show that a collection of n-component vectors of the form $\mathbf{x} = (\mathbf{x}_1, \mathbf{x}_2) = (\mathbf{Bx}_2, \mathbf{x}_2)$, with $\mathbf{x}_1 = \mathbf{Bx}_2$, generates a subspace of E^n whose dimension is the number of components in $\mathbf{x}_2$ (it is assumed that no restriction is placed on the values the components of $\mathbf{x}_2$ may assume). Use this result to prove that the nullity of $\mathbf{A} = n - k$ provided that $r(\mathbf{A}) = k$ and $\mathbf{A}$ is an $m \times n$ matrix.

5–8. Show that a necessary and sufficient condition for a set of m vectors $\mathbf{x}_1, \ldots, \mathbf{x}_m$ to be linearly dependent is that the Gram determinant of these vectors vanish. The Gram determinant $|\mathbf{G}|$ is defined as:

$$|\mathbf{G}| = \begin{vmatrix} \mathbf{x}'_1\mathbf{x}_1 & \cdots & \mathbf{x}'_1\mathbf{x}_m \\ \vdots & & \vdots \\ \mathbf{x}'_m\mathbf{x}_1 & \cdots & \mathbf{x}'_m\mathbf{x}_m \end{vmatrix}.$$

*5–9. Consider the set of m homogeneous equations in n unknowns $\mathbf{Ax} = \mathbf{0}$. Let $\mathbf{a}^j$ ($j = 1, \ldots, m$) be the rows of $\mathbf{A}$. Then the set of equations can be written $\mathbf{a}^j\mathbf{x} = 0$ ($j = 1, \ldots, m$), that is, a solution $\mathbf{x}$ is orthogonal to every row of $\mathbf{A}$. Using this expression and the concept of an orthogonal basis, show that if $r(\mathbf{A}) = n$, then $\mathbf{x} = \mathbf{0}$ is the only solution.

*5–10. Prove that a necessary and sufficient condition for the existence of a solution to $\mathbf{Ax} = \mathbf{b}$ is that $\mathbf{b}$ lie in the subspace spanned by the columns of $\mathbf{A}$. Show that a necessary condition for the existence of a solution to $\mathbf{Ax} = \mathbf{b}$ is that $\mathbf{y}'\mathbf{b} = 0$ for all $\mathbf{y}$ such that $\mathbf{A}'\mathbf{y} = \mathbf{0}$. Can you prove that this condition is also sufficient?

In Problems 5–11 through 5–16, determine whether a solution exists. If there is a solution, is it unique? Find a solution to every set for which solutions (a solution) exist:

5–11. $3x_1 + 2x_2 = 7,$
$\quad\;\, x_1 + x_2 = 7.$

5–12. $x_1 + 2x_2 + x_3 = 1,$
$\quad\;\, 2x_1 + 4x_2 + 5x_3 = 3.$

5–13. $2x_1 + 8x_2 + 7x_3 = 0,$
$\quad\;\, x_1 + 2x_2 + 4x_3 = 0,$
$\quad\;\, 2x_1 + 4x_2 + 6x_3 = 17.$

5–14. $2x_1 + 8x_2 + 7x_3 = 1,$
$\quad\;\, x_1 + 2x_2 + 4x_3 = 0,$
$\quad\;\, 2x_1 + 4x_2 + 6x_3 = 0.$

5–15. $3x_1 + 7x_2 + 4x_3 = 0,$
$\quad\;\, x_1 + 2x_2 + x_3 = 0.$

5–16. $2x_1 + 3x_2 = 7,$
$\quad\;\, 4x_1 + 6x_2 = 3,$
$\quad\;\, x_1 + 17x_2 = 0.$

In Problems 5–17 through 5–19, find a fundamental system of solutions:

5–17. $\quad x_1 + x_2 + x_3 + x_4 = 0,$
$\quad\quad\; 2x_1 + 3x_2 + x_3 + 5x_4 = 0.$

5–18. $x_1 + 2x_2 + x_3 + 4x_4 = 0,$
$\quad\;\, x_1 + x_2 + 5x_3 + x_4 = 0,$
$\quad\;\, 2x_1 + 2x_2 + x_3 + x_4 = 0.$

---

* Starred problems require use of starred material in previous chapters or material with which all readers may not be familiar.

5-19. $x_1 + x_2 = 0$,
$2x_1 + 3x_2 = 0$,
$x_1 + 7x_2 = 0$.

In Problems 5-20 through 5-22, find all existing basic solutions. Do all possible basic solutions actually exist?

5-20. $x_1 + 2x_2 + 3x_3 + 4x_4 = 7$,
$2x_1 + x_2 + x_3 + 2x_4 = 3$.

5-21. $8x_1 + 6x_2 + 13x_3 + x_4 + x_5 = 6$,
$9x_1 + x_2 + 2x_3 + 6x_4 + 10x_5 = 11$.

5-22. $2x_1 + 3x_2 + 4x_3 + x_4 = 2$,
$x_1 + x_2 + 7x_3 + x_4 = 6$,
$3x_1 + 2x_2 + x_3 + 5x_4 = 8$.

5-23. Consider the vectors:

$$\mathbf{a}_1 = \begin{bmatrix} 1 \\ 0 \end{bmatrix}, \quad \mathbf{a}_2 = \begin{bmatrix} 3 \\ 4 \end{bmatrix}, \quad \mathbf{a}_3 = \begin{bmatrix} -3 \\ -4 \end{bmatrix}, \quad \mathbf{a}_4 = \begin{bmatrix} -1 \\ 2 \end{bmatrix}, \quad \mathbf{a}_5 = \begin{bmatrix} -1 \\ -4 \end{bmatrix}, \quad \mathbf{a}_6 = \begin{bmatrix} 1 \\ 4 \end{bmatrix},$$

$$\mathbf{b}_1 = \begin{bmatrix} 5 \\ 6 \end{bmatrix}, \quad \mathbf{b}_2 = \begin{bmatrix} 2 \\ -4 \end{bmatrix}, \quad \mathbf{b}_3 = \begin{bmatrix} 1 \\ 4 \end{bmatrix}.$$

Plot these vectors. Do all possible basic solutions exist to the following sets of equations? How many do exist? Are they nondegenerate?

$$(\mathbf{a}_2, \mathbf{a}_3, \mathbf{a}_4, \mathbf{a}_6)\mathbf{x} = \mathbf{b}_1, \quad (\mathbf{a}_2, \mathbf{a}_4, \mathbf{a}_5, \mathbf{a}_6)\mathbf{x} = \mathbf{b}_2,$$
$$(\mathbf{a}_2, \mathbf{a}_4, \mathbf{a}_6)\mathbf{x} = \mathbf{b}_3, \quad (\mathbf{a}_1, \mathbf{a}_2, \mathbf{a}_4, \mathbf{a}_5)\mathbf{x} = \mathbf{b}_1.$$

5-24. Show that the Gauss-Jordan reduction scheme may be usefully applied to systems $\mathbf{Ax} = \mathbf{b}$, where $\mathbf{A}$ is $m \times n$ and $m < n$. Illustrate this for equations

$$3x_1 + 2x_2 + x_3 + 4x_4 + 6x_5 = 2,$$
$$4x_1 + x_2 + x_3 + 5x_4 + 7x_5 = 10,$$
$$x_1 + 9x_2 + 3x_3 + x_4 + x_5 = 7.$$

*5-25. Recall that any solution $\mathbf{x}$ of the set of inhomogeneous linear equations $\mathbf{Ax} = \mathbf{b}$ can be written $\mathbf{x} = \mathbf{x}_1 + \mathbf{y}$, where $\mathbf{y}$ is a solution to the homogeneous set of equations $\mathbf{Ay} = \mathbf{0}$ and $\mathbf{x}_1$ is any solution to $\mathbf{Ax} = \mathbf{b}$. Furthermore, we can write $\mathbf{y} = \sum c_i \mathbf{y}_i$ if the $\mathbf{y}_i$ form a basis for the null space of $\mathbf{A}$. Compare this with the corresponding results for inhomogeneous linear differential equations of nth order.

5-26. Consider a set of m simultaneous linear equations in n unknowns, $\mathbf{Ax} = \mathbf{b}$, with $r(\mathbf{A}) = m$. Suppose that a solution to this set of equations has exactly m nonzero variables. Furthermore, assume that when all variables other than these m variables are set to zero, the resulting set of equations uniquely determines the m variables. Show that the solution is a basic solution, that is, the columns of **A** corresponding to the m nonzero variables are linearly independent.

5-27. Consider a set of linearly independent vectors $x_1, \ldots, x_q$ in E^n, $q < n$. Suppose that the vector y_1 is linearly independent of the set of vectors x_j; furthermore, the vectors y_2 through y_m are linearly independent of the x_j. Assume in addition that the y_i are linearly independent of each other. Prove that $x_1, \ldots, x_q, y_1, \ldots, y_m$ do not necessarily form a linearly independent set. Give a counterexample and illustrate it geometrically. This is equivalent to showing that $\sum_{i=1}^m \lambda_i y_i$ is not necessarily linearly independent of the x_j, although each y_i is linearly independent of the x_j and the y_i form a linearly independent set.

5-28. Show how to routinize the gaussian reduction technique for solving a set of equations $Ax = b$. In particular, demonstrate that it is never necessary to use the variables explicitly. The reduction can be carried out by using only the augmented matrix $A_b = (A, b)$, that is, A_b is reduced to row echelon form. If the row echelon matrix is (H, g), express the value of the variables in terms of the elements of this matrix. Discuss the use of the sum check, introduced in Problem 4-43, for the purpose of avoiding mistakes in solving equations by the gaussian reduction method.

5-29. Show that the Gauss-Jordan method for solving a set of equations $Ax = b$ can be used by making appropriate transformations on the matrix $A_b = (A, b)$; it is never necessary to introduce the variables explicitly. Sketch the structure of the matrix obtained after stages 1, 2, and 3. Demonstrate that if the elements of the matrix at the start of stage s are denoted by u_{ij}, then the elements $\hat{u}_{ij}$ at the end of stage s are given by

$$\hat{u}_{sj} = \frac{u_{sj}}{u_{ss}}, \quad \text{all } j; \quad \hat{u}_{ij} = u_{ij} - \frac{u_{sj}}{u_{ss}} u_{is}, \quad \text{all } i \neq s, \text{all } j.$$

Show that a sum check can be used to guard against numerical mistakes. Solve Problem 5-2, using the technique suggested in the present problem.

*5-30. We know that a set of m vectors $x_1, \ldots, x_m$ from E^n, $m \leq n$, will be linearly dependent if the rank of $X = (x_1, \ldots, x_m)$ is less than m, that is, if there is no minor of order m in X, which is different from 0. Problem 5-8 showed that the vectors are linearly dependent if their Gram determinant vanishes. What is the connection between these two conditions? Hint: $G = X'X$. Use the theorem on expanding the determinant of the product of two rectangular matrices.

PROBLEMS INVOLVING COMPLEX NUMBERS

5-31. List the important results obtained in this chapter. Show that all these results hold when, in the system of equations $Ax = b$, the elements of $A_b = (A, b)$ are complex numbers.

5-32. Solve the following set of equations by Cramer's rule:

$$(-5 + 2i)x_1 - 3ix_2 + 4x_3 = 7 - i,$$
$$(2 + i)x_1 + (4 - 3i)x_2 + (9 - i)x_3 = 2,$$
$$7ix_1 + 5x_2 + (1 + 2i)x_3 = 4 + 6i.$$

5-33. Solve the following set of equations by the Gauss-Jordan method:

$$(4+i)x_1 + (2-i)x_2 + (7+4i)x_3 = 2i,$$
$$5x_1 + (3+5i)x_2 + (8-9i)x_3 = 6-i,$$
$$3ix_1 + (9-7i)x_2 + (2+i)x_3 = 8.$$

5-34. Find all basic solutions to the set of equations:

$$(5-6i)x_1 + 2ix_2 + 5x_3 + (4+3i)x_4 = 16,$$
$$(3+2i)x_1 + (6-4i)x_2 + (-3+2i)x_3 + (8-i)x_4 = 4i.$$

CHAPTER 6

CONVEX SETS AND n-DIMENSIONAL GEOMETRY

"We can make several things clearer, but we cannot make anything clear."

Frank P. Ramsey.

6–1 Sets. In the preceding chapters, we have on various occasions used the notion of a set, for example: *a set of n component vectors*. The notion of a set is so basic that it is somewhat difficult to define it in terms of more fundamental ideas. The following expressions are synonymous: (1) a set of elements, (2) a collection of objects, (3) a number of things. A set consists of a finite or infinite number of *elements*. The concept of a set is clearly one of great generality; it is also very useful. Set theory was first introduced into mathematics by the German mathematician Georg Cantor at the end of the 19th century. Since then it has developed into a most important branch of mathematics.

The main topic of this chapter is convex set theory. For many years, only a handful of men working in the field of pure mathematics were interested in convex sets. Recently, however, the theory has found important applications in economics, linear programming, game theory, and statistical decision theory. This has stimulated interest in the subject, and within the last fifteen years a great deal of work has been done in developing the theory and applications of convex sets. Before turning to the theory of convex sets, we shall first study briefly some general topics of set theory and then develop the fundamentals of point sets.

Sets will be denoted by capital letters, for example, A, B. The elements of the set will be denoted by a_i, b_i, etc. Braces { } enclose the elements belonging to a set. Thus, the set A can be written

$$A = \{a_i\}. \tag{6-1}$$

EQUALITY: *Two sets A, B are equal, $A = B$, if they contain the same elements, that is, if every element of A is also an element of B, and conversely.*

The notation $a_i \in A$ indicates that a_i is an element of A; $b_i \notin A$ means that b_i is not an element of A.

SUBSET: *A subset B of a set A is a set all of whose elements are in A. However, not all elements of A need to be in subset B. B is a proper subset of A if A contains at least one element which is not in B.*

The notation $B \subset A$ or $A \supset B$ indicates that B is a subset of A.

INTERSECTION: *The intersection of sets A, B, written $A \cap B$, is the set $C = \{c_i\}$ containing all elements common to A, B, that is, $c_i \in A$, $c_i \in B$.*

If there are m sets $A_1, \ldots, A_m$, then $A_1 \cap A_2 \cap \ldots \cap A_m$ is the set of elements C common to $A_1, \ldots, A_m$. We can write C as

$$C = \bigcap_{i=1}^{m} A_i. \tag{6-2}$$

EXAMPLE: If $A = \{1, 2, 3, 5, 8, 15, 21\}$ and $B = \{4, 2, 5, 15, 37, 52\}$, then
$$C = A \cap B = \{2, 5, 15\},$$
since these elements are common to A and B.

UNION: *The union of two sets, written $A \cup B$, is the set $C = \{c_i\}$ containing all elements in either A, or B, or both.*

If there are m sets $A_1, \ldots, A_m$, then $A_1 \cup A_2 \ldots \cup A_m$ is the set of elements C in at least one of $A_1, \ldots, A_m$. C can be written as

$$C = \bigcup_{i=1}^{m} A_i. \tag{6-3}$$

The symbols $\cap$, $\cup$ are sometimes read "cap" and "cup," respectively.

6–2 Point sets. *Point sets are sets whose elements are points or vectors in E^n.* Since we shall approach our subject from a geometrical standpoint, n-tuples will frequently be referred to as points in E^n rather than vectors in E^n; recall, though, that there is no difference between a point and a vector.* Point sets may contain either a finite or an infinite number of points. However, the point sets to be considered here will usually contain an infinite number of points.

Point sets are often defined by some property or properties which the set of points satisfy. In E^2, for example, let us consider the set of points X lying inside a circle of unit radius with center at the origin, that is, the set of points satisfying the inequality

$$x_1^2 + x_2^2 < 1.$$

A convenient representation for the set X is

$$X = \{[x_1, x_2] | x_1^2 + x_2^2 < 1\}.$$

---

* However, here we prefer to continue to use the term "component" rather than "coordinate" for any element in an n-tuple.

In general, the notation
$$X = \{\mathbf{x}|\mathbf{P}(\mathbf{x})\} \tag{6-4}$$
will indicate that the set of points $X = \{\mathbf{x}\}$ has the property (or properties) $\mathbf{P}(\mathbf{x})$. In the above example the property $\mathbf{P}$ is the inequality $x_1^2 + x_2^2 < 1$.

If there are no points with the property $\mathbf{P}$, then the set (6-4) contains no elements and is called empty or vacuous. An empty set will be denoted by 0. When discussing the properties of sets, we shall always assume that there is at least one element in the set unless otherwise stated.

EXAMPLES: (1) $X = \{[x_1, x_2]|x_1^2 + x_2^2 \leq 1\}$ is the set of points lying inside and on the circumference of the circle of radius unity with its center at the origin (compare with the preceding example).

(2) $X = \{[x_1, x_2]|2x_1 + 3x_2 = 4\}$ is the set consisting of all points on the line $2x_1 + 3x_2 = 4$.

(3) $X = \{[x_1, x_2]|x_1 > 0, x_2 > 0, x_1 < 1, x_2 < 1\}$ is the set of points inside the square with corners [0, 0], [1, 0], [1, 1], [0, 1].

(4) $X = \{[x_1, x_2]|x_1^2 + x_2^2 > 1, x_1^2 + x_2^2 < 4\}$ is the set of points inside the annulus with its center at the origin, an outer radius of 2, and an inner radius of 1.

The notion of a point set enables us to illustrate geometrically the concepts of the union and intersection of sets. Let us define two sets A and B by
$$A = \{[x_1, x_2]|x_1^2 + x_2^2 \leq 1\},$$
and
$$B = \{[x_1, x_2]|(x_1 - 1)^2 + x_2^2 \leq 1\}.$$

$A \cap B$ is the shaded region shown in Fig. 6–1; $A \cup B$ is the shaded region shown in Fig. 6–2. A, of course, is the set of points inside and on the circle of radius unity with center at the origin; B is the set of points inside and on the circle of radius unity with center at $x_1 = 1$, $x_2 = 0$.

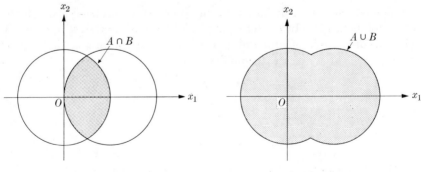

FIGURE 6–1 FIGURE 6–2

According to Section 2–6, the distance between two points **x**, **a** is given by
$$|\mathbf{x} - \mathbf{a}| = [(x_1 - a_1)^2 + \cdots + (x_n - a_n)^2]^{1/2}. \tag{6-5}$$

If **a** is a fixed point and **x** a variable point, then
$$X = \{\mathbf{x} | \, |\mathbf{x} - \mathbf{a}| = \epsilon\} \tag{6-6}$$

is the set of points **x** which are at a given distance ϵ from **a**. In E^2, X becomes

$$[(x_1 - a_1)^2 + (x_2 - a_2)^2]^{1/2} = \epsilon \quad \text{or} \quad (x_1 - a_1)^2 + (x_2 - a_2)^2 = \epsilon^2;$$

this is the equation of a circle of radius ϵ with its center at $[a_1, a_2]$. Any points on the circle are a distance ϵ from the center. In E^3, Eq. (6-6) represents the points on a sphere. In E^n, the relation $|\mathbf{x} - \mathbf{a}| = \epsilon$ defines a hypersphere.

HYPERSPHERE: *A hypersphere in E^n with center at **a** and radius $\epsilon > 0$ is defined as the set of points*

$$X = \{\mathbf{x} | \, |\mathbf{x} - \mathbf{a}| = \epsilon\}. \tag{6-7}$$

Hence, the equation of a hypersphere in E^n is
$$|\mathbf{x} - \mathbf{a}| = \epsilon,$$
or
$$\sum_{i=1}^{n} (x_i - a_i)^2 = \epsilon^2. \tag{6-8}$$

In the preceding paragraph, we generalized the concept of a circle in E^2 and a sphere in E^3 to what we called a hypersphere in E^n. Note that if the equations for a circle in E^2 or a sphere in E^3 are written in vector notation, they both become $|\mathbf{x} - \mathbf{a}| = \epsilon$, which is also the vector form of a hypersphere in E^n. This gives us a key to a very useful procedure for generalizing ideas applicable for E^2 and E^3 to E^n: The vector form obtained for relations in E^2 and E^3 will be a suitable definition in E^n.

We shall frequently be interested in points which are "close" to point **a**, that is, which can be considered to be "inside" some hypersphere with its center at **a**. In E^2, E^3, the inside of a circle or sphere can be represented in vector form by the inequality $|\mathbf{x} - \mathbf{a}| < \epsilon$. This immediately suggests an appropriate definition for E^n.

INSIDE: *The inside of a hypersphere with center at **a** and radius $\epsilon > 0$ is the set of points*
$$X = \{\mathbf{x} | \, |\mathbf{x} - \mathbf{a}| < \epsilon\}. \tag{6-9}$$

ϵ NEIGHBORHOOD: *An ϵ neighborhood about the point **a** is defined as the set of points expressed by (6–9), that is, the set of points inside the hypersphere with center at **a** and radius ϵ > 0.*

When discussing ϵ neighborhoods, we shall often assume that ϵ is small. In general, however, ϵ can be any positive number whatever.

In two and three dimensions, the meaning of the terms "interior" and "boundary" point of a set is intuitively clear. This intuition fails us in E^n, and therefore we need analytic definitions to determine for any given point whether it can be considered an interior or a boundary point of a set. The following definitions are obvious generalizations from E^2 and E^3.

INTERIOR POINT: *A point **a** is an interior point of the set A if there exists an ϵ neighborhood about **a** which contains only points of the set A.*

It may be true that for the neighborhood to contain only points in the set, ϵ will have to be very small. However, it is immaterial how small ϵ is, as long as ϵ > 0. An interior point **a** must be an element of the set because every ϵ neighborhood of **a** contains **a**.

BOUNDARY POINT: *A point **a** is a boundary point of the set A if every ϵ neighborhood about **a** (regardless of how small ϵ > 0 may be) contains points which are in the set and points which are not in the set.*

Note that a boundary point does not have to be an element of the set A. In Fig. 6–3, $\mathbf{a}_1$ is an interior point and $\mathbf{a}_2$ is a boundary point of A.

The concepts of interior and boundary points lead to the notion of open and closed sets, since the boundary points may or may not be elements of the set.

OPEN SET: *A set A is an open set if it contains only interior points.*

CLOSED SET: *A set A is a closed set if it contains all its boundary points.*

EXAMPLES: (1) The set

$$X = \{[x_1, x_2] | x_1^2 + x_2^2 < 1\}$$

is open because points on the circumference of the circle are not included. For any point $[x_1, x_2]$ in the set, it is possible to find an ϵ such that all points $[x_1', x_2']$ given by $(x_1' - x_1)^2 + (x_2' - x_2)^2 < \epsilon^2$ are in the set. A suitable value for ϵ is one-half the distance from $[x_1, x_2]$ to the point on the circumference lying on a radial line through $[x_1, x_2]$. Hence, every point is an interior point and the set is open.

(2) The set

$$X = \{[x_1, x_2] | x_1^2 + x_2^2 \leq 1\}$$

is closed since every point on the circumference is a boundary point,

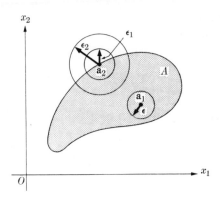

FIGURE 6–3

and all boundary points are included in the set. For any point on the circumference, all neighborhoods about the point contain points which are in the set and points which are not.

It should be noted that open and closed sets are not mutually exclusive or collectively exhaustive concepts. There are sets which are neither open nor closed, such as for example:

$$X = \{[x_1, x_2] | x_1^2 + x_2^2 \leq 1 \text{ for } x_2 \geq 0; x_1^2 + x_2^2 < 1 \text{ for } x_2 < 0\}.$$

Below the x_1-axis, the circumference of the circle is not included; above the x_1-axis, it is. Thus, not all the boundary points are in the set, and it is not closed. However, not every point is an interior point since some boundary points are in the set, and hence the set is not open.

Some sets can be considered to be both open and closed. For example, the set containing every point in E^n is open since an ϵ neighborhood about any point contains only points of E^n. However, E^n has no boundary points, and therefore the set contains all its boundary points and is closed.

COMPLEMENT: *The complement of any set A in E^n, written $\hat{A}$, is the set of all points in E^n not in A.*

Note that $A \cup \hat{A} = E^n$, $A \cap \hat{A} = 0$; furthermore, if **a** is a boundary point of A, it is a boundary point of $\hat{A}$, and vice versa. Hence, if A is closed, $\hat{A}$ is open, and conversely.

EXAMPLE: The complement of the set

$$X = \{[x_1, x_2] | x_1^2 + x_2^2 \leq 1\}$$

is

$$\hat{X} = \{[x_1, x_2] | x_1^2 + x_2^2 > 1\}.$$

Some sets have the property of being "bounded," that is, the components of the points in the set cannot become arbitrarily large or small. We shall define three forms of boundedness:

STRICTLY BOUNDED: *A set A is strictly bounded if there exists a positive number r such that for every $\mathbf{a} \in A$, $|\mathbf{a}| < r$.*

A strictly bounded set lies inside a hypersphere of radius r with its center at the origin.

BOUNDED FROM ABOVE: *The set A is bounded from above if there exists an $\mathbf{r}$ with each component finite such that for all $\mathbf{a} \in A$*

$$\mathbf{a} \leq \mathbf{r}.$$

A set which is bounded from above has an upper limit on each component of every point in the set.

BOUNDED FROM BELOW: *The set A is bounded from below if there exists an $\mathbf{r}$, with each component finite, such that for each $\mathbf{a} \in A$*

$$\mathbf{r} \leq \mathbf{a}.$$

A set which is bounded from below has a lower limit on each component of every point in the set.

EXAMPLES: (1) The set $X = \{[x_1, x_2] \,|\, (x_1 - 3)^2 + (x_2 - 4)^2 \leq 4\}$ is strictly bounded, since every point in the set lies inside the circle of radius 10 with center at the origin.

(2) The set $X = \{[x_1, x_2] \,|\, x_1 \geq x_2, x_1 \geq 0, x_2 \geq 0\}$ is bounded from below by the origin. It is not strictly bounded, however, because x_1, x_2 can become arbitrarily large positive numbers.

6–3 Lines and hyperplanes. We shall find considerable use for the notion of a line in E^n. A suitable definition will be, as usual, the vector form obtained by formulating the equation for a line in E^2 and E^3 in vector terms. In Fig. 6–4, consider the two points $\mathbf{x}_1$, $\mathbf{x}_2$ and the line passing through them. The vector $\mathbf{x}_2 - \mathbf{x}_1$ is parallel to the line passing through $\mathbf{x}_1$, $\mathbf{x}_2$. Any point $\mathbf{x}$ on the line passing through $\mathbf{x}_1$, $\mathbf{x}_2$ can be written

$$\mathbf{x} = \mathbf{x}_1 + \lambda(\mathbf{x}_2 - \mathbf{x}_1) = \lambda \mathbf{x}_2 + (1 - \lambda)\mathbf{x}_1 \qquad (6\text{–}10)$$

for some scalar λ (parallelogram law for addition of vectors). Thus Eq. (6–10) is the vector form for the line through $\mathbf{x}_1$, $\mathbf{x}_2$ in E^2. A similar analysis shows that (6–10) represents a line in E^3. If Eq. (6–10) is written in component form, the ordinary parametric equations for a line are obtained. Thus, Eq. (6–10) is used to define a line in E^n.

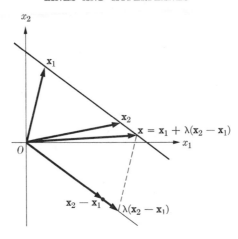

FIGURE 6–4

LINE: *The line passing through points $\mathbf{x}_1$, $\mathbf{x}_2$ ($\mathbf{x}_1 \neq \mathbf{x}_2$) in E^n is defined to be the set of points*

$$X = \{\mathbf{x} | \mathbf{x} = \lambda \mathbf{x}_2 + (1 - \lambda)\mathbf{x}_1, \text{ all real } \lambda\}, \qquad (6\text{--}11)$$

The vector equation (6–10) traces a line in E^n as λ takes on all possible values.

From Fig. 6–4 we see that any point on the line segment joining $\mathbf{x}_1$, $\mathbf{x}_2$ can be written

$$\mathbf{x} = \mathbf{x}_1 + \lambda(\mathbf{x}_2 - \mathbf{x}_1) = \lambda \mathbf{x}_2 + (1 - \lambda)\mathbf{x}_1, \quad 0 \leq \lambda \leq 1. \qquad (6\text{--}12)$$

Thus we make the following general definition:

LINE SEGMENT: *The line segment joining points $\mathbf{x}_1$, $\mathbf{x}_2$ in E^n is defined to be the set of points*

$$X = \{\mathbf{x} | \mathbf{x} = \lambda \mathbf{x}_2 + (1 - \lambda)\mathbf{x}_1, 0 \leq \lambda \leq 1\}. \qquad (6\text{--}13)$$

Many sets have the property of being "connected"; that is, their elements do not exist in groups which are completely isolated from other points in the set by points not belonging to the set. For example, the set defined by

$$X = \{[x_1, x_2] | x_1^2 + x_2^2 < 1 \text{ or } (x_1 - 5)^2 + (x_2 - 6)^2 < 1\}$$

is not connected. Some points of the set are inside the circle of radius unity with center at the origin, and others inside the circle with radius unity and center [5, 6]. These two circles are isolated from each other.

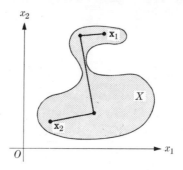

FIGURE 6-5

There is no path leading from one circle to the other while remaining entirely within the set.

CONNECTED SET: *An open set is connected if any two points in the set can be joined by a polygonal path lying entirely within the set.*

A polygonal path joining $\mathbf{x}_1$, $\mathbf{x}_2$ can be defined as follows: Choose any R points $\mathbf{y}_1 = \mathbf{x}_1$, $\mathbf{y}_2, \ldots, \mathbf{y}_{R-1}$, $\mathbf{y}_R = \mathbf{x}_2$. Form the line segments joining $\mathbf{y}_1$ and $\mathbf{y}_2$, $\mathbf{y}_2$ and $\mathbf{y}_3, \ldots, \mathbf{y}_{R-1}$ and $\mathbf{y}_R$. The set of points defined by these $R - 1$ line segments is a polygonal path. Intuitively, a polygonal path is merely a broken line. A connected set and a typical polygonal path are shown in Fig. 6-5. The simplest polygonal path connecting two points is the line segment joining these points.

A set does not need to be open to be connected. However, with closed sets, using the polygonal path as the criterion of our definition may land us in difficulties. For example: From the preceding discussion we would assume that the set of points on the circle $x_1^2 + x_2^2 = 1$ represents a connected set. This set is closed (every point is a boundary point), and there is no polygonal path lying entirely within the set which connects two different points on the circle. Hence, for this closed set, we must generalize the notion of the allowed path connecting any two points. The path must be what we think of in two and three dimensions as an arbitrary continuous curve. We shall not make this generalization. For many closed connected sets, it is possible to find a polygonal path connecting any two points and therefore to demonstrate that the set is connected.

REGION: *A region in E^n is a connected set of points in E^n.*

A region may or may not contain some or all boundary points of the set.

In E^2, $c_1x_1 + c_2x_2 = z$ (c_1, c_2, z constant) represents a straight line; in E^3, $c_1x_1 + c_2x_2 + c_3x_3 = z$ is the equation for a plane. If, in E^2, we write $\mathbf{c} = (c_1, c_2)$, $\mathbf{x} = [x_1, x_2]$; and, in E^3, $\mathbf{c} = (c_1, c_2, c_3)$, $\mathbf{x} = [x_1, x_2, x_3]$, then the line in E^2 or the plane in E^3 can be written as the scalar product

$$\mathbf{cx} = z. \tag{6-14}$$

The equivalent in E^n of a plane in E^3 or a line in E^2 is a hyperplane; it will be defined by Eq. (6–14).

HYPERPLANE: *A hyperplane in E^n is defined to be the set of points*

$$X = \{\mathbf{x} | \mathbf{cx} = z\}, \tag{6-15}$$

with $\mathbf{c} \neq \mathbf{0}$ being a given n-component row vector and z a given scalar.

If the equation for a hyperplane is written out, we obtain

$$\mathbf{cx} = c_1 x_1 + c_2 x_2 + \cdots + c_n x_n = z, \tag{6-16}$$

and any $\mathbf{x}$ satisfying (6–16) *lies on* the hyperplane.

A hyperplane passes through the origin if and only if $z = 0$. When $z = 0$, Eq. (6–16) becomes

$$\mathbf{cx} = 0; \tag{6-17}$$

we see that $\mathbf{c}$ is orthogonal to every vector $\mathbf{x}$ on the hyperplane, and hence we can say that $\mathbf{c}$ is normal to the hyperplane. If $z \neq 0$, and $\mathbf{x}_1$, $\mathbf{x}_2$ are any two distinct points lying on the hyperplane, then

$$\mathbf{cx}_1 - \mathbf{cx}_2 = \mathbf{c}(\mathbf{x}_1 - \mathbf{x}_2) = z - z = 0,$$

and $\mathbf{c}$ is orthogonal to every vector $\mathbf{x}_1 - \mathbf{x}_2$ ($\mathbf{x}_1$, $\mathbf{x}_2$ being on the hyperplane). In E^2 and E^3, vector $\mathbf{x}_1 - \mathbf{x}_2$ is parallel to the line or plane, respectively. Thus, even with $z \neq 0$, we can say that $\mathbf{c}$ is normal to the hyperplane.

If $\mathbf{cx} = z$ is multiplied by the scalar $\lambda \neq 0$, we have $(\lambda \mathbf{c})\mathbf{x} = \lambda z$. The same hyperplane is defined by either $\lambda \mathbf{c}$ and λz or $\mathbf{c}$, z. Hence if $\mathbf{c}$ is normal to a hyperplane, so is $\lambda \mathbf{c}$. Let us suppose that $\lambda = 1/|\mathbf{c}|$. Then if (6–16) becomes

$$\mathbf{n} = \frac{\mathbf{c}}{|\mathbf{c}|}, \qquad b = \frac{z}{|\mathbf{c}|},$$

$$\mathbf{nx} = b, \tag{6-18}$$

and $|\mathbf{n}| = 1$. The vector $\mathbf{n}$ is called a unit normal to the hyperplane. There are two unit normals to any hyperplane; the other is given by $\mathbf{n} = -\mathbf{c}/|\mathbf{c}|$. The above discussion can be summarized in the following definition for E^n:

NORMALS: *Given the hyperplane $\mathbf{cx} = z$ in E^n ($\mathbf{c} \neq \mathbf{0}$), then $\mathbf{c}$ is a vector normal to the hyperplane. Any vector $\lambda \mathbf{c}$ is also normal to the hyperplane ($\lambda \neq 0$). The two vectors of unit length $\mathbf{c}/|\mathbf{c}|$, $-\mathbf{c}/|\mathbf{c}|$ are the unit normals to the hyperplane.*

In E^2 and E^3, the concept of a normal to a hyperplane can be clearly illustrated (see Fig. 6-6): Let **n** be a vector of unit length lying along the line normal to $c_1 x_1 + c_2 x_2 = z$. Since **n** points towards the line, moving the line **cx** $= z$ parallel to itself in the direction of **n** increases its distance from the origin. It will be noted that for any **x** on the line **cx** $= z$, $b = |\mathbf{x}| \cos \theta$ or

$$\mathbf{nx} = b = |\mathbf{x}| \cos \theta; \tag{6-19}$$

this is the equation of the line. If we multiply Eq. (6-19) by a λ such that $z = \lambda b$, we see that $\lambda \mathbf{n} = \mathbf{c}$, and $\lambda = \pm|\mathbf{c}|$. Hence **c** is normal to the line. The same reasoning applies to a plane in E^3. This discussion also demonstrates that, in E^2 or E^3, b is the distance of the line or plane from the origin. *Similarly, in E^n, $|z|/|\mathbf{c}|$ ($|z|$ is the absolute value of z) is the distance of the hyperplane from the origin.*

If we move any line in E^2 parallel to itself in the direction of **n** (see Fig. 6-6), b and the distance of the line from the origin increase. Let us consider the line **cx** $= z$: If $z > 0$, then **c** points in the same direction as **n**; hence, moving the line parallel to itself in the direction of **c** increases z. However, if $z < 0$, **c** points in the direction opposite to **n**; moving the line parallel to itself in the direction of **c** decreases the absolute value of z (moves the line closer to the origin), but increases its algebraic value. Thus both alternatives result in an algebraic increase of z.

EXAMPLE: Consider $2x_1 + 3x_2 = -6$ in Fig. 6-7. A normal vector to the line is $\mathbf{c} = (2, 3)$. Line $2x_1 + 3x_2 = 0$, obtained by moving the first line parallel to itself in the direction of **c**, has indeed a greater algebraic value of z. The slope is, of course, the same for both lines.

The intuitive concept of moving a line or plane parallel to itself in order to increase the algebraic value of z can easily be generalized to E^n. First,

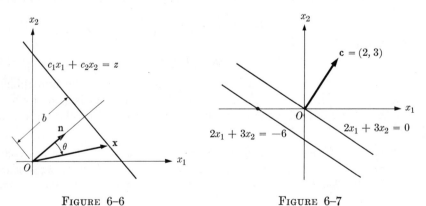

FIGURE 6-6   FIGURE 6-7

we define the notion of two parallel hyperplanes by direct extension from E^2 and E^3.

PARALLEL HYPERPLANES: *Two hyperplanes are parallel if they have the same unit normal.*

Thus the hyperplanes $c_1x = z_1$, $c_2x = z_2$ are parallel if $c_1 = \lambda c_2$, $\lambda \neq 0$.

To understand the concept of moving a hyperplane parallel to itself let us consider the hyperplane with normal c passing through x_0, that is, $cx = z_0$ where $cx_0 = z_0$. Let us find the value of z for a hyperplane with normal c passing through the point $x_1 = x_0 + \lambda c'$, $\lambda > 0$ (we use c' since c is assumed to be a row vector). When $x_1 = x_0 + \lambda c'$, $\lambda > 0$, we can say that in going from x_0 to x_1, we move in the direction of c (illustrate this geometrically). The hyperplane through x_1 is $cx = z_1$, where $cx_1 = z_1$. But

$$cx_1 = c(x_0 + \lambda c') = z_0 + \lambda |c|^2.$$

Therefore,

$$z_1 = z_0 + \lambda |c|^2, \tag{6-20}$$

and

$$z_1 > z_0, \quad \text{since } \lambda > 0, \ |c| \neq 0.$$

Hence z_1 is algebraically greater than z_0. Furthermore, if x_0 is *any* point on the hyperplane $cx = z_0$, then $x_1 = x_0 + \lambda c'$ lies on the hyperplane $cx = z_1$. Thus we say the hyperplane $cx = z_0$ has been moved parallel to itself in the direction of c to yield $cx = z_1$. Points lying on the hyperplane $cx = z_1$ satisfy the inequality

$$cx > z_0, \tag{6-21}$$

that is, moving a hyperplane parallel to itself in the direction of c moves it in the "greater than" direction.

Note: In linear programming the function to be optimized, $z = \sum c_j x_j$, is a hyperplane for a given value of z. Since the c_j are constants, the hyperplanes corresponding to different values of z have the same normal and are therefore parallel. Hence, to maximize z, we move this hyperplane parallel to itself over the region representing the feasible solutions until z is made algebraically as large as possible. If z is positive, we move the hyperplane in the direction of c until it is as far away from the origin as possible. If z is negative, then the hyperplane is moved as close to the origin as possible.

We have noted previously that any two different points in E^n can be used to define a line in E^n. In order to determine a hyperplane, it is necessary to specify the n components of c and z, that is, $n + 1$ parameters. These are determined only up to a multiplicative constant, however;

that is, λc_i ($i = 1, \ldots, n$), λz yield the same hyperplane as the c_i and z. Thus, there are only n independent parameters, and hence n points in E^n will be needed to determine a hyperplane; however, not any arbitrary set of n points will provide a unique definition. Only a set of n points $\mathbf{x}_1, \ldots, \mathbf{x}_n$ which can be numbered so that the $n - 1$ vectors $\mathbf{x}_1 - \mathbf{x}_n$, $\mathbf{x}_2 - \mathbf{x}_n, \ldots, \mathbf{x}_{n-1} - \mathbf{x}_n$ are linearly independent, will describe a unique hyperplane. (The n points $\mathbf{x}_1, \ldots, \mathbf{x}_n$ may or may not form a set of n linearly independent vectors; it is the set of differences $\mathbf{x}_1 - \mathbf{x}_n$, etc., which is of importance.)

To prove this let us consider the set of equations

$$\mathbf{c}\mathbf{x}_i - z = 0, \qquad i = 1, \ldots, n. \tag{6-22}$$

This expression represents a set of n homogeneous linear equations in $n + 1$ unknowns, the c_i and z. If the nth equation is used to eliminate z, we obtain a new set of $n - 1$ homogeneous equations in n unknowns, the c_i:

$$\mathbf{c}(\mathbf{x}_i - \mathbf{x}_n) = 0, \qquad i = 1, \ldots, n - 1. \tag{6-23}$$

If the vectors $\mathbf{x}_i - \mathbf{x}_n$ are linearly independent, the matrix of the coefficients $\|x_{ji} - x_{jn}\|'$ has rank $n - 1$ and nullity 1. Thus, there is a solution to (6-23) with not all $c_i = 0$, and with only one degree of freedom, that is, the c_i are determined up to a multiplicative constant. Then the nth equation of (6-22) uniquely determines z for any $\mathbf{c}$. We have proved that n points in E^n for which the vectors $\mathbf{x}_1 - \mathbf{x}_n, \ldots, \mathbf{x}_{n-1} - \mathbf{x}_n$ are linearly independent uniquely define a hyperplane in E^n. If the rank of $\|x_{ji} - x_{jn}\|$ is $n - k$, then, according to Section 5-6, we can determine k linearly independent vectors $\mathbf{c}$ which satisfy (6-23); in this case the chosen points lie in the intersection of k hyperplanes. Furthermore, the k hyperplanes are not unique if $k \geq 2$; there exists an infinite number of sets of k hyperplanes whose intersection contains the n points.

Let us consider any point $\mathbf{x}$ in E^n such that $\mathbf{x} - \mathbf{x}_n$ can be written as a linear combination of the $n - 1$ linearly independent vectors $\mathbf{x}_1 - \mathbf{x}_n, \ldots, \mathbf{x}_{n-1} - \mathbf{x}_n$. Then $\mathbf{x}$ lies on the hyperplane determined by the points $\mathbf{x}_1, \ldots, \mathbf{x}_n$ since, by Eq. (6-23), $\mathbf{c}(\mathbf{x} - \mathbf{x}_n) = 0$ or $\mathbf{c}\mathbf{x} = \mathbf{c}\mathbf{x}_n = z$. Furthermore, any point $\mathbf{x}$ on the hyperplane has the property that $\mathbf{x} - \mathbf{x}_n$ can be written as a linear combination of $\mathbf{x}_1 - \mathbf{x}_n, \ldots, \mathbf{x}_{n-1} - \mathbf{x}_n$. If this were not true, $\mathbf{x} - \mathbf{x}_n$ would be linearly independent of $\mathbf{x}_1 - \mathbf{x}_n, \ldots, \mathbf{x}_{n-1} - \mathbf{x}_n$; then if the equation $\mathbf{c}(\mathbf{x} - \mathbf{x}_n) = 0$ were annexed to (6-23), the rank of the matrix of the coefficients would be n, and the only solution would be $\mathbf{c} = \mathbf{0}$. Hence, if we choose any n points on a hyperplane with the property that $\mathbf{x}_1 - \mathbf{x}_n, \ldots, \mathbf{x}_{n-1} - \mathbf{x}_n$ are linearly independent, then any other point $\mathbf{x}$ on the hyperplane is such that $\mathbf{x} - \mathbf{x}_n$ can be written as a linear combination of $\mathbf{x}_1 - \mathbf{x}_n, \ldots, \mathbf{x}_{n-1} - \mathbf{x}_n$ or, equivalently, $\mathbf{x}$ can be expressed as a linear combination of $\mathbf{x}_1, \ldots, \mathbf{x}_n$.

We can choose $\mathbf{x}_n = \mathbf{0}$ for a hyperplane through the origin. Then any $n - 1$ linearly independent points $\mathbf{x}_1, \ldots, \mathbf{x}_{n-1}$ on the hyperplane can be used to determine $\mathbf{c}$, and every other point on the hyperplane can be written as a linear combination of $\mathbf{x}_1, \ldots, \mathbf{x}_{n-1}$.

A hyperplane $\mathbf{cx} = z$ in E^n divides all of E^n into three mutually exclusive and collectively exhaustive sets. These are

$$(1) \quad X_1 = \{\mathbf{x} | \mathbf{cx} < z\}, \tag{6-24a}$$

$$(2) \quad X_2 = \{\mathbf{x} | \mathbf{cx} = z\}, \tag{6-24b}$$

$$(3) \quad X_3 = \{\mathbf{x} | \mathbf{cx} > z\}. \tag{6-24c}$$

OPEN HALF-SPACES: *The sets $X_1 = \{\mathbf{x}|\mathbf{cx} < z\}$ and $X_3 = \{\mathbf{x}|\mathbf{cx} > z\}$ are called open half-spaces.*

In E^2, E^3 a half-space is all of E^2 or E^3 lying on one side of a line or plane, respectively.

CLOSED HALF-SPACES: *The sets $X_4 = \{\mathbf{x}|\mathbf{cx} \leq z\}$ and $X_5 = \{\mathbf{x}|\mathbf{cx} \geq z\}$ are called closed half-spaces.*

Note: $X_4 \cap X_5 = X_2$, that is, if $\mathbf{x}$ is on $\mathbf{cx} = z$, it is in both closed half-spaces. However, no point can be in more than one of the sets X_1, X_2, X_3, and every point is in one of these sets. Furthermore $X_1 \cap X_3 = 0$.

It is easy to see that hyperplanes are closed sets. Choose any point $\mathbf{x}_0$ on the hyperplane $\mathbf{cx} = z$. Form an ϵ neighborhood about the point $\mathbf{x}_0$ and consider the point $\mathbf{x}_1 = \mathbf{x}_0 + (\epsilon/2)(\mathbf{c}'/|\mathbf{c}|)$. The point $\mathbf{x}_1$ is in the ϵ neighborhood since $|\mathbf{x}_1 - \mathbf{x}_0| = \epsilon/2 < \epsilon$. However,

$$\mathbf{cx}_1 = \mathbf{cx}_0 + \frac{\epsilon}{2}|\mathbf{c}| = z_0 + \frac{\epsilon}{2}|\mathbf{c}| > z_0,$$

and $\mathbf{x}_1$ is not on the hyperplane. This holds true for every $\epsilon > 0$. Therefore, every point on a hyperplane is a boundary point. A hyperplane has no interior points. Furthermore, there are no boundary points for a hyperplane which are not on the hyperplane. A point in either open half-space cannot be a boundary point for the hyperplane since it is always possible to find an ϵ neighborhood about such a point, with all points in the neighborhood being in the open half-space. To show this explicitly, let us suppose $\mathbf{cx}_1 = z_1 < z$. Take $\epsilon = (z - z_1)/2|\mathbf{c}|$. For any point $\mathbf{x}$ in this ϵ neighborhood of $\mathbf{x}_1$,

$$\mathbf{cx} = \mathbf{cx}_1 + \mathbf{c}(\mathbf{x} - \mathbf{x}_1) \leq z_1 + |\mathbf{c}(\mathbf{x} - \mathbf{x}_1)|,$$

and by the Schwarz inequality,

$$\mathbf{cx} \leq z_1 + |\mathbf{c}||\mathbf{x} - \mathbf{x}_1| \leq z_1 + |\mathbf{c}|\epsilon = z_1 + \frac{z - z_1}{2} = \frac{z_1 + z}{2} < z.$$

Hence every point in this ϵ neighborhood of $\mathbf{x}_1$ is in the half-space and $\mathbf{x}_1$ cannot be a boundary point of the hyperplane. Consequently, a hyperplane contains all its boundary points and is a closed set.

A closed half-space is also a closed set. The boundary points of the half-space X_4 or X_5 are all the points in the set $X_2 = \{\mathbf{x}|\mathbf{cx} = z\}$, because every ϵ neighborhood of any point on $\mathbf{cx} = z$ contains points in both closed half-spaces, that is, every point on $\mathbf{cx} = z$ is a boundary point of either closed half-space. The preceding discussion has also demonstrated that for any point in an open half-space there exists an ϵ neighborhood which contains only points in the open half-space. Hence, no such point can be a boundary point of the closed half-space. However, a closed half-space contains all the points $\mathbf{cx} = z$, that is, all its boundary points, and is a closed set.

6–4 Convex sets.

CONVEX SET: *A set X is convex if for any points $\mathbf{x}_1$, $\mathbf{x}_2$ in the set, the line segment joining these points is also in the set.*

This definition implies that if $\mathbf{x}_1, \mathbf{x}_2 \in X$, then every point

$$\mathbf{x} = \lambda \mathbf{x}_2 + (1 - \lambda)\mathbf{x}_1, \qquad 0 \leq \lambda \leq 1,$$

must also be in the set. It is immediately obvious that a convex set is connected because the polygonal path joining the points is the line joining the points. Hence, a convex set is a region. By convention, we say that any set containing only one point is convex.

The expression $(1 - \lambda)\mathbf{x}_1 + \lambda \mathbf{x}_2$, $0 \leq \lambda \leq 1$, is often referred to as a convex combination of $\mathbf{x}_1$, $\mathbf{x}_2$ (for a given λ). A set is convex if every convex combination of any two points in the set is also in the set.

Intuitively, a convex set cannot have any "holes" in it, that is, it is "solid," and not "re-entrant," i.e., its boundaries are always "flat" or "bent away" from the set. These intuitive ideas, of course, are all rigorously expressed in the definition of a convex set.

EXTREME POINT: *A point $\mathbf{x}$ is an extreme point of a convex set if and only if there do not exist points $\mathbf{x}_1$, $\mathbf{x}_2 (\mathbf{x}_1 \neq \mathbf{x}_2)$ in the set such that*

$$\mathbf{x} = (1 - \lambda)\mathbf{x}_1 + \lambda \mathbf{x}_2, \qquad 0 < \lambda < 1. \tag{6-25}$$

Note that strict inequalities are imposed on λ. The definition stipulates that an extreme point cannot be "between" any other two points of the set, that is, it cannot be on the line segment joining the points $(0 < \lambda < 1)$. Clearly, an extreme point is a boundary point of the set. To prove this, let us suppose that $\mathbf{x}_0$ is any interior point of X. Then there is an $\epsilon > 0$ such that every point in this ϵ neighborhood of $\mathbf{x}_0$ is in the set.

Let $\mathbf{x}_1 \neq \mathbf{x}_0$ be a point in the ϵ neighborhood. Consider the point (illustrate this geometrically)

$$\mathbf{x}_2 = -\mathbf{x}_1 + 2\mathbf{x}_0, \qquad |\mathbf{x}_2 - \mathbf{x}_0| = |\mathbf{x}_1 - \mathbf{x}_0|;$$

then $\mathbf{x}_2$ is in the ϵ neighborhood. Furthermore,

$$\mathbf{x}_0 = \tfrac{1}{2}(\mathbf{x}_1 + \mathbf{x}_2),$$

and hence $\mathbf{x}_0$ is not an extreme point.

Not all boundary points of a convex set are necessarily extreme points. Some boundary points may lie between two other boundary points.

If a convex set contains only a single point, this point will be considered an extreme point.

EXAMPLES: (1) A triangle and its interior form a convex set. The vertices of the triangle are its only extreme points, since they do not lie between any other two points of the set. The other points on the sides of the triangle are not extreme points because they lie between the vertices.

(2) The set

$$X = \{[x_1, x_2] | x_1^2 + x_2^2 \leq 1\}$$

is convex. Every point on the circumference is an extreme point.

(3) The set in Fig. 6–8 is not convex, since the line joining $\mathbf{x}_1$ and $\mathbf{x}_2$ does not lie entirely within the set. The set is re-entrant.

(4) The four sets in Fig. 6–9 are convex, and the extreme points are the vertices. Point $\mathbf{x}_1$ is not an extreme point because it can be represented as a convex combination of $\mathbf{x}_2$, $\mathbf{x}_3$ with $0 < \lambda < 1$.

A hyperplane is a convex set: If $\mathbf{x}_1$, $\mathbf{x}_2$ are on the hyperplane, that is, $\mathbf{c}\mathbf{x}_1 = z$ and $\mathbf{c}\mathbf{x}_2 = z$, then $\mathbf{x} = \lambda \mathbf{x}_2 + (1 - \lambda)\mathbf{x}_1$ is on the hyperplane, since

$$\mathbf{c}\mathbf{x} = \mathbf{c}[\lambda \mathbf{x}_2 + (1 - \lambda)\mathbf{x}_1] = \lambda \mathbf{c}\mathbf{x}_2 + (1 - \lambda)\mathbf{c}\mathbf{x}_1 = \lambda z + (1 - \lambda)z = z.$$

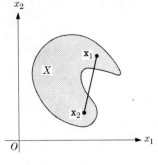

FIGURE 6–8

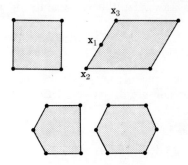

FIGURE 6–9

Similarly, *a closed half-space is also a convex set.* Suppose $\mathbf{x}_1$, $\mathbf{x}_2$ are in the closed half-space $\mathbf{cx} \leq z$; if $\mathbf{x} = \lambda \mathbf{x}_2 + (1 - \lambda)\mathbf{x}_1$ $(0 \leq \lambda \leq 1)$, then

$$\mathbf{cx} = \lambda \mathbf{cx}_2 + (1 - \lambda)\mathbf{cx}_1 \leq \lambda z + (1 - \lambda)z = z,$$

and $\mathbf{x}$ is in the half-space. The same arguments will prove that *an open half-space is a convex set.*

The intersection of two convex sets is also a convex set. Given the convex sets X_1, X_2, let $\mathbf{x}_1$, $\mathbf{x}_2$ be any two points in $X_3 = X_1 \cap X_2$ (if there is only one point in X_3, then X_3 is automatically convex). Thus,

$$\lambda \mathbf{x}_2 + (1 - \lambda)\mathbf{x}_1 \in X_1 \quad \text{for} \quad 0 \leq \lambda \leq 1,$$

$$\lambda \mathbf{x}_2 + (1 - \lambda)\mathbf{x}_1 \in X_2 \quad \text{for} \quad 0 \leq \lambda \leq 1.$$

Hence

$$\lambda \mathbf{x}_2 + (1 - \lambda)\mathbf{x}_1 \in X_1 \cap X_2 = X_3,$$

and X_3 is convex. In Problem 6–15 the reader will be required to show that if X_i $(i = 1, \ldots, m)$ are convex, then $X = \cap_{i=1}^{m} X_i$ is also convex.

If X_1, X_2 are closed sets, then $X_3 = X_1 \cap X_2$ is also closed. We see immediately that an interior point of both X_1 and X_2 will also be an interior point of X_3. Similarly, a point which is not in X_1 and/or not in X_2 cannot be a boundary point of X_3. Therefore, every boundary point of X_3 is a boundary point of X_1 or X_2. However, X_1, X_2, and hence X_3 contain all their boundary points; thus X_3 is closed. The same is true for the intersection of any finite number of closed sets (proof to be furnished in Problem 6–16).

We have shown that hyperplanes and half-spaces are convex sets. Since the intersection of any finite number of convex sets is also convex, *the intersection of a finite number of hyperplanes, or half-spaces, or of both is also a convex set.* Furthermore, *the intersection of a finite number of hyperplanes, or closed half-spaces, or of both is a closed convex set.*

This has immediate application to linear programming. In Section 1–4, we saw that the constraints on a linear programming problem can be written

$$\sum_{j=1}^{r} a_{ij}x_j \{\leq \; = \; \geq\} b_i, \qquad i = 1, \ldots, m, \qquad (6\text{–}26)$$

where one of the $\geq$, $=$, $\leq$ signs holds for each i. In addition, there are the non-negativity restrictions

$$x_j \geq 0, \qquad j = 1, \ldots, r. \qquad (6\text{–}27)$$

If we define the row vector by

$$\mathbf{a}^i = (a_{i1}, \ldots, a_{ir}), \qquad (6\text{–}28)$$

Eq. (6–26) becomes

$$\mathbf{a}^i\mathbf{x}\{\leq\ =\ \geq\}b_i, \qquad i = 1, \ldots, m. \tag{6–29}$$

Each one of the constraints requires that the allowable $\mathbf{x}$ be in some given closed half-space in E^r, or, if the strict equality holds, lie on some given hyperplane.

The non-negativity restrictions can also be written in the form of (6–29). If we write

$$\mathbf{a}^{m+j} = \mathbf{e}'_j, \tag{6–30}$$

then $x_j \geq 0$ becomes

$$\mathbf{a}^{m+j}\mathbf{x} = \mathbf{e}'_j\mathbf{x} \geq 0, \qquad j = 1, \ldots, r. \tag{6–31}$$

Each of the non-negativity restrictions also requires that the allowable $\mathbf{x}$ be in some closed half-space. *The region of E^n defined by $x_j \geq 0$, $j = 1, \ldots, n$, is called the non-negative orthant of E^n.*

Now we can see that any feasible solution $\mathbf{x}$ must simultaneously be an element of each of the following sets [(6–29) and (6–31)]:

$$X_i = \{\mathbf{x} | \mathbf{a}^i\mathbf{x}(\leq\ =\ \geq)b_i\}, \qquad i = 1, \ldots, m + r, \tag{6–32}$$

where $b_i = 0$, $i = m + 1, \ldots, m + r$. This is equivalent to saying that the set of feasible solutions X is the intersection of the sets X_i:

$$X = \bigcap_{i=1}^{m+r} X_i. \tag{6–33}$$

Therefore, *the set of feasible solutions to a linear programming problem (if a feasible solution exists) is a closed convex set.*

The preceding analysis also shows that the set of solutions to a system of m linear equations in n unknowns, $\mathbf{Ax} = \mathbf{b}$, is a closed convex set. To see this, note that the set of equations can be written

$$\mathbf{a}^i\mathbf{x} = b_i, \qquad i = 1, \ldots, m, \tag{6–34}$$

where $\mathbf{a}^i$ is the ith row of $\mathbf{A}$. The set of points which satisfies the ith equation comprises the hyperplane $\mathbf{a}^i\mathbf{x} = b_i$. The set of points which simultaneously satisfies all m equations is the intersection of the m hyperplanes (6–34); it is therefore a closed convex set provided that a solution exists. Furthermore, the set of non-negative solutions to $\mathbf{Ax} = \mathbf{b}$, that is, the set of solutions with $\mathbf{x} \geq \mathbf{0}$, is also a closed convex set.

We have already defined a convex combination of two points $\mathbf{x}_1, \mathbf{x}_2$ to be $\lambda \mathbf{x}_2 + (1 - \lambda)\mathbf{x}_1$, $0 \leq \lambda \leq 1$. This definition can easily be generalized to the concept of a convex combination of m points.

CONVEX COMBINATION: *A convex combination of a finite number of points* $\mathbf{x}_1, \ldots, \mathbf{x}_m$ *is defined as a point*

$$\mathbf{x} = \sum_{i=1}^{m} \mu_i \mathbf{x}_i, \quad \mu_i \geq 0, \quad i = 1, \ldots, m, \quad \sum_{i=1}^{m} \mu_i = 1. \quad (6\text{-}35)$$

A convex combination of m points can be interpreted physically. Suppose that we associate a mass m_i with the point $\mathbf{x}_i$. The center of mass of the m points is then given by

$$\mathbf{x} = \frac{1}{M} \sum_{i=1}^{m} m_i \mathbf{x}_i, \quad M = \sum_{i=1}^{m} m_i. \quad (6\text{-}36)$$

If $\mu_i = m_i/M$, the center of mass is obtained from (6-35). Hence a convex combination of m points can be thought of as the center of mass of the points, with the mass assigned to $\mathbf{x}_i$ being a fraction μ_i of the total mass.

The set of all convex combinations of a finite number of points $\mathbf{x}_1, \ldots, \mathbf{x}_m$ *is a convex set*, that is, the set

$$X = \left\{ \mathbf{x} \,\middle|\, \mathbf{x} = \sum_{i=1}^{m} \mu_i \mathbf{x}_i, \text{ all } \mu_i \geq 0, \sum_{i=1}^{m} \mu_i = 1 \right\} \quad (6\text{-}37)$$

is convex. To prove this, let $\mathbf{v}$, $\mathbf{w}$ be any points such that

$$\mathbf{v} = \sum_{i=1}^{m} \mu_i' \mathbf{x}_i, \quad \mu_i' \geq 0, \quad \sum \mu_i' = 1,$$

$$\mathbf{w} = \sum_{i=1}^{m} \mu_i'' \mathbf{x}_i, \quad \mu_i'' \geq 0, \quad \sum \mu_i'' = 1.$$

The set will be convex if $\lambda \mathbf{w} + (1 - \lambda)\mathbf{v}$ is also in the set for any λ ($0 \leq \lambda \leq 1$). Now

$$\lambda \mathbf{w} + (1 - \lambda)\mathbf{v} = \sum_{i=1}^{m} [\lambda \mu_i'' + (1 - \lambda)\mu_i']\mathbf{x}_i. \quad (6\text{-}38)$$

But

$$\lambda \mu_i'' + (1 - \lambda)\mu_i' \geq 0,$$

and

$$\sum_{i=1}^{m} [\lambda \mu_i'' + (1 - \lambda)\mu_i'] = \lambda \sum \mu_i'' + (1 - \lambda) \sum \mu_i' = 1.$$

Thus $\lambda \mathbf{w} + (1 - \lambda)\mathbf{v}$ is also a convex combination of the $\mathbf{x}_i$, and the set is convex.

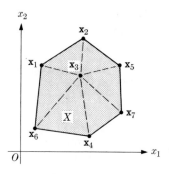

FIGURE 6–10

EXAMPLE: Figure 6–10 shows the set of all convex combinations of the points $x_1, \ldots, x_7$. In E^2 we find the set of all convex combinations of m points by connecting all the points with straight lines. The resulting polygon and its interior is the desired set.

6–5 The convex hull. Given any set A which is not convex, it is possible to imbed A in another set X which is convex such that every point in A is also in X.* We often wish to find the "smallest" convex set containing A. This smallest convex set which contains A is called the *convex hull* of A.

EXAMPLES: (1) The convex hull of the set $A = \{[x_1, x_2] | x_1^2 + x_2^2 = 1\}$ is $X = \{[x_1, x_2] | x_1^2 + x_2^2 \leq 1\}$. The convex hull of the points on the circumference of a circle is the circumference plus the interior of the circle. This is the smallest convex set containing the circumference.

(2) The convex hull of two points x_1, x_2 is the set of all convex combinations of these points $X = \{x | x = \lambda x_2 + (1 - \lambda) x_1, \text{ all } \lambda, 0 \leq \lambda \leq 1\}$. This is the smallest convex set containing x_1, x_2.

So far we have not defined clearly the meaning of the term "smallest" in E^n. To avoid intuitive interpretations, mathematicians define the convex hull as follows:

CONVEX HULL: *The convex hull of a set A is the intersection of all convex sets which contain A.*

The intersection of all convex sets containing A must be the smallest convex set which contains A. Hence, our definition is merely a more rigorous and elegant formulation of the intuitive idea of "smallest." It should be recalled that the intersection of convex sets is also convex.

---

* Note that E^n is a convex set and hence X always exists.

The convex hull of a finite number of points $\mathbf{x}_1, \ldots, \mathbf{x}_m$ *is the set of all convex combinations of* $\mathbf{x}_1, \ldots, \mathbf{x}_m$. This theorem states that the convex hull of $\mathbf{x}_1, \ldots, \mathbf{x}_m$ is the set

$$X = \left\{ \mathbf{x} \mid \mathbf{x} = \sum_{i=1}^{m} \mu_i \mathbf{x}_i, \text{ all } \mu_i \geq 0, \sum_{i=1}^{m} \mu_i = 1 \right\}. \tag{6-39}$$

In the preceding section we have shown that X is convex. The proof will be made by induction on the number of points, that is, on m. We must show that every convex set containing $\mathbf{x}_1, \ldots, \mathbf{x}_m$ also contains X. Clearly the theorem is true for $m = 1$ since there is only one point and $\mu_1 = 1$. We have already defined a set with one point as convex. The theorem is equally obvious for $m = 2$, but we need not make use of this directly. Now let us suppose that the theorem is true for $m - 1$, that is, the convex hull of $\mathbf{x}_1, \ldots, \mathbf{x}_{m-1}$ is the set

$$X_1 = \left\{ \mathbf{x} \mid \mathbf{x} = \sum_{i=1}^{m-1} \beta_i \mathbf{x}_i, \text{ all } \beta_i \geq 0, \sum_{i=1}^{m-1} \beta_i = 1 \right\}. \tag{6-40}$$

Then, consider the convex hull X of $\mathbf{x}_1, \ldots, \mathbf{x}_m$. Obviously, $\mathbf{x}_m$ must be an element of the convex hull. Similarly, every point in X_1 must be an element of X because X_1 is by assumption the convex hull of $\mathbf{x}_1, \ldots, \mathbf{x}_{m-1}$. In addition, X must contain all points on the line segments joining points in X_1 to $\mathbf{x}_m$, that is, all points

$$\mathbf{x} = \lambda \sum_{i=1}^{m-1} \beta_i \mathbf{x}_i + (1 - \lambda) \mathbf{x}_m, \qquad 0 \leq \lambda \leq 1. \tag{6-41}$$

If
$$\mu_i = \lambda \beta_i, \quad i = 1, \ldots, m - 1, \quad \mu_m = (1 - \lambda),$$

then all $\mu_i \geq 0$ and

$$\sum_{i=1}^{m} \mu_i = \sum_{i=1}^{m-1} \lambda \beta_i + (1 - \lambda) = \lambda + (1 - \lambda) = 1.$$

Furthermore, since each β_i and λ can vary between 0 and 1, each μ_i can assume any value between 0 and 1, the only restriction being $\sum \mu_i = 1$. Hence, the set defined by (6–41) is the set of all convex combinations of $\mathbf{x}_1, \ldots, \mathbf{x}_m$ and represents the convex hull of $\mathbf{x}_1, \ldots, \mathbf{x}_m$ if X_1 is the convex hull of $\mathbf{x}_1, \ldots, \mathbf{x}_{m-1}$ since it is the smallest convex set containing X_1 and $\mathbf{x}_m$. By induction, the convex hull of m points is the set of all convex combinations of the m points.

Note that the convex hull of m points is also a closed set. In Problem 6–31, the reader will be asked to provide the argument in detail.

We shall find it convenient to denote convex hulls of m points by a special name. The following definition is a direct generalization from E^2, E^3, where the convex hull of m points is a polyhedron.

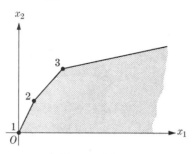

FIGURE 6–11

CONVEX POLYHEDRON: *The convex hull of a finite number of points is called the convex polyhedron spanned by these points.*

It is obvious that the convex polyhedron spanned by m points cannot have more than m extreme points since it is the set of all convex combinations of the m points. All elements of the set, except $x_1, \ldots, x_m$, lie between other points of the set. Thus only $x_1, \ldots, x_m$ can be extreme points. Each point $x_1, \ldots, x_m$ will not necessarily be an extreme point. One or more of these points may be interior points of the convex polyhedron (see Fig. 6–10).

This discussion suggests that any point in a convex polyhedron can be represented as a convex combination of the extreme points of the polyhedron, i.e., any x can be written

$$x = \sum \mu_i x_i^*, \quad \mu_i \geq 0, \quad \sum \mu_i = 1, \qquad (6\text{-}42)$$

where the x_i^* are extreme points. We shall prove this statement later. However, not every convex set with a finite number of extreme points has the property that any point in the set can be represented as a convex combination of the extreme points. For example: It is not true that any point in the convex set shown in Fig. 6–11 can be represented as a convex combination of the extreme points 1, 2, 3. Intuitively, we can see the reason for this: The set is unbounded. We shall prove in one of the following sections that any strictly bounded closed convex set with a finite number of extreme points is the convex hull of the extreme points.

SIMPLEX: *The convex hull of any set of $n+1$ points from E^n which do not lie on a hyperplane in E^n is called a simplex.*

A simplex is a special case of a convex polyhedron. Since the $n+1$ points do not lie on a hyperplane, n points must be linearly independent, for otherwise all the points would lie on a hyperplane passing through the origin. In E^2 a triangle and its interior form a simplex. The three points which generate the simplex are the vertices of the triangle.

*6–6 Theorems on separating hyperplanes. In this and the following two sections, four important theorems are proved which are of significance for a wide range of problems in linear programming, decision theory, and game theory.

THEOREM I: *Given any closed convex set X, a point $\mathbf{y}$ either belongs to the set X or there exists a hyperplane which contains $\mathbf{y}$ such that all of X is contained in one open half-space produced by that hyperplane.*

This is the theorem of the separating hyperplanes. Before considering the proof, note that the two parts of the theorem are mutually exclusive. If $\mathbf{y}$ belongs to the hyperplane, then it does not belong to the open half-space.

We shall begin proving the theorem by assuming that $\mathbf{y}$ does not belong to X. We then find the point $\mathbf{w}$ in X such that for all $\mathbf{u}$ in X

$$|\mathbf{w} - \mathbf{y}| = \min |\mathbf{u} - \mathbf{y}|. \tag{6-43}$$

The point $\mathbf{w}$ is the point in X closest to $\mathbf{y}$ ("closest" means "shortest distance"). Since the set is closed, we know that the minimum distance is actually assumed† for some $\mathbf{w}$. There can be only one such point $\mathbf{w}$, for if there were two, the point halfway between them would be in X and also closer‡ to $\mathbf{y}$.

Next, let us consider any $\mathbf{u} \in X$. Then the point

$$(1 - \lambda)\mathbf{w} + \lambda\mathbf{u}, \qquad 0 \leq \lambda \leq 1,$$

---

* The results of Sections 6–6, 6–7, and 6–8 are important. The theorems are geometrically quite obvious, but the proofs are rather tedious. It is sufficient to read the theorems and to study their geometrical interpretation. If desired, the material in the starred sections may be omitted entirely without loss of continuity.

† We have not proved that the distance will actually assume its minimum value for a point in the set. This proof (which exceeds the scope of this text) could be based on the theorem of Weierstrass, which states that a continuous function defined over a closed bounded set (X need not be bounded, but only a part of X "near" to $\mathbf{y}$ must be considered) actually takes on its minimum at some point in the set. Intuitively, however, the result is fairly obvious.

‡ Suppose that there are two points $\mathbf{w}_1, \mathbf{w}_2$ both of which have the same minimum distance from $\mathbf{y}$. Then, by the triangle inequality, we obtain

$$|\tfrac{1}{2}(\mathbf{w}_1 + \mathbf{w}_2) - \mathbf{y}| = \tfrac{1}{2}|(\mathbf{w}_1 - \mathbf{y}) + (\mathbf{w}_2 - \mathbf{y})| < \tfrac{1}{2}(|\mathbf{w}_1 - \mathbf{y}| + |\mathbf{w}_2 - \mathbf{y}|).$$

The strict inequality holds if $\mathbf{w}_1 - \mathbf{y} \neq \lambda(\mathbf{w}_2 - \mathbf{y})$; this is the case here, since $|\mathbf{w}_1 - \mathbf{y}| = |\mathbf{w}_2 - \mathbf{y}|$, and $\mathbf{w}_1 \neq \mathbf{w}_2$. Therefore,

$$|\tfrac{1}{2}(\mathbf{w}_1 + \mathbf{w}_2) - \mathbf{y}| < |\mathbf{w}_1 - \mathbf{y}|.$$

is in X. But by (6–43)

$$|(1-\lambda)\mathbf{w} + \lambda\mathbf{u} - \mathbf{y}|^2 \geq |\mathbf{w} - \mathbf{y}|^2, \qquad 0 \leq \lambda \leq 1, \qquad (6\text{–}44)$$

or

$$|(\mathbf{w} - \mathbf{y}) + \lambda(\mathbf{u} - \mathbf{w})|^2 \geq |\mathbf{w} - \mathbf{y}|^2.$$

Expanding the preceding expression, we arrive at

$$|\mathbf{w} - \mathbf{y}|^2 + 2\lambda(\mathbf{w} - \mathbf{y})'(\mathbf{u} - \mathbf{w}) + \lambda^2|\mathbf{u} - \mathbf{w}|^2 \geq |\mathbf{w} - \mathbf{y}|^2,$$

or

$$2\lambda(\mathbf{w} - \mathbf{y})'(\mathbf{u} - \mathbf{w}) + \lambda^2|\mathbf{u} - \mathbf{w}|^2 \geq 0. \qquad (6\text{–}45)$$

Take $\lambda > 0$. Dividing (6–45) by λ yields

$$2(\mathbf{w} - \mathbf{y})'(\mathbf{u} - \mathbf{w}) + \lambda|\mathbf{u} - \mathbf{w}|^2 \geq 0.$$

Let λ tend to zero. In the limit we have

$$(\mathbf{w} - \mathbf{y})'(\mathbf{u} - \mathbf{w}) \geq 0; \qquad (6\text{–}46)$$

however,

$$\mathbf{u} - \mathbf{w} = \mathbf{u} - \mathbf{y} - (\mathbf{w} - \mathbf{y}). \qquad (6\text{–}47)$$

Substituting (6–47) into (6–46), we have

$$(\mathbf{w} - \mathbf{y})'(\mathbf{u} - \mathbf{y}) \geq |\mathbf{w} - \mathbf{y}|^2.$$

But $|\mathbf{w} - \mathbf{y}|^2 > 0$ because $\mathbf{w} \in X$ and $\mathbf{y} \notin X$. Therefore,

$$(\mathbf{w} - \mathbf{y})'(\mathbf{u} - \mathbf{y}) > 0, \qquad (6\text{–}48)$$

or

$$(\mathbf{w} - \mathbf{y})'\mathbf{u} > (\mathbf{w} - \mathbf{y})'\mathbf{y}. \qquad (6\text{–}49)$$

Define

$$\mathbf{c} = (\mathbf{w} - \mathbf{y})', \qquad z = (\mathbf{w} - \mathbf{y})'\mathbf{y} = \mathbf{cy}. \qquad (6\text{–}50)$$

Consider the hyperplane $\mathbf{cx} = z$. Since $\mathbf{cy} = z$, $\mathbf{y}$ is on the hyperplane. However, according to (6–49), any $\mathbf{u} \in X$ satisfies

$$\mathbf{cu} > z. \qquad (6\text{–}51)$$

Thus any point in X is in the half-space* $\mathbf{cx} > z$. The theorem is proved.

The geometrical interpretation of the theorem in E^2, E^3 is very simple. Examination of Fig. 6–12 shows that when $\mathbf{y}$ is not in X, we take for the

---

* The direction of the inequality in $\mathbf{cx} > z$ is immaterial. If both sides are multiplied by (-1), we obtain $(-\mathbf{c})\mathbf{x} < -z$. The hyperplane $(-\mathbf{c})\mathbf{x} = -z$ also contains $\mathbf{y}$.

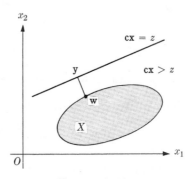

FIGURE 6–12

hyperplane of the theorem the line through **y** perpendicular to the line representing the shortest distance from **y** to X. Thus for E^2 and E^3, the validity of the theorem is obvious. For E^n we have proved it rigorously by algebraic methods.

Theorem I states that for any $\mathbf{y} \notin X$ (regardless of how close **y** is to the set X), there is a hyperplane $\mathbf{cx} = z$ containing **y** such that $\mathbf{cu} > z$ for all $\mathbf{u} \in X$. This fact suggests that if **w** is any boundary point of X, there is a hyperplane $\mathbf{cx} = z$ containing **w** such that $\mathbf{cu} \geq z$ for all $\mathbf{u} \in X$. This is true, as we shall prove shortly. The hyperplane containing the boundary point is called a supporting hyperplane to the convex set X.

SUPPORTING HYPERPLANE: *Given a boundary point* **w** *of a convex set* X; *then* $\mathbf{cx} = z$ *is called a supporting hyperplane at* **w** *if* $\mathbf{cw} = z$ *and if all of* X *lies in one closed half-space produced by the hyperplane, that is,* $\mathbf{cu} \geq z$ *for all* $\mathbf{u} \in X$ *or* $\mathbf{cu} \leq z$ *for all* $\mathbf{u} \in X$.

THEOREM II: *If* **w** *is a boundary point of a closed convex set, then there is at least one supporting hyperplane at* **w**.

Theorem I did not actually show that the point **w** closest to **y** was a boundary point of X. Clearly, this has to be the case; otherwise there would exist an $\mathbf{s} \in X$ on the line joining **w**, **y**, with

$$\mathbf{s} = \lambda \mathbf{w} + (1 - \lambda)\mathbf{y}, \quad 0 < \lambda < 1,$$

such that

$$|\mathbf{s} - \mathbf{y}| = \lambda |\mathbf{w} - \mathbf{y}| < |\mathbf{w} - \mathbf{y}|,$$

and **w** would not be closest to **y**. Thus **w** is a boundary point of X.

Using (6–50) in (6–46), we have

$$\mathbf{c}(\mathbf{u} - \mathbf{w}) \geq 0,$$

or

$$\mathbf{cu} \geq \mathbf{cw}. \tag{6–52}$$

If we consider the hyperplane

$$\mathbf{cx} = \mathbf{cw} = z, \qquad (6\text{–}53)$$

then $\mathbf{w}$ is on this hyperplane, and by (6–52), $\mathbf{cu} \geq z$ for all $\mathbf{u} \in X$. Thus, boundary point $\mathbf{w}$ has a supporting hyperplane. This fact, however, does not prove Theorem II since in Theorem I, $\mathbf{w}$ was not an arbitrary boundary point but was determined by the choice of $\mathbf{y}$. Theorem II could be proved if, for any given boundary point $\mathbf{w}$, we could find a point $\mathbf{y} \notin X$ such that $\mathbf{w}$ was the closest point in the set to $\mathbf{y}$. It is clear how to do this geometrically in E^2 and E^3.

We construct a normal to the set at $\mathbf{w}$, and any point on this normal, but outside the set, will satisfy the condition. Such an approach cannot be easily generalized to E^n (in fact it is difficult to formalize in E^2 and E^3) and hence we shall proceed in a slightly different way.

Select any boundary point $\mathbf{w}$ of X. Consider an ϵ neighborhood about $\mathbf{w}$. For any $\epsilon > 0$, however small, there are points inside the hypersphere which are not in X. Choose a given ϵ (ϵ_k) and select from the neighborhood a $\mathbf{y}_k$ which is not in X. Then a boundary point $\mathbf{w}_k$ of X will be the closest point in X to $\mathbf{y}_k$. We have shown that there is a supporting hyperplane $\mathbf{c}_k\mathbf{x} = \mathbf{c}_k\mathbf{w}_k = z_k$ at $\mathbf{w}_k$. Next, we choose a sequence of ϵ_k such that $\epsilon_k \to 0$ as $k \to \infty$. By the triangle inequality,

$$|\mathbf{w}_k - \mathbf{w}| = |\mathbf{w}_k - \mathbf{y}_k + \mathbf{y}_k - \mathbf{w}| \leq |\mathbf{w}_k - \mathbf{y}_k| + |\mathbf{y}_k - \mathbf{w}|. \qquad (6\text{–}54)$$

The choice of $\mathbf{y}_k$ requires that $|\mathbf{y}_k - \mathbf{w}| < \epsilon_k$, and therefore $|\mathbf{y}_k - \mathbf{w}| \to 0$ as $k \to \infty$. Since $|\mathbf{w}_k - \mathbf{y}_k|$ is the shortest distance between $\mathbf{y}_k$ and X, $|\mathbf{w}_k - \mathbf{y}_k| \leq |\mathbf{y}_k - \mathbf{w}|$, and hence $|\mathbf{y}_k - \mathbf{w}_k| \to 0$ as $k \to \infty$. We conclude that

$$|\mathbf{w}_k - \mathbf{w}| \to 0 \quad \text{as} \quad k \to \infty, \quad \text{or} \quad \mathbf{w}_k \to \mathbf{w}.$$

For each $\mathbf{w}_k$ there is a supporting hyperplane $\mathbf{c}_k\mathbf{x} = z_k$. Dividing by $|\mathbf{c}_k|$, we obtain

$$\mathbf{n}_k\mathbf{x} = b_k, \qquad \mathbf{n}_k = \mathbf{c}_k/|\mathbf{c}_k|, \qquad b_k = z_k/|\mathbf{c}_k|, \qquad (6\text{–}55)$$

and $|\mathbf{n}_k| = 1$. By the Schwarz inequality (Section 2–6),

$$|b_k| \leq |\mathbf{n}_k||\mathbf{w}_k| = |\mathbf{w}_k|, \qquad (6\text{–}56)$$

because $\mathbf{w}_k$ is on the hyperplane (6–55). For sufficiently large k there exists an $r > 0$ such that $|\mathbf{w}_k| < r$ independent of k. This follows since

$$|\mathbf{w}_k| = |\mathbf{w}_k - \mathbf{w} + \mathbf{w}| \leq |\mathbf{w}| + |\mathbf{w}_k - \mathbf{w}|,$$

and $|\mathbf{w}_k - \mathbf{w}| \to 0$. An r which satisfies the above requirement is

$r = 2|\mathbf{w}|$. Equation (6-56) can then be written in the form of

$$-r < -|\mathbf{w}_k| \leq b_k \leq |\mathbf{w}_k| < r. \tag{6-57}$$

Hence the b_k form a bounded infinite sequence. Similarly, the components of $\mathbf{n}_k$ form bounded infinite sequences because $|\mathbf{n}_k| = 1$. One theorem on sequences states *that bounded infinite sequences possess at least one limit point*. Hence, there is a subsequence of points $\mathbf{y}_k$ for which $\mathbf{n}_k \to \mathbf{n}$, and $b_k \to b$ as $\mathbf{w}_k \to \mathbf{w}$. Furthermore, for every k, $\mathbf{n}_k \mathbf{w}_k - b_k = 0$, and in the limit, $\mathbf{nw} = b$. Thus we have shown that for $\mathbf{w}$ there is a hyperplane $\mathbf{nx} = b$ for which

$$\mathbf{nw} = b, \quad \mathbf{nu} \geq b, \quad \text{all } \mathbf{u} \in X;$$

and this is a supporting hyperplane. Theorem II has been proved.

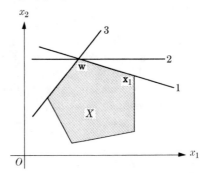

FIGURE 6-13

Unfortunately, in proving the theorem, we had to use the ideas of limits and bounded sequences. We have not discussed these subjects, which may be unfamiliar to the reader. However, these concepts had to be introduced since we are not sure that $\mathbf{n}_k$, b_k approach unique values when we chose the sequence of points $\mathbf{y}_k$ in some specific way. For arbitrary sequences $\mathbf{y}_k$, it is not always true that $\mathbf{n}_k$, b_k approach unique values. The theorem on bounded sequences assures us that there is a sequence of $\mathbf{y}_k$ for which $\mathbf{n}_k$ and b_k do approach unique limits. However, the supporting hyperplane at $\mathbf{w}$ need not be unique. Figure 6-13 illustrates that any one of the hyperplanes 1, 2, 3 is a supporting hyperplane at $\mathbf{w}$. Hence, an arbitrary sequence of points $\mathbf{y}_k$ could not be used to obtain unique values.

It should be clear to the reader that Theorem II holds even if X is not closed, i.e., there is at least one supporting hyperplane at a boundary point $\mathbf{w}$, regardless of whether or not $\mathbf{w} \in X$.

*6–7 A basic result in linear programming. We have already shown that the set of feasible solutions to a linear programming problem is a closed convex set, and that the function to be optimized is a hyperplane. This hyperplane is moved parallel to itself over the convex set of feasible solutions until z is made as large as possible (if z is being maximized) while still having at least one point **x** on the hyperplane in the convex set of feasible solutions.

Hence, if a given hyperplane corresponds to the optimal value of z, then no point on the hyperplane can be an interior point of the convex set† of feasible solutions X. To see this, let us suppose that $z = \mathbf{cx}$ is an optimal hyperplane and that one of its points $\mathbf{x}_0$ is an interior point of the set. We select an $\epsilon > 0$ such that every point in this ϵ neighborhood of $\mathbf{x}_0$ is in X. The point $\mathbf{x}_1 = \mathbf{x}_0 + (\epsilon/2)(\mathbf{c}'/|\mathbf{c}|)$ is in X, and $\mathbf{cx}_1 = z + (\epsilon/2)|\mathbf{c}| > z$. This contradicts the fact that z is the maximum value. Hence every point of X on the optimal hyperplane must be a boundary point. Therefore, if $\mathbf{x}_0$ is an optimal solution to the linear programming problem, then $z = \mathbf{cx}_0$ and $\mathbf{cu} \leq z$ for all $\mathbf{u} \in X$ (we assume, of course, that z is being maximized). Thus, an optimal hyperplane is a supporting hyperplane to the convex set of feasible solutions at an optimal solution $\mathbf{x}_0$.

THEOREM III: *A closed convex set which is bounded from below has extreme points in every supporting hyperplane.*

The convex set of feasible solutions to a linear programming problem is closed and bounded from below by **0** because $x_j \geq 0$ for all j. Hence, the theorem states that if there is an optimal solution, at least one of the extreme points of the convex set of feasible solutions will be an optimal solution. In E^n, as in E^2, E^3, the convex set of feasible solutions will have only a finite number of extreme points.‡ Hence, if we had the means of selecting the extreme points of the convex set of feasible solutions, only a finite number of points would have to be examined to find an optimal solution to the problem. And indeed, it is possible to determine analytically the extreme points. This is the basis of the simplex method. We move from one extreme point to a new one (having a value of z at least as large as the preceding one) until an optimal solution is found.

We shall now prove Theorem III. The hyperplane $\mathbf{cx} = z$ will be assumed to be a supporting hyperplane at $\mathbf{x}_0$ to the closed convex set X which is bounded from below. The intersection of X and the set

---

† It may turn out that there is no maximum value of z, that is, z can be made arbitrarily large. Then this result does not hold. However, when we speak of an optimal solution to a linear programming problem, we shall imply that z has a finite maximum or finite minimum.

‡ In Problem 6–33 the reader will be asked to prove the theorem for E^n.

$S = \{\mathbf{x} | \mathbf{c}\mathbf{x} = z\}$ will be denoted by T. The intersection is not empty because $\mathbf{x}_0 \in T$; furthermore, since X and S are closed convex sets, so is T; T is also bounded from below since X is.

We shall show that any extreme point of T is also an extreme point of X. If $\mathbf{t}$ is any point in T, and if

$$\mathbf{t} = \lambda \mathbf{x}_2 + (1 - \lambda)\mathbf{x}_1, \quad 0 < \lambda < 1,$$

where $\mathbf{x}_1, \mathbf{x}_2 \in X$, then $\mathbf{x}_1, \mathbf{x}_2 \in T$. This follows from the fact that

$$\mathbf{c}\mathbf{t} = \lambda \mathbf{c}\mathbf{x}_2 + (1 - \lambda)\mathbf{c}\mathbf{x}_1 = z, \tag{6-58}$$

and $\mathbf{c}\mathbf{x}_2 \geq z$, $\mathbf{c}\mathbf{x}_1 \geq z$ because $\mathbf{c}\mathbf{x} = z$ is a supporting hyperplane. Noting that $\lambda, (1 - \lambda) > 0$, we see that (6–58) will hold if and only if $\mathbf{c}\mathbf{x}_2 = z$, $\mathbf{c}\mathbf{x}_1 = z$, that is, if and only if $\mathbf{x}_1, \mathbf{x}_2 \in T$. Thus an extreme point of T cannot be represented as a convex combination of any two points in X with $0 < \lambda < 1$. Hence an extreme point of T is an extreme point of X.

We still have to prove that T actually has an extreme point; this will be accomplished by finding an extreme point. Out of all the points in T, choose the one with the smallest (algebraic) first component. There is at least one such point since T is closed and bounded from below.

If there is more than one point with a smallest first component, choose the point or points with the smallest first and second components. If again there is more than one point with the smallest first and second components, find the point or points with the smallest first, second, and third components, etc. Finally, a unique point will be obtained since only one point can have all its components of minimum algebraic value.

The unique point $\mathbf{t}^*$ determined by the above process is an extreme point. If $\mathbf{t}^*$ were not an extreme point, we could write

$$\mathbf{t}^* = \lambda \mathbf{t}_1 + (1 - \lambda)\mathbf{t}_2, \quad 0 < \lambda < 1; \quad \mathbf{t}_1 \neq \mathbf{t}_2 \in T. \tag{6-59}$$

Suppose the unique $\mathbf{t}^*$ was determined on minimizing the jth component. If t_{j1}, t_{j2} are the jth components of $\mathbf{t}_1, \mathbf{t}_2$, then the jth component of (6–59) is

$$t_j^* = \lambda t_{j1} + (1 - \lambda) t_{j2}, \quad 0 < \lambda < 1. \tag{6-60}$$

Furthermore (why?)

$$t_i^* = t_{i1} = t_{i2} \quad (i = 1, \ldots, j - 1).$$

But then (6–60) requires that $t_j^* = t_{j1} = t_{j2}$, for otherwise $t_j^* > \min[t_{j_1}, t_{j_2}]$. However, this result contradicts the fact that there is only one point with this t_j^* when all components $1, \ldots, j - 1$ are at their minimum values. Consequently, $\mathbf{t}^*$ cannot be represented as a convex combination of any two other points in T $(0 < \lambda < 1)$. Hence $\mathbf{t}^*$ is an extreme point and the theorem is proved.† The above proof also demonstrates that a strictly bounded convex set has extreme points in every supporting hyperplane.

*6–8 Convex hull of extreme points.

THEOREM IV: *If a closed, strictly bounded convex set X has a finite number of extreme points, any point in the set can be written as a convex combination of the extreme points, that is, set X is the convex hull of its extreme points.*

We were led to suspect in Section 6–5 that a result of this sort might be true. We are now able to prove it.

Take the extreme points of X to be $\mathbf{y}_1, \ldots, \mathbf{y}_m$; S will be defined as the convex hull of the extreme points,

$$S = \left\{ \mathbf{y} \mid \mathbf{y} = \sum_{i=1}^{m} \mu_i \mathbf{y}_i, \text{ all } \mu_i \geq 0, \sum \mu_i = 1 \right\}. \quad (6\text{–}61)$$

Suppose that there is a point $\mathbf{v}_0 \in X$ and $\mathbf{v}_0 \notin S$. Then by Theorem I (on separating hyperplanes) there is a hyperplane $\mathbf{cx} = z$ containing $\mathbf{v}_0$ such that $\mathbf{cy} > z$ for all $\mathbf{y} \in S$. In addition, there is a $\mathbf{w} \in S$ which is closest to $\mathbf{v}_0$ (S is closed). Furthermore, we can write $\mathbf{c}' = \mathbf{w} - \mathbf{v}_0$. Consider a set of points $\mathbf{x}_k$ not necessarily in X such that

$$\mathbf{x}_k = \mathbf{v}_0 - \lambda_k(\mathbf{w} - \mathbf{v}_0); \quad \lambda_k \geq 0. \quad (6\text{–}62)$$

Then

$$|\mathbf{w} - \mathbf{x}_k| = (1 + \lambda_k)|\mathbf{w} - \mathbf{v}_0| \geq |\mathbf{w} - \mathbf{v}_0|.$$

The hyperplane with normal $\mathbf{c}$ which passes through $\mathbf{x}_k$ may be written $\mathbf{cx} = z_k$, and

$$\mathbf{cx} = \mathbf{cx}_k = \mathbf{cv}_0 - \lambda_k |\mathbf{c}|^2 = z - \lambda_k |\mathbf{c}|^2 = z_k < z.$$

For any $\mathbf{y} \in S$, $\mathbf{cy} > z$, and hence $\mathbf{cy} > z_k$. Increase λ_k to the largest possible value for which the hyperplane $\mathbf{cx} = z_k$ will contain a point of X. There will be a largest λ_k since X is closed and strictly bounded. Call z^* the z_k for this hyperplane. Hence a $\mathbf{v}^* \in X$ exists such that $\mathbf{cv}^* = z^*$, and $\mathbf{cv} \geq z^*$ for all $\mathbf{v} \in X$. It follows that $\mathbf{cx} = z^*$ is a supporting hyperplane to X. However, it contains no extreme points of X, since they are all in S, and every point $\mathbf{y} \in S$ satisfies $\mathbf{cy} > z^*$. This contradicts the

---

† The same line of reasoning will reveal that if a linear programming problem has a feasible solution, at least one feasible solution will be a boundary point, and there will be at least one extreme point (see Problem 6–23).

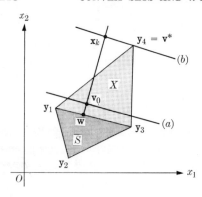

FIGURE 6–14   FIGURE 6–15

fact that there are extreme points in every supporting hyperplane. Hence every point of X must be in S, and X is the convex hull of its extreme points.

Figure 6–14 interprets our procedure geometrically. We assume that the points y_1, y_2, y_3 are the only extreme points of the convex set X. They generate the convex hull S. Point v_0 is not in S. A hyperplane (a) is passed through v_0 so that all of S lies in one half-space produced by the hyperplane. Then the hyperplane is moved parallel to itself until (b) is reached; if it were moved any further, no point of X would be on the hyperplane. Hence (b) is a supporting hyperplane which presumably contains no extreme points of X. This is obviously not true (see Fig. 6–14). The contradiction is, of course, a result of our forgetfulness: We did not count extreme point y_4 in our original determination of the extreme points of X.

EXAMPLE: Suppose that we wish to write any point w inside a triangle as a convex combination of the vertices (extreme points):

$$\mathbf{w} = \sum_{i=1}^{3} \mu_i \mathbf{x}_i, \qquad \mu_i \geq 0, \qquad \sum \mu_i = 1.$$

The situation is illustrated in Fig. 6–15: First, draw a line from $\mathbf{x}_2$ through $\mathbf{w}$. It will intersect the opposite side of the triangle at $\mathbf{v}$. Then

$$\mathbf{w} = \lambda_1 \mathbf{x}_2 + (1 - \lambda_1)\mathbf{v}, \qquad 0 \leq \lambda_1 \leq 1;$$

but

$$\mathbf{v} = \lambda_2 \mathbf{x}_1 + (1 - \lambda_2)\mathbf{x}_3, \qquad 0 \leq \lambda_2 \leq 1;$$

thus

$$\mathbf{w} = \lambda_1 \mathbf{x}_2 + (1 - \lambda_1)\lambda_2 \mathbf{x}_1 + (1 - \lambda_1)(1 - \lambda_2)\mathbf{x}_3.$$

Let $\mu_1 = \lambda_2(1 - \lambda_1)$, $\mu_2 = \lambda_1$, $\mu_3 = (1 - \lambda_1)(1 - \lambda_2)$.

Clearly, $\mu_i \geq 0$ and

$$\mu_1 + \mu_2 + \mu_3 = \lambda_2(1 - \lambda_1) + \lambda_1 + (1 - \lambda_1) - (1 - \lambda_1)\lambda_2 = 1.$$

The desired expression for $\mathbf{w}$ is $\mathbf{w} = \sum_{i=1}^{3} \mu_i \mathbf{x}_i$.

From the definition of a convex polyhedron and from Theorem IV it follows that every closed, strictly bounded convex set with a finite number of extreme points is a convex polyhedron.

Convex sets with only a finite number of extreme points are essentially convex polyhedrons. However, they may be of the type shown in Fig. 6-11, that is, they may not be strictly bounded. For convex sets with a finite number of extreme points, it is useful to introduce the concepts of an *edge* and of adjacent extreme points.

EDGE: *Let $\mathbf{x}_1^*$, $\mathbf{x}_2^*$ be distinct extreme points of the convex set X. The line segment joining them is called an edge of the convex set if it is the intersection of X with a supporting hyperplane. If $\mathbf{x}_1^*$ is an extreme point of X, and if there exists another point $\bar{\mathbf{x}}$ in X such that $\mathbf{x} = \mathbf{x}_1^* + \lambda(\bar{\mathbf{x}} - \mathbf{x}_1^*)$ is in X for every $\lambda \geq 0$, and if, in addition, the set $L = \{\mathbf{x} | \mathbf{x} = \mathbf{x}_1^* + \lambda(\bar{\mathbf{x}} - \mathbf{x}_1^*),$ all $\lambda \geq 0\}$ is the intersection of X with a supporting hyperplane, then the set L is said to be an edge of X which extends to infinity.*

ADJACENT EXTREME POINTS: *Two distinct extreme points $\mathbf{x}_1^*$, $\mathbf{x}_2^*$ of the convex set X are called adjacent if the line segment joining them is an edge of the convex set.*

These definitions conform to our conception of an edge and of adjacent extreme points in E^2, E^3. In Fig. 6-13, the line segment joining the extreme points $\mathbf{w}$, $\mathbf{x}_1$ is the intersection of X with hyperplane 1, and hence is an edge of the set. Since the line joining them is an edge of the set, $\mathbf{w}$, $\mathbf{x}_1$ are adjacent extreme points.

6-9 Introduction to convex cones. In the preceding sections, we have examined some of the properties of a special class of convex sets, the convex polyhedrons. Now another class of convex sets, i.e., convex cones, will be studied which are useful in the theory of linear programming and in linear economic models.

CONE: *A cone C is a set of points with the following property: If $\mathbf{x}$ is in the set, so is $\mu \mathbf{x}$ for all $\mu \geq 0$.*

The cone "generated" by a set of points $X = \{\mathbf{x}\}$ is the set

$$C = \{\mathbf{y} | \mathbf{y} = \mu \mathbf{x}, \text{ all } \mu \geq 0 \text{ and all } \mathbf{x} \in X\}. \tag{6-63}$$

220   CONVEX SETS AND n-DIMENSIONAL GEOMETRY   [CHAP. 6

Note that a cone is never a strictly bounded set (except in the trivial case where **0** is the only element in the cone). However, a cone may be bounded from above or from below. In E^2 and E^3, a cone as "a set of points" is often identical with the usual geometrical concept of a cone.

VERTEX: *The point* **0** *is an element of any cone and is called the vertex of the cone.*

EXAMPLE: Figure 6–16 shows a cone in E^3 generated by the set of points
$$X = \{[x_1, x_2, x_3] | x_1^2 + x_2^2 \leq 1,\ x_3 = 1\}.$$

NEGATIVE: *The negative C^- of a cone $C = \{\mathbf{u}\}$ is the set of points $C^- = \{-\mathbf{u}\}$.*

C^- is clearly a cone if C is.

SUM: *The sum of two cones $C_1 = \{\mathbf{u}\},\ C_2 = \{\mathbf{v}\}$, written $C_1 + C_2$, is the set of all points* $\mathbf{u} + \mathbf{v},\ \mathbf{u} \in C_1,\ \mathbf{v} \in C_2$.

The sum $C_1 + C_2$ is a cone because $\mu(\mathbf{u} + \mathbf{v}) = \mu\mathbf{u} + \mu\mathbf{v}$, and $\mu\mathbf{u} \in C_1$ if $\mathbf{u} \in C_1$, $\mu\mathbf{v} \in C_2$ if $\mathbf{v} \in C_2$, so that $\mu(\mathbf{u} + \mathbf{v}) \in C_1 + C_2$ for all $\mu \geq 0$.

POLAR CONE: *If $C = \{\mathbf{u}\}$ is a cone, then C^+, the cone polar to C, is the collection of points $\{\mathbf{v}\}$ such that $\mathbf{v}'\mathbf{u} \geq 0$ for each $\mathbf{v}$ in the set and all $\mathbf{u} \in C$.*

Obviously, C^+ is a cone since if $\mathbf{v}'\mathbf{u} \geq 0$ for all $\mathbf{u} \in C$, then $\mu\mathbf{v}'\mathbf{u} \geq 0$ for all $\mu \geq 0$. Intuitively, a polar cone is the collection of all vectors which form a nonobtuse angle with all the vectors in C. Note that each $\mathbf{v} \in C^+$ *must* form a nonobtuse angle with *every* vector in C.

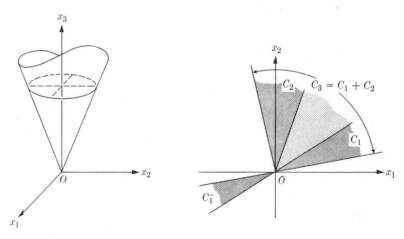

FIGURE 6–16   FIGURE 6–17

EXAMPLES: (1) Figure 6–17 shows the given cones C_1, C_2 as well as $C_3 = C_1 + C_2$ and C_1^-.
(2) Figure 6–18 shows C^+, the cone polar to the given cone C.
We can easily prove that for any cones C_1, C_2,

$$(C_1 + C_2)^+ = C_1^+ \cap C_2^+, \tag{6-64}$$

and if $C_1 \subset C_2$, then $C_2^+ \subset C_1^+$. The details of the proofs are to be supplied in Problems 6–38 and 6–39.

CONVEX CONE: *A cone is a convex cone if it is a convex set.*

All the cones in our examples have been convex cones. The cone in Fig. 6–19 is not convex since it consists of two separate parts.

A set of points is a convex cone if and only if the sum $\mathbf{v}_1 + \mathbf{v}_2$ *is in the set when* $\mathbf{v}_1$, $\mathbf{v}_2$ *are, and if* $\mu\mathbf{v}$ *is in the set when* $\mathbf{v}$ *is for any* $\mu \geq 0$. To see that the conditions are necessary, note that if C is a convex cone, then $\mu\mathbf{v}(\mu \geq 0) \in C$ if $\mathbf{v} \in C$ by the definition of a cone. To see that $\mathbf{v}_1 + \mathbf{v}_2$ must be in C if $\mathbf{v}_1$, $\mathbf{v}_2 \in C$, note that since C is a cone, we can write

$$\mathbf{v}_1 = \lambda\boldsymbol{\omega}_1, \quad 0 \leq \lambda \leq 1, \quad \boldsymbol{\omega}_1 \in C, \quad \mathbf{v}_2 = (1 - \lambda)\boldsymbol{\omega}_2, \quad \boldsymbol{\omega}_2 \in C.$$

However, because C is convex, $\lambda\boldsymbol{\omega}_1 + (1 - \lambda)\boldsymbol{\omega}_2 \in C$, and thus $\mathbf{v}_1 + \mathbf{v}_2$ must be in C. To see that the conditions are sufficient, let us suppose that the sum $\mathbf{v}_1 + \mathbf{v}_2$ is in the set if $\mathbf{v}_1$, $\mathbf{v}_2$ are, and $\mu\mathbf{v}$ is in the set if $\mathbf{v}$ is for all $\mu \geq 0$. The second condition indicates clearly that the set is a cone. The cone will be convex if $\lambda\boldsymbol{\omega}_1 + (1 - \lambda)\boldsymbol{\omega}_2$, $0 \leq \lambda \leq 1$, is in the set when $\boldsymbol{\omega}_1$, $\boldsymbol{\omega}_2$ are in the set. The definitions $\mathbf{v}_1 = \lambda\boldsymbol{\omega}_1$, $\mathbf{v}_2 = (1 - \lambda)\boldsymbol{\omega}_2$, and the fact that $\mathbf{v}_1 + \mathbf{v}_2$ is in the set, determine that the set is convex and therefore a convex cone.

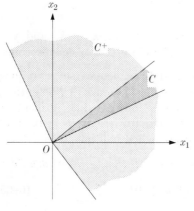

FIGURE 6–18

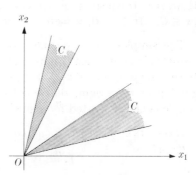

FIGURE 6–19

This set of conditions for convex cones looks very similar to those used for defining a subspace of E^n. The only difference is that, for convex cones, the scalar $\mu \geq 0$. Because of this restriction, a convex cone is not, in general, a subspace of E^n, although it can be a subspace in certain cases.

The sum $C_1 + C_2$ of two convex cones is also convex. Let $\mathbf{x}_1, \mathbf{x}_2$ be any two points in $C_1 + C_2$. By definition of the sum,

$$\mathbf{x}_1 = \mathbf{u}_1 + \mathbf{v}_1, \qquad \mathbf{x}_2 = \mathbf{u}_2 + \mathbf{v}_2, \qquad \mathbf{u}_1, \mathbf{u}_2 \in C_1, \qquad \mathbf{v}_1, \mathbf{v}_2 \in C_2.$$

Then for any point $\mathbf{x}$ which is a convex combination of $\mathbf{x}_1, \mathbf{x}_2$, we obtain

$$\mathbf{x} = \lambda \mathbf{x}_1 + (1-\lambda)\mathbf{x}_2 = \lambda \mathbf{u}_1 + (1-\lambda)\mathbf{u}_2 + \lambda \mathbf{v}_1 + (1-\lambda)\mathbf{v}_2.$$

But $\lambda \mathbf{u}_1 + (1-\lambda)\mathbf{u}_2 \in C_1$, $\lambda \mathbf{v}_1 + (1-\lambda)\mathbf{v}_2 \in C_2$. Hence $\mathbf{x} \in C_1 + C_2$, and $C_1 + C_2$ is a convex cone. In Problem 6–40 the reader will be required to show that if C_i ($i = 1, \ldots, m$) are convex, $\sum_{i=1}^{m} C_i$ is convex too.

The cone generated by a convex set is a convex cone. Given the convex set $X = \{\mathbf{x}\}$, we wish to prove that $C = \{\mathbf{y}|\mathbf{y} = \mu\mathbf{x}, \mu \geq 0, \text{ all } \mathbf{x} \in X\}$ is a convex cone. Clearly, C is a cone. To show that C is convex, we must demonstrate that if $\mathbf{y}_1, \mathbf{y}_2 \in C$, then any convex combination of $\mathbf{y}_1, \mathbf{y}_2 \in C$. Note that $\mathbf{y}_1 = \mu_1 \mathbf{x}_1$, $\mathbf{y}_2 = \mu_2 \mathbf{x}_2$, $\mathbf{x}_1, \mathbf{x}_2 \in X$. Thus we wish to show that any point $\mathbf{y}$,

$$\mathbf{y} = \lambda \mu_1 \mathbf{x}_1 + (1-\lambda)\mu_2 \mathbf{x}_2 \in C, \qquad 0 \leq \lambda \leq 1,$$

is in the set. Write

$$\zeta = \lambda \mu_1 + (1-\lambda)\mu_2.$$

Hence if

$$\mathbf{y} = \zeta[\alpha \mathbf{x}_1 + (1-\alpha)\mathbf{x}_2], \qquad \zeta \neq 0,$$

then

$$\alpha = \frac{\lambda \mu_1}{\zeta}; \qquad 0 \leq \alpha \leq 1.$$

However, because X is convex, $\mathbf{x} = \alpha \mathbf{x}_1 + (1-\alpha)\mathbf{x}_2 \in X$, and $\mathbf{y} = \zeta \mathbf{x} \in C$. If $\zeta = 0$, $\mathbf{y} = \mathbf{0}$, and $\mathbf{0} \in C$. Therefore C is a convex cone.

The simplest convex cones are generated by a convex set containing a single element (a set containing a single element is convex by definition). If the single element $\mathbf{x} \neq \mathbf{0}$, then the convex cone is a line segment beginning at the origin, that is, the set of all multiples $\mu\mathbf{x}$, $\mu \geq 0$. Such simple cones generated by a single point are called half-lines.

HALF-LINE: *Given a single point $\mathbf{a} \neq \mathbf{0}$, a half-line or ray is defined as the set*

$$L = \{\mathbf{y}|\mathbf{y} = \mu\mathbf{a}, \text{ all } \mu \geq 0\}. \tag{6–65}$$

The symbol L will always refer to a half-line.

The polar cone H of a half-line L is a closed half-space containing $\mathbf{0}$ on its bounding hyperplane because H is the set

$$H = \{\mathbf{v}|\mathbf{a}'\mathbf{v} \geq 0\}. \tag{6-66}$$

This follows because H is the collection of points $\mathbf{v}$ with $\mathbf{v}'\mathbf{y} \geq 0$ for all $\mathbf{y} \in L$. However, any $\mathbf{y}$ can be written $\mu\mathbf{a}$, $\mu \geq 0$ and $\mu\mathbf{v}'\mathbf{a} \geq 0$ if $\mathbf{v}'\mathbf{a} \geq 0$. Thus H is the collection of points $\mathbf{v}$ for which $\mathbf{v}'\mathbf{a} = \mathbf{a}'\mathbf{v} \geq 0$, which is a closed half-space.

CONTAINING HALF-SPACE: *Given a cone C. Let $\mathbf{a} \neq \mathbf{0}$ be an element of C^+. Then the set*

$$H_s = \{\mathbf{x}|\mathbf{a}'\mathbf{x} \geq 0\} \tag{6-67}$$

is a containing half-space for C.

Any containing half-space includes all of cone C. The boundary hyperplane P_s of any containing half-space passes through the origin:

$$P_s = \{\mathbf{x}|\mathbf{a}'\mathbf{x} = 0\}. \tag{6-68}$$

EXAMPLE: A cone C and a typical containing half-space are illustrated in Fig. 6–20.

ORTHOGONAL CONE: *Given a cone C in E^n. The cone $C^\perp$ which is the set of all vectors $\mathbf{u}$ in E^n, such that each $\mathbf{u}$ is orthogonal to every vector $\mathbf{v} \in C$, that is, the set*

$$C^\perp = \{\mathbf{u}|\mathbf{u}'\mathbf{v} = 0, \text{ all } \mathbf{v} \in C\} \tag{6-69}$$

is called the orthogonal cone to C.

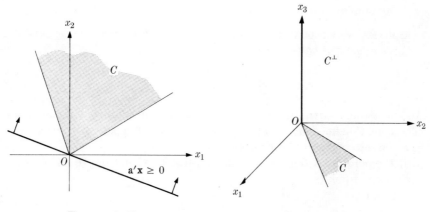

FIGURE 6–20   FIGURE 6–21

$C^\perp$ is a cone since if $\mathbf{u}'\mathbf{v} = 0$, then $\mu \mathbf{u}'\mathbf{v} = 0$ for all $\mu \geq 0$. $C^\perp$ is not only a cone, it is also a subspace of E^n. This follows because if $\mathbf{u} \in C^\perp$, $\mathbf{u}'\mathbf{v} = 0$ and $\mu \mathbf{u}'\mathbf{v} = 0$ for all μ; therefore $\mu \mathbf{u}$ is in $C^\perp$ for any μ. Furthermore, if $\mathbf{u}_1, \mathbf{u}_2 \in C^\perp$, then $\mathbf{u}_1 + \mathbf{u}_2 \in C^\perp$ since $\mathbf{u}_1'\mathbf{v} = 0$ and $\mathbf{u}_2'\mathbf{v} = 0$ imply that $(\mathbf{u}_1 + \mathbf{u}_2)'\mathbf{v} = 0$. In some cases, the set $C^\perp$ will contain only the element $\mathbf{0}$. For example in E^3, $\mathbf{u} = \mathbf{0}$ is the only vector for which $\mathbf{u}'\mathbf{v} = 0$ when the $\mathbf{v}$ are the elements of the cone shown in Fig. 6–16.

EXAMPLE: In E^3 (see Fig. 6–21), cone $C^\perp$ (orthogonal to C) is all of the x_3-axis.

Given any half-line L, the orthogonal cone P to this half-line is a hyperplane. If $\mathbf{a} \neq \mathbf{0}$ generates the half-line, P is the set

$$P = \{\mathbf{x} | \mathbf{a}'\mathbf{x} = 0\}, \tag{6-70}$$

which is a hyperplane through the origin.

DIMENSION OF A CONE: *The dimension of a cone C is defined as the maximum number of linearly independent vectors in C.*

The dimension of C is the dimension of the "smallest subspace of E^n" which contains C, that is, the dimension of the intersection of all subspaces containing C.

It is easy to prove that the smallest subspace containing a *convex* cone C is $C + C^-$. The details of the proof will have to be supplied in Problem 6–41.

EXAMPLE: The dimension of the cone (Fig. 6–22) is 2 because there are two linearly independent vectors in the cone. Thus the smallest subspace of E^2 containing the cone is E^2 itself.

6–10 Convex polyhedral cones.

CONVEX POLYHEDRAL CONE: *A convex polyhedral cone C is the sum of a finite number of half-lines,*

$$C = \sum_{i=1}^{r} L_i. \tag{6-71}$$

In this definition the term "sum" is used in the sense of sums of cones. The cone C defined by (6–71) is convex because the sum of convex cones is a convex cone and half-lines are convex cones.

If the point $\mathbf{a}_i \neq \mathbf{0}$ generates the half-line L_i, then from (6–71) C is the collection of points

$$\mathbf{y} = \sum_{i=1}^{r} \mu_i \mathbf{a}_i, \quad \text{all } \mu_i \geq 0, \quad i = 1, \ldots, r. \tag{6-72}$$

EXAMPLE: Figure 6–23 shows the convex polyhedral cone generated by the half-lines 1, 2, 3. Note that any cross section of the polyhedral cone is a convex polyhedron.

The cone C generated by a convex polyhedron is a convex polyhedral cone. Let the set X be a convex polyhedron. Then any point $\mathbf{x}$ in X can be written as a convex combination of the extreme points $\mathbf{x}_i^*$ (assumed to be r in number):

$$\mathbf{x} = \sum_{i=1}^{r} \mu_i \mathbf{x}_i^*, \quad \mu_i \geq 0, \quad \sum \mu_i = 1.$$

Cone C is the collection of points $\alpha \mathbf{x}$, all $\alpha \geq 0$ and all $\mathbf{x} \in X$. However,

$$\alpha \mathbf{x} = \sum_{i=1}^{r} \mu_i \alpha \mathbf{x}_i^* = \sum_{i=1}^{r} \lambda_i \mathbf{x}_i^*, \quad \mu_i \alpha = \lambda_i \geq 0. \tag{6–73}$$

Each extreme point $\mathbf{x}_i^*$ generates a half-line L_i; we see by (6–72) and (6–73) that

$$C = \sum_{i=1}^{r} L_i;$$

thus C is a convex polyhedral cone.

If $\mathbf{A}$ is an $n \times r$ matrix $\mathbf{A} = (\mathbf{a}_1, \ldots, \mathbf{a}_r)$, then the set of points

$$\mathbf{y} = \mathbf{A}\mathbf{x} = \sum_{i=1}^{r} x_i \mathbf{a}_i, \quad \text{all } \mathbf{x} \geq \mathbf{0}, \tag{6–74}$$

is a convex polyhedral cone in E^n. This follows immediately from (6–72). The columns $\mathbf{a}_i$ of $\mathbf{A}$ generate the half-lines whose sum yields the polyhedral cone. The fact that $C = \{\mathbf{y}\}$, with $\mathbf{y}$ given by (6–74), is a poly-

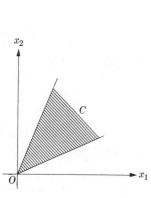

FIGURE 6–22

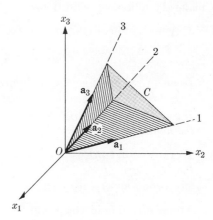

FIGURE 6–23

hedral cone indicates that there is a non-negative solution $\mathbf{x} \geq \mathbf{0}$ to the set of simultaneous linear equations

$$\mathbf{Ax} = \mathbf{b}$$

if and only if $\mathbf{b}$ is an element of the convex polyhedral cone generated by the columns of $\mathbf{A}$.

Any given finite number of points $\mathbf{a}_1, \ldots, \mathbf{a}_r$ from E^n can be thought of as generating a convex polyhedral cone C. Each $\mathbf{a}_i$ generates a half-line, and the cone C is the sum of these r half-lines. Similarly, we can imagine that any convex polyhedral cone in E^n has been generated by a finite number of points $\mathbf{a}_1, \ldots, \mathbf{a}_r$ from E^n. We only need to choose one nonzero point on each half-line whose sum yields the cone. If $r \geq n$, and if in addition there are n points in the set $\mathbf{a}_1, \ldots, \mathbf{a}_r$ which are linearly independent, then the points $\mathbf{a}_1, \ldots, \mathbf{a}_r$ generate a cone of dimension n, that is, no subspace of E^n containing C has dimension $< n$. The cone shown in Fig. 6–23, for example, has dimension 3, the dimension of E^3; it is generated by three linearly independent points.

Suppose that we have in E^n an n-dimensional cone C generated by $\mathbf{a}_1, \ldots, \mathbf{a}_r$. Out of the set $\mathbf{a}_1, \ldots, \mathbf{a}_r$ let us choose any $n - 1$ linearly independent points $\mathbf{b}_1, \ldots, \mathbf{b}_{n-1}$. These points determine a unique hyperplane $\mathbf{cx} = 0$ through the origin in E^n because $\mathbf{cb}_1 = 0, \ldots, \mathbf{cb}_{n-1} = 0$ form a set of $n - 1$ homogeneous linear equations in the n unknowns, the c_i. Furthermore, since $\mathbf{b}_1, \ldots, \mathbf{b}_{n-1}$ are linearly independent, there is a nonvanishing determinant of order $n - 1$ in the matrix $(\mathbf{b}_1, \ldots, \mathbf{b}_{n-1})$; hence the c_i are determined up to a common multiplicative constant. Thus the hyperplane is uniquely determined. It may or may not be true that all of the cone C will lie in one of the closed half-spaces $\mathbf{cx} \geq 0$ or $\mathbf{cx} \leq 0$. If C does lie in one closed half-space produced by $\mathbf{cx} = 0$, we make the following definitions:

EXTREME SUPPORTING HALF-SPACE: *The set of points from E^n,*

$$H_F = \{\mathbf{v} | \mathbf{cv} \geq 0\}, \tag{6–75}$$

is an extreme supporting half-space for the n-dimensional convex polyhedral cone C generated by the points $\mathbf{a}_1, \ldots, \mathbf{a}_r$ if C lies in the half-space H_F and $n - 1$ linearly independent points from the set $\mathbf{a}_1, \ldots, \mathbf{a}_r$ lie on the hyperplane $\mathbf{cv} = 0$.

EXTREME SUPPORTING HYPERPLANE: *The hyperplane $\mathbf{cv} = 0$ which forms the boundary of the extreme supporting half-space is called an extreme supporting hyperplane for the convex polyhedral cone C.*

The reader should be careful to note the difference between an extreme supporting hyperplane for a convex polyhedral cone and a supporting hyperplane for any convex set as defined earlier. An extreme supporting hyperplane must have $n - 1$ linearly independent points of the cone lying on it, while a supporting hyperplane need not have more than a single point in common with the cone.

It is clear that the intersection F of an extreme supporting hyperplane $\mathbf{cv} = 0$ and the polyhedral cone C yield a collection of points which are boundary points of C; that is, every point of C on the extreme supporting hyperplane $\mathbf{cv} = 0$ is a boundary point of C since C lies in the half-space $\mathbf{cv} \geq 0$. Furthermore, the intersection F is itself a convex polyhedral cone generated by the points from $\mathbf{a}_1, \ldots, \mathbf{a}_r$ which lie on the hyperplane $\mathbf{cv} = 0$. This proof is trivial, and the details will have to be worked out in Problem 6–43. Since the set F is a subset of C and a convex polyhedral cone, we call F a subcone of C. The subcone F lies in the hyperplane $\mathbf{cv} = 0$. Since there are precisely $n - 1$ linearly independent points in F, F has dimension $n - 1$. In Fig. 6–23 any two of the vectors $\mathbf{a}_1, \mathbf{a}_2, \mathbf{a}_3$ uniquely determine a plane through the origin such that the cone C is contained in one half-space produced by the plane. Furthermore, any two points from $\mathbf{a}_1, \mathbf{a}_2, \mathbf{a}_3$ are linearly independent, and hence the resulting hyperplane is an extreme supporting hyperplane. The intersection of C with any one of the three extreme supporting hyperplanes yields a face of the cone C. This face is a cone—the cone F discussed above. *In general, we call the $(n - 1)$-dimensional convex polyhedral cone F which is the intersection of an n-dimensional polyhedral cone C in E^n with an extreme supporting hyperplane a facet or face of the cone C.*

An n-dimensional convex polyhedral cone in E^n generated by $\mathbf{a}_1, \ldots, \mathbf{a}_r$ can have only a finite number of extreme supporting hyperplanes since there are at most $r!/(n - 1)!(r - n + 1)!$ sets of $n - 1$ linearly independent points in the set $\mathbf{a}_1, \ldots, \mathbf{a}_r$. Not every n-dimensional convex polyhedral cone in E^n needs to have an extreme supporting hyperplane. The convex polyhedral cone may be all of E^n and thus cannot lie in any half-space of E^n. For example, in E^2 the convex polyhedral cone generated by the points $(0, 1)$, $(1, 0)$, $(0, -1)$, $(-1, 0)$ is all of E^2 and hence does not have an extreme supporting half-space. *An n-dimensional convex polyhedral cone which contains every point of E^n is called solid.*

If an n-dimensional convex polyhedral cone is not all of E^n, that is, if it is not solid, then it does have an extreme supporting hyperplane.

6–11 Linear transformations of regions. Consider a linear transformation represented by the $m \times n$ matrix $\mathbf{A}$ which takes E^n into a subspace of E^m. Frequently, we wish to know how the transformation affects some region in E^n, that is, what image the region will assume in E^m.

One of the most important properties of a linear transformation is that it always takes lines into lines or a line into a point. It also takes hyperplanes in E^n into hyperplanes in E^m, or a hyperplane in E^n into the intersection of two or more hyperplanes in E^m, or a hyperplane in E^n into all of E^m. We have noted previously that a line in E^n is the set of points

$$\mathbf{x} = \lambda \mathbf{x}_1 + (1 - \lambda)\mathbf{x}_2, \quad \text{all } \lambda, \quad \mathbf{x}_1 \neq \mathbf{x}_2. \tag{6-76}$$

The set of points $\mathbf{y} = \mathbf{A}\mathbf{x}$ for $\mathbf{x}$ given by (6-76) is

$$\mathbf{y} = \lambda \mathbf{A}\mathbf{x}_1 + (1 - \lambda)\mathbf{A}\mathbf{x}_2 = \lambda \mathbf{y}_1 + (1 - \lambda)\mathbf{y}_2, \tag{6-77}$$

where $\mathbf{y}_1 = \mathbf{A}\mathbf{x}_1$, $\mathbf{y}_2 = \mathbf{A}\mathbf{x}_2$. If $\mathbf{y}_1 \neq \mathbf{y}_2$, we obtain a line in E^m. When $\mathbf{y}_1 = \mathbf{y}_2$, the line in E^n is transformed into a point in E^m. If $\mathbf{A}$ is a nonsingular nth-order matrix, $\mathbf{y}_1$ will differ from $\mathbf{y}_2$ (when $\mathbf{x}_1 \neq \mathbf{x}_2$) so that a nonsingular transformation takes a line in E^n into another line in E^n.

To determine the effects of a linear transformation on a hyperplane $\mathbf{c}\mathbf{x} = z$ in E^n, we choose n points $\mathbf{x}_1, \ldots, \mathbf{x}_n$ on the hyperplane such that the vectors $\mathbf{x}_1 - \mathbf{x}_n, \ldots, \mathbf{x}_{n-1} - \mathbf{x}_n$ are linearly independent. Then the hyperplane is the set of all points $\mathbf{x}$ for which we can write

$$\mathbf{x} - \mathbf{x}_n = \sum_{i=1}^{n-1} \lambda_i (\mathbf{x}_i - \mathbf{x}_n). \tag{6-78}$$

If we set $\mathbf{y} = \mathbf{A}\mathbf{x}$, $\mathbf{y}_i = \mathbf{A}\mathbf{x}_i$, $i = 1, \ldots, n$, then the set of points in E^m which is the image of the hyperplane can be written

$$\mathbf{y} - \mathbf{y}_n = \sum_{i=1}^{n-1} \lambda_i (\mathbf{y}_i - \mathbf{y}_n). \tag{6-79}$$

If m of the $\mathbf{y}_i - \mathbf{y}_n$ are linearly independent, the image of the hyperplane is all of E^m. When $m - 1$ of the $\mathbf{y}_i - \mathbf{y}_n$ are linearly independent, the image of the hyperplane in E^n is a hyperplane in E^m. If less than $m - 1$ of the $\mathbf{y}_i - \mathbf{y}_n$ are linearly independent, the image of the hyperplane in E^n lies in the intersection of two or more hyperplanes in E^m. In the event that $\mathbf{A}$ is an nth-order nonsingular matrix, the transformation takes a hyperplane in E^n into another hyperplane in E^n. If $\mathbf{x} = \mathbf{A}^{-1}\mathbf{y}$, then $\mathbf{c}\mathbf{x} = z$ becomes $\mathbf{c}\mathbf{A}^{-1}\mathbf{y} = z$. This hyperplane has a normal $\mathbf{c}\mathbf{A}^{-1}$, whereas the original hyperplane had a normal $\mathbf{c}$.

The linear transformation represented by the $m \times n$ matrix $\mathbf{A}$ takes a cone in E^n into a cone in E^m because if $C = \{\mathbf{x}\}$, then when $\mathbf{y} = \mathbf{A}\mathbf{x}$, it follows that $\mu \mathbf{y} = \mu \mathbf{A}\mathbf{x} = \mathbf{A}(\mu \mathbf{x})$, $\mu \geq 0$, is an element of the image of C since $\mu \mathbf{x} \in C$. Thus if $\mathbf{y}$ is in the image of C, so is $\mu \mathbf{y}$ for $\mu \geq 0$. Hence the image of C is also a cone. In general, a linear transformation takes any convex set into a convex set. (To be proved in Problem 6-18.)

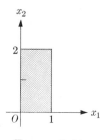
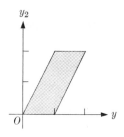

FIGURE 6–24   FIGURE 6–25

EXAMPLE: Let us study what the linear transformation

$$\begin{bmatrix} y_1 \\ y_2 \end{bmatrix} = \begin{bmatrix} 1 & \frac{1}{2} \\ 0 & 1 \end{bmatrix} \begin{bmatrix} x_1 \\ x_2 \end{bmatrix}$$

does to the rectangular region of the $x_1 x_2$-plane shown in Fig. 6–24. Because the transformation is nonsingular, lines will be taken into lines. Hence it is only necessary to examine what happens to the corners of the region. Once the images of the corners have been determined, the image of the rectangular region is found by joining the images of the corners with straight lines. The origin goes into the origin and hence this corner remains unchanged. Corner $\mathbf{x} = [0, 2]$ becomes $\mathbf{y} = [1, 2]$, $\mathbf{x} = [1, 2]$ becomes $\mathbf{y} = [2, 2]$, and $\mathbf{x} = [1, 0]$ becomes $\mathbf{y} = [1, 0]$. Thus Fig. 6–25 shows the image of the rectangular region shown in Fig. 6–24. The region has been sheared by the linear transformation.

All linear transformations of E^n into a subspace of E^m take the origin of E^n into the origin of E^m since $\mathbf{y} = \mathbf{A0} = \mathbf{0}$. This is equivalent to the statement that a linear transformation will never translate a region. A more general transformation of the form $\mathbf{y} = \mathbf{Ax} + \mathbf{b}$, $\mathbf{b} \neq \mathbf{0}$, takes the origin of E^n into the point $\mathbf{b}$ of E^m. It performs a so-called affine transformation. Affine transformations are not linear. (What is the relation between affine transformations and affine subspaces defined in Section 5–7?)

REFERENCES

The following papers discuss various aspects of convex sets and convex cones. However, they are no easy reading matter for a person with little background in the subject.

1. T. BONNESEN, and W. FENCHEL, "Theorie der konvexen Körper," *Ergebnisse der Mathematik und ihrer Grenzgebiete*. Berlin: Verlag Julius Springer, 1934, Vol. 3, No. 1.

2. M. GLEZERMAN, and L. PONTRYAGIN, *Intersections in Manifolds*. Translation No. 50, New York: American Mathematical Society, 1951.

3. J. G. Kemeny, H. Mirkil, J. L. Snell, G. L. Thompson. *Finite Mathematical Structures*. Englewood Cliffs: Prentice-Hall, 1959, Chapters 2 and 5.

4. T. C. Koopmans, Ed., *Activity Analysis of Production and Allocation*. New York: John Wiley and Sons, 1951. Chapter XVII. "Convex Polyhedral Cones and Linear Inequalities" by David Gale; Chapter XVIII. "Theory of Convex Polyhedral Cones" by Murray Gerstenhaber.

5. H. W. Kuhn, and A. W. Tucker, Eds. *Linear Inequalities and Related Systems*. Princeton: Princeton University Press, 1956; *Paper 2*. "Polyhedral Convex Cones" by A. J. Goldman and A. W. Tucker; *Paper 3*. "Resolution and Separation Theorems for Polyhedral Convex Sets" by A. J. Goldman; *Paper 13*. "Integral Boundary Points of Convex Polyhedra" by A. J. Hoffman and J. G. Kruskal; *Paper 15*. "Neighboring Vertices on a Convex Polyhedron" by David Gale.

6. H. W. Kuhn, and A. W. Tucker, Eds. *Contributions to the Theory of Games*, Vol. I, Princeton: Princeton University Press, 1950; *Paper 1*. "The Elementary Theory of Convex Polyhedra" by H. Weyl; translated by H. W. Kuhn from "Elementare Theorie der Konvexen Polyeder," *Commentarii Mathematici Helvetici* **7**: 290–306, 1935.

7. J. von Neumann, and O. Morgenstern, *Theory of Games and Economic Behavior*. Princeton: Princeton University Press, 1944. A brief discussion of convex set theory is presented in Chapter 3.

Problems

6–1. Show graphically the regions represented by the following point sets:

(a) $X = \{[x_1, x_2] | x_1^2 + x_2^2 \geq 1\}$;
(b) $X = \{[x_1, x_2] | x_1^2 + 3x_2^2 \leq 6\}$;
(c) $X = \{[x_1, x_2] | x_1 \geq 2, x_2 \leq 4\}$.

6–2. Show graphically $A \cap B$ and $A \cup B$ in the following cases:

(a) $A = \{[x_1, x_2] | x_1 > 2\}$, $B = \{[x_1, x_2] | x_1 < 3\}$;
(b) $A = \{[x_1, x_2] | x_1^2 + x_2^2 \leq 4\}$, $B = \{[x_1, x_2] | x_1^2 + x_2 \leq 1\}$;
(c) $A = \{[x_1, x_2] | x_1 \geq 0, x_2 \geq 0, x_1 \leq 1, x_2 \leq 1\}$,
$B = \{[x_1, x_2] | (x_1 - 1)^2 + x_2^2 \leq 1\}$.

6–3. Illustrate graphically $A_1 \cap A_2 \cap A_3$, $A_1 \cup A_2 \cup A_3$:

$$A_1 = \{[x_1, x_2] | x_1 - x_2 \geq 0\},$$
$$A_2 = \{[x_1, x_2] | x_1 x_2 \leq 1\},$$
$$A_3 = \{[x_1, x_2] | 3x_1 + 2x_2 \geq 0\}.$$

6–4. Show that a line in E^n is a closed set.

6-5. Give the complements of the following sets and indicate whether the complements are open or closed sets (or neither).

(a) $\mathbf{cx} < z$;
(b) $|\mathbf{x} - \mathbf{a}| < \epsilon$;
(c) $|\mathbf{x} - \mathbf{a}| = \epsilon$;
(d) E^n.

6-6. Draw the line $6x_1 + 3x_2 = 4$ and a vector normal to it. Find the line passing through $(2, 2)$ with the same normal. Can this line be obtained from $6x_1 + 3x_2 = 4$ by moving it parallel to itself in the direction of the normal?

6-7. Express the point $(0.5, 1)$ as a convex combination of $(2, 0)$, $(0, \frac{4}{3})$.

6-8. Let $\mathbf{x}$ be a point on the line joining $\mathbf{x}_1$ and $\mathbf{x}_2$; if $\mathbf{x}$ is a fraction μ of the distance from $\mathbf{x}_1$ to $\mathbf{x}_2$, write $\mathbf{x}$ as a convex combination of $\mathbf{x}_1$, $\mathbf{x}_2$.

6-9. Given three points $\mathbf{x}_1$, $\mathbf{x}_2$, $\mathbf{x}_3$ in E^n, how can we quickly ascertain whether they lie on the same line? How can we decide whether $\mathbf{x}_3$ lies between $\mathbf{x}_1$, $\mathbf{x}_2$? Hint: Consider $\mathbf{x}_1 - \mathbf{x}_2$ and $\mathbf{x}_1 - \mathbf{x}_3$.

6-10. Given the hyperplane $3x_1 + 2x_2 + 4x_3 + 6x_4 = 7$. In which half-space is the point $\mathbf{x} = [6, 1, 7, 2]$?

6-11. Consider the hyperplane $\mathbf{cx} = 0$ in E^n. Prove that this hyperplane is a subspace of dimension $n - 1$. Conversely, prove that any subspace of dimension $n - 1$ in E^n is a hyperplane through the origin.

6-12. Prove that E^n and any subspace of E^n are convex sets.

6-13. Is a set convex if, given any two points $\mathbf{x}_1$, $\mathbf{x}_2$ in the set, the point $\mathbf{x} = \frac{1}{2}(\mathbf{x}_1 + \mathbf{x}_2)$, i.e., the point halfway between these points, is also in the set? Can you give a counterexample?

6-14. Which of the following sets are convex?

(a) $X = \{[x_1, x_2] | 3x_1^2 + 2x_2^2 \leq 6\}$;
(b) $X = \{[x_1, x_2] | x_1 \geq 2, x_1 \leq 3\}$;
(c) $X = \{[x_1, x_2] | x_1 x_2 \leq 1, x_1 \geq 0, x_2 \geq 0\}$;
(d) $X = \{[x_1, x_2] | x_2 - 3 \geq -x_1^2, x_1 \geq 0, x_2 \geq 0\}$.

6-15. Prove that if $X_1, \ldots, X_n$ are convex sets, then $\bigcap_{i=1}^{n} X_i$ is a convex set. Is $\bigcup_{i=1}^{n} X_i$ convex? If not, give a counterexample.

6-16. If $X_1, \ldots, X_n$ in Problem 6-15 are closed, prove that $\bigcap_{i=1}^{n} X_i$ is closed also.

6-17. Sketch the convex polyhedra generated by the following sets of points:

(a) $(0, 0)$, $(1, 0)$, $(0, 1)$, $(1, 1)$;
(b) $(3, 4)$, $(5, 6)$, $(0, 0)$, $(2, 2)$, $(1, 0)$, $(2, 5)$, $(4, 7)$;
(c) $(-1, 2)$, $(3, -4)$, $(4, 4)$, $(0, 0)$, $(6, 5)$, $(7, 1)$.

6-18. Prove that a linear transformation of E^n into a subspace of E^m will take a convex set into a convex set.

6-19. Show that a nonsingular linear transformation takes an extreme point of a convex set into an extreme point. Give an example illustrating that a singular linear transformation can take an extreme point into an interior point.

6–20. Prove that the convex polyhedron generated by a finite number of points $\mathbf{x}_1, \ldots, \mathbf{x}_m$ is strictly bounded if each $\mathbf{x}_i$ is of finite length.

6–21. Prove that a closed convex set which is bounded from above has an extreme point in every supporting hyperplane.

6–22. Prove that a strictly bounded, closed convex set has an extreme point in every supporting hyperplane by showing that the point in the intersection of the hyperplane and the convex set which is farthest removed from the origin is an extreme point.

6–23. Show that any nonempty closed convex set which is bounded from below contains at least one extreme point. If a set contains only one point, this point will be considered an extreme point.

6–24. Prove that a strictly bounded, closed convex set with two points in any supporting hyperplane has at least two extreme points. Hint: According to Problem 6–22, $\mathbf{t}^*$, the point in the intersection of the hyperplane and the convex set farthest from the origin, was an extreme point. Consider the point in the intersection farthest from $\mathbf{t}^*$.

6–25. Consider the intersection of the following half-spaces or hyperplanes:

$$\mathbf{a}^i \mathbf{x} \{ \leq\ =\ \geq \} b_i, \qquad i = 1, \ldots, m.$$

Find the set of points which is a subset of the above set and includes all its boundary points.

6–26. Show that the boundary of the intersection of m half-spaces is connected.

6–27. Show that every point in the intersection of m hyperplanes in E^n is a boundary point of this set and hence that the set is closed.

6–28. Prove that the intersection of m hyperplanes has no extreme points when the variables are unrestricted in sign, unless there is only a single point in the intersection. Hint: Let $\mathbf{x}_1, \mathbf{x}_2$ be any two points in the intersection. Then $\mathbf{x}_3 = \mathbf{x}_1 + \lambda(\mathbf{x}_2 - \mathbf{x}_1)$ is also in the intersection for any real λ. Is $\mathbf{x}_1$ ever an extreme point?

6–29. What happens to the theorem of the separating hyperplanes when the convex set is open? Is the theorem true if the set is not closed? Hint: Consider a set which contains some but not all of its boundary points.

6–30. Given the convex set $X = \{[x_1, x_2] | x_1^2 + x_2^2 \leq 1\}$. Find the equation for the supporting hyperplane at any boundary point (ξ, η), and the equation for the supporting hyperplane at $[\sqrt{2}/2, \sqrt{2}/2]$.

6–31. Prove that the convex hull of a finite number of points is a closed set.

6–32. Consider the triangle with vertices $(0, 0)$, $(2, 0)$, $(1, 1)$. Express the point $(0.5, 0.5)$ as a convex combination of the extreme points. Do the same for point $(0.3, 0.2)$.

6–33. Show that the intersection of a finite number of closed half-spaces (and perhaps some hyperplanes) can have only a finite number of extreme points in E^n. Prove that if the number m of half-spaces is less than n, there is no extreme point at all.

6–34. Prove that the set of solutions to $\mathbf{A}\mathbf{x} = \mathbf{b}$ is a convex set by showing that if $\mathbf{x}_1, \mathbf{x}_2$ are solutions, so is $\lambda \mathbf{x}_1 + (1 - \lambda)\mathbf{x}_2$, $0 \leq \lambda \leq 1$. Demonstrate by the same method that the set of solutions with $\mathbf{x} \geq \mathbf{0}$ is a convex set.

6-35. Prove that a strictly bounded intersection of m closed half-spaces is a convex polyhedron. Hint: It is only necessary to show that there are a finite number of extreme points.

6-36. What difficulties would be involved in solving a linear programming problem if the constraints were open (i.e., have a ">" or "<" sign) rather than closed half-spaces. Would the absolute maximum or minimum actually be taken on by a point in the set of feasible solutions?

6-37. Sketch the half-lines generated by the points $\mathbf{a}_1 = [2, 1]$, $\mathbf{a}_2 = [1, 3]$, $\mathbf{a}_3 = [-1, 2]$. Sketch the cone C which is the sum of these three half-lines. Sketch C^+, C^- $C^\perp$.

6-38. If C_1, C_2 are cones and $C_1 \subset C_2$, prove that $C_2^+ \subset C_1^+$.

6-39. If C_1, C_2 are cones, prove that $(C_1 + C_2)^+ = C_1^+ \cap C_2^+$. Generalize this result and show that

$$\left[\sum_{i=1}^m C_i\right]^+ = \bigcap_{i=1}^m C_i^+.$$

Hint: If $\mathbf{u}_1 \in C_1$, $\mathbf{u}_2 \in C_2$, then any vector $\mathbf{v}$ satisfying $\mathbf{v}'(\mathbf{u}_1 + \mathbf{u}_2) \geq 0$ must also satisfy $\mathbf{v}'\mathbf{u}_1 \geq 0$, $\mathbf{v}'\mathbf{u}_2 \geq 0$, and hence $(C_1 + C_2)^+ \subset C_1^+ \cap C_2^+$.

6-40. Let C_i, $i = 1, \ldots, m$ be cones. Show that $C = \sum_{i=1}^m C_i$ is also a cone. If the C_i are convex cones, show that C is also convex.

6-41. Prove that $C + C^-$ is the intersection of all subspaces containing the convex cone C, that is, $C + C^-$ is the smallest subspace containing C. Is this true if C is not convex? Can you supply a counterexample?

6-42. Sketch the convex polyhedral cone generated by the points $(1, 0, 0)$, $(1, 1, 1)$, $(0, 1, 0)$.

6-43. Prove that the intersection of an n-dimensional convex polyhedral cone C (generated by $\mathbf{a}_1, \ldots, \mathbf{a}_r$) in E^n with an extreme supporting hyperplane is a polyhedral cone F of dimension $n - 1$. Prove that F is the polyhedral cone generated by the points from the set $\mathbf{a}_1, \ldots, \mathbf{a}_r$ lying on the extreme supporting hyperplane.

6-44. Prove that the intersection of k hyperplanes $\mathbf{c}^i\mathbf{x} = 0$ in E^n (the $\mathbf{c}^i$ are linearly independent) is a subspace of dimension $n - k$. Show that any subspace of dimension $n - k$ can be represented as the intersection of k hyperplanes. Is this representation unique? Hint: Review Section 5-6.

6-45. The lineality of a cone C is the dimension of the subspace which is the convex hull of all subspaces in C; hence it is the dimension of the smallest subspace containing all the subspaces in C. This space is the lineality space of C. Prove that if C is convex, the lineality space of C is contained in C. A convex cone with zero lineality is called pointed. Illustrate this geometrically.

6-46. Show that the non-negative orthant of E^n is a convex polyhedral cone.

6-47. Prove: If a convex polyhedral cone C in E^n contains no vector $\mathbf{x} < \mathbf{0}$, then C^+ contains a vector $\mathbf{y} \geq \mathbf{0}$, $\mathbf{y} \neq \mathbf{0}$. Hint: If P is the non-negative orthant, then $C + P$ does not contain P^- and therefore $C + P$ is not all of E^n. Consequently, there is a vector $\mathbf{w} \neq \mathbf{0}$ in $(C + P)^+$. But $(C + P)^+ = C^+ \cap P^+$, and $P^+ = P$, since $\mathbf{e}_1, \ldots, \mathbf{e}_n$ are in P; if $\mathbf{w}'\mathbf{u} \geq 0$ for each $\mathbf{u} \in P$, then each component of $\mathbf{w}$ is non-negative.

6–48. Referring to Problem 6–47, consider the hyperplane $\mathbf{y'x} = 0$. Show that C is contained in $\mathbf{y'x} \geq 0$ and P^- in $\mathbf{y'x} \leq 0$. Thus the hyperplane $\mathbf{y'x} = 0$ separates C and P^-.

6–49. If $\mathbf{A}$ is an $n \times r$ matrix, show that the following sentence is a statement of Problem 6–47 in matrix language: If $\mathbf{Ax} \geq \mathbf{0}$ for all $\mathbf{x} \geq \mathbf{0}$, then there exists $\mathbf{w} \geq \mathbf{0}$ in E^n such that $\mathbf{w'A} \geq \mathbf{0}$.

6–50. Given an $n \times r$ matrix $\mathbf{A}$, demonstrate that there exists either a vector $\mathbf{x} \geq \mathbf{0}, \sum_{i=1}^{r} x_i = 1$, such that $\mathbf{Ax} \leq \mathbf{0}$, or a vector $\mathbf{w} \geq \mathbf{0}, \sum_{i=1}^{n} w_i = 1$, such that $\mathbf{w'A} > \mathbf{0}$. Prove the result by showing that such an $\mathbf{x}$ exists if $\mathbf{0}$ is in the convex polyhedron spanned by the columns of $\mathbf{A}$ and the unit vectors. If $\mathbf{0}$ is not in the convex polyhedron, demonstrate, by means of the theorem on separating hyperplanes, that there is a hyperplane through $\mathbf{0}$ such that each point of the convex polyhedron is in one open half-space. Show that $\mathbf{w}$ is thus a normal to this hyperplane. Can this problem be solved immediately using the results of Problem 6–49? (Von Neumann and Morgenstern call this the theorem of alternatives for matrices.)

6–51. Show that the two alternatives mentioned in Problem 6–50 are mutually exclusive, that is, vectors $\mathbf{x}, \mathbf{w}$ cannot both exist. Hint: Assume $\mathbf{x}$ and $\mathbf{w}$ exist and consider $\mathbf{w'Ax}$.

6–52. Prove that a convex polyhedral cone is a closed set.

6–53. Interpret the solutions to $\mathbf{Ax} = \mathbf{b}$, with $\mathbf{A}$ being an $m \times n$ matrix, as the intersection of m hyperplanes in E^n. What is the geometrical interpretation of inconsistent equations? What is the geometrical interpretation of redundant equations?

6–54. Consider the convex set of solutions to $\mathbf{Ax} = \mathbf{b}, \mathbf{x} \geq \mathbf{0}$, that is, the set of non-negative solutions to $\mathbf{Ax} = \mathbf{b}$. Show that every extreme point of the convex set is a basic solution to $\mathbf{Ax} = \mathbf{b}$ with $\mathbf{x}_B \geq \mathbf{0}$. Also show that every basic solution to $\mathbf{Ax} = \mathbf{b}$ with $\mathbf{x}_B \geq \mathbf{0}$ is an extreme point of the convex set. Hint: Let $\mathbf{x} = [\mathbf{x}_B, \mathbf{0}] \geq \mathbf{0}$ be a non-negative basic solution to $\mathbf{Ax} = \mathbf{b}$. Do there exist any other non-negative solutions $\mathbf{x}_1, \mathbf{x}_2$ such that $\mathbf{x} = \lambda \mathbf{x}_1 + (1 - \lambda) \mathbf{x}_2$, $0 < \lambda < 1$? Consider the last $n - m$ components of this relation. To show that an extreme point is a basic non-negative solution, let $\mathbf{x}^*$ be an extreme point. Now show that the columns of $\mathbf{A}$ associated with positive variables are linearly independent. To do this assume that the positive variables appear in the first k components of $\mathbf{x}^*$. Next assume that $\sum_{j=1}^{k} \lambda_j \mathbf{a}_j = \mathbf{0}$ and at least one $\lambda_j \neq 0$. Let $\boldsymbol{\lambda}$ be the n-component vector $[\lambda_1, \ldots, \lambda_k, 0, \ldots, 0]$. Choose an ϵ so small that $\mathbf{x}_1 = \mathbf{x}^* + \epsilon \boldsymbol{\lambda} \geq \mathbf{0}$ and $\mathbf{x}_2 = \mathbf{x}^* - \epsilon \boldsymbol{\lambda} \geq \mathbf{0}$. How is this done? Then $\mathbf{x}^* = \frac{1}{2}(\mathbf{x}_1 + \mathbf{x}_2)$.

6–55. Show that the general linear programming problem, as defined by Eqs. (1–12), (1–13), can be converted into an equivalent linear programming problem $\mathbf{Ax} = \mathbf{b}, \mathbf{x} \geq \mathbf{0}$, max or min $z = \mathbf{cx}$, where "equivalent" means that both problems have the same set of optimal solutions. Note that in the new form the constraints are a set of simultaneous linear equations. Hint: For any inequality in Eq. (1–13) of the form $\sum_{j=1}^{r} a_{ij} x_j \leq b_i$, define a new non-negative variable x_{r+i} (called a slack variable) by $x_{r+i} = b_i - \sum_{j=1}^{r} a_{ij} x_j$. What is the physical interpretation of a slack variable? For any inequality in Eq. (1–13) of the form $\sum_{j=1}^{r} a_{ij} x_j \geq b_i$, define a new non-negative variable x_{r+i} (called a surplus variable) by $x_{r+i} = \sum_{j=1}^{r} a_{ij}x_j - b_i$. What is the physical interpretation of a surplus variable? What prices should be assigned to slack and surplus variables?

6–56. Show that if the linear programming problem $\mathbf{Ax} = \mathbf{b}$, $\mathbf{x} \geq \mathbf{0}$, max $z = \mathbf{cx}$ has an optimal solution, then at least one of the basic feasible solutions will be optimal.

CHAPTER 7

CHARACTERISTIC VALUE PROBLEMS AND QUADRATIC FORMS*

*"Siftings on siftings in oblivion,
Till change hath broken down
All things . . ."*

Ezra Pound–Hugh Selwyn Mauberley.

7–1 Characteristic value problems. A problem which arises frequently in applications of linear algebra is that of finding values of a scalar parameter λ for which there exist vectors $\mathbf{x} \neq \mathbf{0}$ satisfying

$$\mathbf{A}\mathbf{x} = \lambda \mathbf{x}, \qquad (7\text{–}1)$$

where $\mathbf{A}$ is a given nth-order matrix. Such a problem is called a characteristic value (eigenvalue, or proper value) problem. If $\mathbf{x} \neq \mathbf{0}$ satisfies (7–1) for a given λ, then $\mathbf{A}$ operating on $\mathbf{x}$ yields a vector which is a scalar multiple of $\mathbf{x}$.

Clearly, $\mathbf{x} = \mathbf{0}$ is one solution of (7–1) for any λ; however, this trivial solution is not of interest. We are looking for vectors $\mathbf{x} \neq \mathbf{0}$ which satisfy (7–1). Now (7–1) can be written

$$\mathbf{A}\mathbf{x} = \lambda \mathbf{I} \mathbf{x},$$

or

$$(\mathbf{A} - \lambda \mathbf{I})\mathbf{x} = \mathbf{0}. \qquad (7\text{–}2)$$

If we choose a given λ, then any $\mathbf{x}$ which satisfies (7–1) must satisfy the set of n homogeneous linear equations in n unknowns (7–2). There will be a solution $\mathbf{x} \neq \mathbf{0}$ to (7–2) if and only if

$$|\mathbf{A} - \lambda \mathbf{I}| = 0, \qquad (7\text{–}3)$$

that is, if and only if

$$\begin{vmatrix} a_{11} - \lambda & a_{12} & \cdots & a_{1n} \\ a_{21} & a_{22} - \lambda & \cdots & a_{2n} \\ \vdots & & & \vdots \\ a_{n1} & a_{n2} & \cdots & a_{nn} - \lambda \end{vmatrix} = 0. \qquad (7\text{–}4)$$

---

* This chapter is based on the assumption that the reader has studied Sections 2–11, 2–12, and Section 3–16 or Problem 4–18.

Obviously, this determinant is a polynomial in λ. The highest-order term in λ comes from the product of the diagonal elements. Thus $(-\lambda)^n$ is the highest-order term and $|\mathbf{A} - \lambda\mathbf{I}|$ is an nth-degree polynomial. We can write

$$f(\lambda) = |\mathbf{A} - \lambda\mathbf{I}| = (-\lambda)^n + b_{n-1}(-\lambda)^{n-1} + \cdots + b_1(-\lambda) + b_0. \tag{7-5}$$

Equation (7-4) is called the characteristic, or secular, equation* for the matrix $\mathbf{A}$, *and* $f(\lambda)$ *is called the characteristic polynomial for* $\mathbf{A}$.

From the fundamental theorem of algebra we know that the nth-degree equation $f(\lambda) = 0$ has n roots. Not all these roots need to be different, but if a root is counted a number of times equal to its multiplicity, there are n roots, which may be either real or complex numbers. It follows that there cannot be more than n different values of λ for which $|\mathbf{A} - \lambda\mathbf{I}| = 0$. For values of λ different from the roots of $f(\lambda) = 0$, the only solution to (7-1) is $\mathbf{x} = \mathbf{0}$. If λ is set equal to one of the roots λ_i, then $|\mathbf{A} - \lambda_i\mathbf{I}| = 0$, and there is at least one $\mathbf{x} \neq \mathbf{0}$ which satisfies (7-1). The maximum number of linearly independent vectors $\mathbf{x}$ which satisfy (7-1) when $\lambda = \lambda_i$, will be the nullity of $\mathbf{A} - \lambda_i\mathbf{I}$. *The roots of the characteristic equation, which will be denoted by* λ_i, $i = 1, \ldots, n$, *are called the characteristic values, eigenvalues, proper values, or latent roots of the matrix* $\mathbf{A}$. *The vectors* $\mathbf{x} \neq \mathbf{0}$ *which satisfy (7-1) are called characteristic vectors or eigenvectors of the matrix* $\mathbf{A}$.

We can write the polynomial $f(\lambda)$ in factored form, using the roots of $f(\lambda) = 0$, that is,

$$f(\lambda) = (\lambda_1 - \lambda)(\lambda_2 - \lambda) \cdots (\lambda_n - \lambda). \tag{7-6}$$

Comparison of (7-6) and (7-5) yields the well-known relations between the roots and coefficients of a polynomial (see Problem 7-1 for details):

$$b_{n-1} = \sum_{i=1}^{n} \lambda_i = \lambda_1 + \lambda_2 + \cdots + \lambda_n,$$

$$b_{n-2} = \sum_{j>i} \lambda_i\lambda_j = \lambda_1\lambda_2 + \cdots + \lambda_1\lambda_n + \lambda_2\lambda_3 + \cdots + \lambda_2\lambda_n$$

$$\vdots \qquad\qquad\qquad + \cdots + \lambda_{n-1}\lambda_n, \tag{7-7}$$

$$b_{n-r} = \sum_{k>\cdots>j>i} \lambda_i\lambda_j \cdots \lambda_k \quad \text{(each term is a product of } r \text{ of the } \lambda_i\text{)},$$

$$\vdots$$

$$b_0 = \lambda_1\lambda_2\lambda_3 \cdots \lambda_n.$$

---

* The name "secular equation" arose because Eq. (7-4) appears in the theory of secular perturbations in astronomy.

There are also relations between the b_j and the elements of $\mathbf{A}$; in particular, if we set $\lambda = 0$ in (7-5), we see that $b_0 = |\mathbf{A}|$. The derivation of the other b_j in terms of the a_{ij} is to be supplied in Problem 7-2.

To obtain numerical values for the eigenvalues of $\mathbf{A}$, it is necessary to solve the characteristic equation $f(\lambda) = 0$. This can be a difficult undertaking, especially if the equation is of high degree (say 3 or higher). In fact, a great deal of work is involved even in computing the coefficients b_j of $f(\lambda)$ [see Eq. (7-5)] from the a_{ij} of $\mathbf{A}$. We shall not concentrate on numerical methods for finding the eigenvalues of a matrix; instead we shall emphasize the theoretical development of the subject.

EXAMPLE: The characteristic polynomial for a second-order matrix $\mathbf{A} = \|a_{ij}\|$ is

$$f(\lambda) = \begin{vmatrix} a_{11} - \lambda & a_{12} \\ a_{21} & a_{22} - \lambda \end{vmatrix} = (a_{11} - \lambda)(a_{22} - \lambda) - a_{12}a_{21}$$

$$= (-\lambda)^2 + (a_{11} + a_{22})(-\lambda) + a_{11}a_{22} - a_{12}a_{21}$$

$$= (-\lambda)^2 + b_1(-\lambda) + b_0;$$

hence

$$b_1 = a_{11} + a_{22}, \quad b_0 = a_{11}a_{22} - a_{12}a_{21} = |\mathbf{A}|.$$

The two roots of $f(\lambda) = 0$ are

$$\lambda = \tfrac{1}{2}\{(a_{11} + a_{22}) \pm [(a_{11} + a_{22})^2 - 4|\mathbf{A}|]^{1/2}\}.$$

If λ_1, λ_2 denote these roots, then

$$\lambda_1\lambda_2 = |\mathbf{A}| = b_0, \quad \lambda_1 + \lambda_2 = a_{11} + a_{22} = b_1.$$

For most characteristic value problems of physical interest which have matrices $\mathbf{A}$ whose elements are real numbers, it turns out that $\mathbf{A}$ is also a symmetric matrix. The theory of eigenvalue problems involving symmetric matrices $\mathbf{A}$ is thus very important. Interestingly enough, the theory is much simpler in this case than in that of a nonsymmetric matrix $\mathbf{A}$. However, prior to discussing topics related to characteristic value problems for symmetric matrices, we shall introduce the notion of the similarity of matrices, which will be needed in the following sections.

7-2 Similarity. Suppose that $\mathbf{x}$ is an eigenvector of $\mathbf{A}$ corresponding to the eigenvalue λ and that $\mathbf{P}$ is an nth-order nonsingular matrix. Then the vector $\mathbf{y} = \mathbf{Px}$ will not, in general, be an eigenvector of $\mathbf{A}$ corresponding to the eigenvalue λ since on multiplying the left-hand side of (7-1)

by **P**, we obtain
$$\mathbf{PAx} = \lambda \mathbf{Px}, \tag{7-8}$$

which is not the same as $\mathbf{APx} = \lambda \mathbf{Px}$. However, $\mathbf{x} = \mathbf{P}^{-1}\mathbf{Px}$. Substituting this into (7–8), we obtain
$$\mathbf{PAP}^{-1}\mathbf{Px} = \lambda \mathbf{Px}, \tag{7-9}$$
or
$$\mathbf{PAP}^{-1}\mathbf{y} = \lambda \mathbf{y}.$$

Thus **y** is an eigenvector of the matrix $\mathbf{PAP}^{-1}$, corresponding to the eigenvalue λ. We have shown that if λ is an eigenvalue of **A**, then λ is also an eigenvalue of $\mathbf{PAP}^{-1}$ for any nth-order nonsingular matrix **P**. If $\mathbf{B} = \mathbf{PAP}^{-1}$, then $\mathbf{A} = \mathbf{P}^{-1}\mathbf{BP}$, and $\mathbf{x} = \mathbf{P}^{-1}\mathbf{y}$. This demonstrates that any eigenvalue of **B** must also be an eigenvalue of **A**. Hence the matrices **A**, **B** have identical sets of eigenvalues and are called similar matrices.

SIMILARITY: *If there exists a nonsingular matrix* **P** *such that* $\mathbf{B} = \mathbf{PAP}^{-1}$, *the square matrices* **A** *and* **B** *are said to be similar.*

If $\mathbf{B} = \mathbf{PAP}^{-1}$, we say that **B** is obtained by a similarity transformation* on **A**. A similarity transformation is a special case of an equivalence transformation, defined in Section 4–6. If **B** is similar to **A**, **B** is also equivalent to **A**.

Since similar matrices have the same set of eigenvalues and the characteristic polynomial of a matrix can be written in the form (7–6), it follows that similar matrices have the same characteristic polynomial, that is, for any value of λ, $f_A(\lambda) = f_B(\lambda)$, where $f_A(\lambda)$ and $f_B(\lambda)$ are the characteristic polynomials for **A**, **B**, respectively.

7–3 Characteristic value problems for symmetric matrices. In general, the eigenvalues of a matrix **A** need not be real numbers—they may be complex.† We have been assuming that the elements of **A** are real numbers. It does not follow that the roots of (7–4) will be real numbers since the roots of a polynomial equation with real coefficients may be complex.

---

* Note that if $\mathbf{B} = \mathbf{PAP}^{-1}$, then $\mathbf{B} = \mathbf{R}^{-1}\mathbf{AR}$, where $\mathbf{R} = \mathbf{P}^{-1}$. Thus either $\mathbf{PAP}^{-1}$ or $\mathbf{P}^{-1}\mathbf{AP}$ represents a similarity transformation on **A**. We shall call **B** similar to **A** if $\mathbf{B} = \mathbf{PAP}^{-1}$ for some **P** or $\mathbf{B} = \mathbf{R}^{-1}\mathbf{AR}$ for some **R**. The definitions are equivalent since the substitution $\mathbf{R} = \mathbf{P}^{-1}$ converts one form into the other.

† If an eigenvalue is complex, the components of any eigenvector corresponding to this eigenvalue cannot all be real. The reader should note that everything that was stated in Sections 7–1 and 7–2 is still true since the general theory developed in Chapters 2 through 5 also applies to matrices with complex elements.

However, if **A** *in (7–1) is a symmetric matrix, then we can easily show that the eigenvalues of* **A** *are real.*

To carry out the proof, it is necessary to use several simple properties of complex numbers. (We suggest that the reader refresh his memory by examining the problems on complex numbers at the end of Chapters 2 and 3.) Assume that λ is a real or complex eigenvalue of the symmetric matrix **A**. Then there will be at least one eigenvector (which may have complex components) such that

$$\mathbf{Ax} = \lambda \mathbf{x}. \tag{7–10}$$

Taking the complex conjugate of (7–10) and recalling that the elements of **A** are real, we obtain

$$\mathbf{Ax}^* = \lambda^* \mathbf{x}^*, \tag{7–11}$$

where * denotes the complex conjugate, and $\mathbf{x}^* = [x_1^*, \ldots, x_n^*]$. Multiplying (7–10) by $(\mathbf{x}^*)'$, (7–11) by $\mathbf{x}'$, and subtracting, we have

$$(\mathbf{x}^*)'\mathbf{Ax} - \mathbf{x}'\mathbf{Ax}^* = (\lambda - \lambda^*)\mathbf{x}'\mathbf{x}^*, \tag{7–12}$$

since $(\mathbf{x}^*)'\mathbf{x} = \mathbf{x}'\mathbf{x}^*$. Recall that $\mathbf{A}' = \mathbf{A}$. Furthermore, $\mathbf{x}'\mathbf{Ax}^*$ is a number, and the transpose of a number (matrix of a single element) is itself the number. Thus

$$\mathbf{x}'\mathbf{Ax}^* = (\mathbf{x}'\mathbf{Ax}^*)' = (\mathbf{x}^*)'\mathbf{Ax}.$$

In addition, $\mathbf{x}'\mathbf{x}^* = \sum_{i=1}^{n} x_i x_i^*$ is real and positive since $\mathbf{x} \neq \mathbf{0}$. Consequently, $\lambda = \lambda^*$, and λ is real; that is, if $\lambda = a + bi$, then $\lambda^* = a - bi$, and $\lambda = \lambda^*$ implies $b = 0$.

Each eigenvalue of a symmetric matrix is real; hence the components of the eigenvectors will also be real because a set of homogeneous linear equations with real coefficients yields solutions **x** whose components are real. It is interesting to note that we did not have to use the characteristic equation in order to prove that its roots were real.

When **A** *is symmetric, another important result follows: The eigenvectors corresponding to different eigenvalues are orthogonal;* i.e., if $\mathbf{x}_i$ is an eigenvector corresponding to eigenvalue λ_i, and $\mathbf{x}_j$ is an eigenvector corresponding to eigenvalue λ_j ($\lambda_j \neq \lambda_i$), then $\mathbf{x}_j'\mathbf{x}_i = 0$. To prove this, we observe first that, by assumption,

$$\mathbf{Ax}_i = \lambda_i \mathbf{x}_i \quad \text{and} \quad \mathbf{Ax}_j = \lambda_j \mathbf{x}_j.$$

Thus

$$\mathbf{x}_j'\mathbf{Ax}_i = \lambda_i \mathbf{x}_j'\mathbf{x}_i \quad \text{and} \quad \mathbf{x}_i'\mathbf{Ax}_j = \lambda_j \mathbf{x}_j'\mathbf{x}_i. \tag{7–13}$$

Subtracting and noting that $\mathbf{x}_j'\mathbf{A}\mathbf{x}_i = \mathbf{x}_i'\mathbf{A}\mathbf{x}_j$, we have

$$(\lambda_i - \lambda_j)\mathbf{x}_j'\mathbf{x}_i = 0, \qquad (7\text{–}14)$$

or

$$\mathbf{x}_j'\mathbf{x}_i = 0, \quad \text{since} \quad \lambda_i \neq \lambda_j.$$

If $\mathbf{x} \neq \mathbf{0}$ is an eigenvector of $\mathbf{A}$, then any scalar multiple of $\mathbf{x}$ is also an eigenvector of $\mathbf{A}$, corresponding to the same eigenvalue. In general, the length of an eigenvector is not of interest. For convenience, we shall always assume that the eigenvectors of a symmetric matrix are of unit length; they will be denoted by $\mathbf{u}_i$. When the eigenvectors are of unit length, they are said to have been normalized. A set of two or more normalized eigenvectors $\mathbf{u}_i$ corresponding to different eigenvalues of $\mathbf{A}$ satisfies the equation $\mathbf{u}_i\mathbf{u}_j = \delta_{ij}$. Any set of vectors satisfying such an equation is said to be orthonormal.

EXAMPLE: Find the eigenvalues and eigenvectors of

$$\mathbf{A} = \begin{bmatrix} 2 & \sqrt{2} \\ \sqrt{2} & 1 \end{bmatrix}.$$

The characteristic equation is

$$|\mathbf{A} - \lambda\mathbf{I}| = \begin{vmatrix} 2-\lambda & \sqrt{2} \\ \sqrt{2} & 1-\lambda \end{vmatrix} = \lambda^2 - 3\lambda = 0,$$

whence the eigenvalues are

$$\lambda_1 = 0, \quad \lambda_2 = 3.$$

To determine the eigenvectors corresponding to λ_i, we must solve the set of homogeneous equations $(\mathbf{A} - \lambda_i\mathbf{I})\mathbf{x} = \mathbf{0}$. Let us take $\lambda_1 = 0$ first; then the set of equations becomes

$$2x_1 + \sqrt{2}x_2 = 0,$$
$$\sqrt{2}x_1 + x_2 = 0,$$

or

$$x_1 = -\frac{1}{\sqrt{2}}x_2.$$

If we wish to find an eigenvector of unit length, we must require that $x_1^2 + x_2^2 = 1$. Thus $(3/2)x_2^2 = 1$; choosing the positive square root, we obtain

$$x_2 = \sqrt{\frac{2}{3}}, \quad x_1 = -\frac{1}{\sqrt{3}},$$

and the eigenvector of unit length corresponding to λ_1 is

$$\mathbf{u}_1 = \left[-\frac{1}{\sqrt{3}}, \sqrt{\frac{2}{3}}\right].$$

Note that $\mathbf{u}_1$ is not completely specified by the requirement of $|\mathbf{u}_1| = 1$. The vector $-\mathbf{u}_1$ is also an eigenvector of length 1. However, $\mathbf{u}_1$ and $-\mathbf{u}_1$ are not linearly independent. Only one linearly independent eigenvector corresponds to λ_1.

For $\lambda_2 = 3$, the set of equations becomes

$$-x_1 + \sqrt{2}\, x_2 = 0, \qquad \sqrt{2}\, x_1 - 2x_2 = 0,$$

or

$$x_1 = \sqrt{2}\, x_2.$$

If $x_1^2 + x_2^2 = 1$, then $3x_2^2 = 1$; taking the positive square root, we have

$$x_2 = \frac{1}{\sqrt{3}}, \qquad x_1 = \sqrt{\frac{2}{3}},$$

and

$$\mathbf{u}_2 = \left[\sqrt{\frac{2}{3}}, \frac{1}{\sqrt{3}}\right].$$

Again, there is only one linearly independent eigenvector which corresponds to λ_2. It can be easily checked that $\mathbf{u}_2'\mathbf{u}_1 = 0$, in agreement with the theory.

7–4 Additional properties of the eigenvectors of a symmetric matrix. Suppose that all the eigenvalues of an nth-order symmetric matrix are different. Then there exists a set of n vectors $\mathbf{u}_i$ (one for each eigenvalue λ_i) such that

$$\mathbf{u}_j'\mathbf{u}_i = \delta_{ij} \qquad \text{(all } i,j = 1, \ldots, n\text{)}, \tag{7-15}$$

where δ_{ij} is the Kronecker delta. This follows because eigenvectors corresponding to different eigenvalues are orthogonal. Thus the $\mathbf{u}_i$ $(i = 1, \ldots, n)$ form an orthonormal basis for E^n. Hence, when the eigenvalues are all different, no more than one linearly independent eigenvector can correspond to a given eigenvalue. If there were two linearly independent eigenvectors corresponding to a given eigenvalue, both would have to be orthogonal to $n - 1$ orthonormal eigenvectors belonging to the other eigenvalues. However, we know from Section 2–11 that if two nonzero vectors from E^n are orthogonal to an orthonormal set of $n - 1$ vectors, then one is a scalar multiple of the other, and they are not linearly independent. Thus a contradiction is obtained.

Let us now consider the case where the eigenvalues of **A** are not all distinct. We shall show: (1) If an eigenvalue λ_j of the nth-order symmetric matrix **A** has multiplicity $k \geq 2$, there exist k orthonormal (and linearly independent) eigenvectors with eigenvalue λ_j. In fact, there exist an infinite number of sets of k orthonormal eigenvectors corresponding to λ_j. (2) There cannot be more than k linearly independent eigenvectors with the same eigenvalue λ_j; hence, if an eigenvalue has multiplicity k, the eigenvectors with eigenvalue λ_j span a subspace of E^n of dimension k. Then if the sets of eigenvectors corresponding to all the different eigenvalues are combined, it is possible to obtain an orthonormal basis for E^n.

To prove that there exist k linearly independent eigenvectors corresponding to an eigenvalue λ_j of multiplicity k, we must show that the nullity of $\mathbf{A} - \lambda_j \mathbf{I}$ is greater than, or equal to, k. To do this, we begin by noting that there will be at least one eigenvector with eigenvalue λ_j, say $\mathbf{u}_j$. From Section 2–11 we know that there exist $n-1$ vectors $\mathbf{v}_1, \ldots, \mathbf{v}_{n-1}$ such that the set $\mathbf{u}_j, \mathbf{v}_1, \ldots, \mathbf{v}_{n-1}$ is an orthonormal basis for E^n. Consider the matrix

$$\mathbf{Q}_1 = (\mathbf{u}_j, \mathbf{v}_1, \ldots, \mathbf{v}_{n-1}); \tag{7-16}$$

then

$$\mathbf{A}\mathbf{Q}_1 = (\mathbf{A}\mathbf{u}_j, \mathbf{A}\mathbf{v}_1, \ldots, \mathbf{A}\mathbf{v}_{n-1}) = (\lambda_j \mathbf{u}_j, \mathbf{A}\mathbf{v}_1, \ldots, \mathbf{A}\mathbf{v}_{n-1}), \tag{7-17}$$

and

$$\mathbf{Q}_1' \mathbf{A} \mathbf{Q}_1 = \begin{bmatrix} \lambda_j & \mathbf{u}_j' \mathbf{A} \mathbf{v}_1 & \cdots & \mathbf{u}_j' \mathbf{A} \mathbf{v}_{n-1} \\ \lambda_j \mathbf{v}_1' \mathbf{u}_j & \mathbf{v}_1' \mathbf{A} \mathbf{v}_1 & \cdots & \mathbf{v}_1' \mathbf{A} \mathbf{v}_{n-1} \\ \vdots & & & \vdots \\ \lambda_j \mathbf{v}_{n-1}' \mathbf{u}_j & \mathbf{v}_{n-1}' \mathbf{A} \mathbf{v}_1 & \cdots & \mathbf{v}_{n-1}' \mathbf{A} \mathbf{v}_{n-1} \end{bmatrix}. \tag{7-18}$$

However,

$$\mathbf{v}_i' \mathbf{u}_j = 0 \quad (i = 1, \ldots, n-1)$$

and

$$\mathbf{u}_j' \mathbf{A} \mathbf{v}_i = (\mathbf{u}_j' \mathbf{A} \mathbf{v}_i)' = \mathbf{v}_i' \mathbf{A} \mathbf{u}_j = \lambda_j \mathbf{v}_i' \mathbf{u}_j = 0 \quad (i = 1, \ldots, n-1).$$

Therefore, if

$$\alpha_{is} = \mathbf{v}_i' \mathbf{A} \mathbf{v}_s \quad (i, s = 1, \ldots, n-1) \quad \text{and} \quad \boldsymbol{\alpha} = \|\alpha_{is}\|,$$

then

$$\mathbf{A}_1 = \mathbf{Q}_1' \mathbf{A} \mathbf{Q}_1 = \begin{bmatrix} \lambda_j & \mathbf{0} \\ \mathbf{0} & \boldsymbol{\alpha} \end{bmatrix}, \tag{7-19}$$

and $\boldsymbol{\alpha}$ is a symmetric matrix of order $n-1$.

Now observe that

$$\mathbf{Q}'_1\mathbf{Q}_1 = \mathbf{I} \quad \text{or} \quad \mathbf{Q}_1^{-1} = \mathbf{Q}'_1; \tag{7-20}$$

that is, the inverse of $\mathbf{Q}_1$ is equal to its transpose. Hence $\mathbf{A}_1$ is similar to $\mathbf{A}$, and $\mathbf{A}_1$, $\mathbf{A}$ have the same set of eigenvalues. From (7-19)

$$|\mathbf{A}_1 - \lambda \mathbf{I}_n| = (\lambda_j - \lambda)|\boldsymbol{\alpha} - \lambda \mathbf{I}_{n-1}|. \tag{7-21}$$

Thus if λ_j is an eigenvalue of $\mathbf{A}$ with multiplicity $k \geq 2$, it must be true that $|\boldsymbol{\alpha} - \lambda_j \mathbf{I}_{n-1}| = 0$. Hence all minors of

$$\mathbf{A}_1 - \lambda_j \mathbf{I}_n = \begin{bmatrix} 0 & 0 \\ 0 & \boldsymbol{\alpha} - \lambda_j \mathbf{I}_{n-1} \end{bmatrix} \tag{7-22}$$

of order $n - 1$ vanish, and the nullity of $\mathbf{A}_1 - \lambda_j \mathbf{I}$ is ≥ 2. Since $r(\mathbf{A} - \lambda_j \mathbf{I}) = r(\mathbf{A}_1 - \lambda_j \mathbf{I})$, the nullity of $\mathbf{A} - \lambda_j \mathbf{I}$ is ≥ 2. Consequently, there exists an eigenvector $\hat{\mathbf{u}}_j$ of $\mathbf{A}$ with eigenvalue λ_j which is linearly independent of, and orthogonal to, $\mathbf{u}_j$.

If the multiplicity $k = 2$, our development is completed. If $k \geq 3$, the above procedure is repeated: There exist $n - 2$ vectors $\hat{\mathbf{v}}_1, \ldots, \hat{\mathbf{v}}_{n-2}$ such that $\mathbf{u}_j, \hat{\mathbf{u}}_j, \hat{\mathbf{v}}_1, \ldots, \hat{\mathbf{v}}_{n-2}$ is an orthonormal basis for E^n. If $\mathbf{Q}_2 = (\mathbf{u}_j, \hat{\mathbf{u}}_j, \hat{\mathbf{v}}_1, \ldots, \hat{\mathbf{v}}_{n-2})$, then

$$\mathbf{Q}_2^{-1} = \mathbf{Q}'_2, \tag{7-23}$$

and

$$\mathbf{A}_2 = \mathbf{Q}'_2 \mathbf{A} \mathbf{Q}_2 = \begin{bmatrix} \lambda_j & 0 & 0 \\ 0 & \lambda_j & 0 \\ 0 & 0 & \boldsymbol{\beta} \end{bmatrix}, \tag{7-24}$$

where $\boldsymbol{\beta} = \|\hat{\mathbf{v}}'_i \mathbf{A} \hat{\mathbf{v}}_s\|$ is a symmetric matrix of order $n - 2$. Then $|\mathbf{A}_2 - \lambda \mathbf{I}| = (\lambda_j - \lambda)^2 |\boldsymbol{\beta} - \lambda \mathbf{I}_{n-2}|$, and since $k \geq 3$, $|\boldsymbol{\beta} - \lambda_j \mathbf{I}_{n-2}| = 0$, and all minors of $\mathbf{A}_2 - \lambda_j \mathbf{I}$ of order $n - 2$ vanish. Hence the nullity of $\mathbf{A} - \lambda_j \mathbf{I}$ is ≥ 3, and there exists an eigenvector with eigenvalue λ_j which is orthogonal to $\mathbf{u}_j, \hat{\mathbf{u}}_j$. In this way one shows that if the multiplicity of λ_j is k, there exist at least k orthonormal eigenvectors with eigenvalue λ_j.

Now it is also true that there cannot be more than k orthonormal eigenvectors with eigenvalue λ_j if λ_j has multiplicity k. This follows because each eigenvector corresponding to λ_j is orthogonal to every other eigenvector corresponding to an eigenvalue different from λ_j. The foregoing results have shown that if we sum over all different eigenvalues, we obtain an orthonormal set containing at least n vectors. However, there cannot be more than n orthonormal vectors in E^n, and hence the eigenvectors of an eigenvalue of multiplicity k span a k-dimensional subspace of E^n.

In summary, we have proved:

(1) The eigenvectors of an nth-order symmetric matrix $\mathbf{A}$ span E^n.

(2) There exists at least one orthonormal set of eigenvectors of $\mathbf{A}$ which span E^n.

(3) If an eigenvalue λ_j has multiplicity k, there will be exactly k eigenvectors with eigenvalue λ_j in any set of n orthonormal eigenvectors of $\mathbf{A}$.

(4) If an eigenvalue λ_j has multiplicity k, the eigenvectors corresponding to λ_j span a subspace of E^n, of dimension k.

(5) If one or more eigenvalues have multiplicity $k \geq 2$, there will be an infinite number of different sets of orthonormal eigenvectors of $\mathbf{A}$ which span E^n, corresponding to the different ways of selecting orthonormal sets to span the subspaces with dimension $k \geq 2$.

EXAMPLE: Find a set of three orthonormal eigenvectors for the symmetric matrix

$$\mathbf{A} = \begin{bmatrix} 3 & 0 & 0 \\ 0 & 4 & \sqrt{3} \\ 0 & \sqrt{3} & 6 \end{bmatrix}.$$

The characteristic polynomial is

$$f(\lambda) = (3 - \lambda)[(4 - \lambda)(6 - \lambda) - 3] = (3 - \lambda)(\lambda^2 - 10\lambda + 21),$$

and the eigenvalues are

$$\lambda = 3 \quad \text{(twice)}, \quad 7.$$

The eigenvalue 3 has multiplicity 2. Write $\lambda_1 = 7$, $\lambda_2 = 3$.

There will be only one linearly independent eigenvector with eigenvalue 7; it is a solution to $(\mathbf{A} - 7\mathbf{I})\mathbf{x} = \mathbf{0}$, that is,

$$-4x_1 = 0,$$
$$-3x_2 + \sqrt{3}\, x_3 = 0,$$
$$\sqrt{3}\, x_2 - x_3 = 0,$$

whence

$$x_1 = 0, \qquad x_2 = \frac{1}{\sqrt{3}} x_3.$$

If $\sum x_i^2 = 1$, then $(4/3)x_3^2 = 1$; choosing the positive square root, we obtain

$$x_3 = \frac{\sqrt{3}}{2}, \qquad x_2 = \frac{1}{2}.$$

Thus
$$\mathbf{u}_1 = \left[0, \frac{1}{2}, \frac{\sqrt{3}}{2}\right]$$
is an eigenvector of unit length with eigenvalue 7. The only other eigenvector of unit length with eigenvalue 7 is $-\mathbf{u}_1$.

Since $\lambda_2 = 3$ has multiplicity 2, there should be two orthonormal eigenvectors with eigenvalue λ_2. In fact, there should exist an infinite number of sets of two orthonormal eigenvectors with eigenvalue 3. Any eigenvector corresponding to λ_2 must satisfy the set of equations $(\mathbf{A} - 3\mathbf{I})\mathbf{x} = \mathbf{0}$, that is,
$$x_2 + \sqrt{3}\, x_3 = 0, \qquad \sqrt{3}\, x_2 + 3x_3 = 0,$$
or
$$x_2 = -\sqrt{3}\, x_3, \qquad x_1 \text{ arbitrary.} \tag{7-25}$$

If $\sum x_i^2 = 1$, then $x_1^2 + 4x_3^2 = 1$.

Let us choose $x_1 = 0$: Then, taking the positive square root, we obtain
$$x_3 = \frac{1}{2}, \qquad x_2 = -\frac{\sqrt{3}}{2};$$
hence
$$\mathbf{u}_2 = \left[0, -\frac{\sqrt{3}}{2}, \frac{1}{2}\right]$$
is an eigenvector with eigenvalue λ_2.

We now wish to find another eigenvector with eigenvalue λ_2 which is orthogonal to $\mathbf{u}_2$. We can do this by annexing
$$\mathbf{u}_2' \mathbf{x} = -\frac{\sqrt{3}}{2} x_2 + \frac{1}{2} x_3 = 0 \tag{7-26}$$
to the above set of equations. This procedure will automatically ensure that the new solution is orthogonal to $\mathbf{u}_2$. Equation (7-26) requires that $x_2 = (1/\sqrt{3})x_3$ which, together with (7-25), implies that $x_2 = x_3 = 0$. However, we want $\sum x_i^2 = 1$, and thus $x_1 = \pm 1$. Selecting the positive value, we see that
$$\hat{\mathbf{u}}_2 = [1, 0, 0]$$
is an eigenvector with eigenvalue λ_2 which is orthogonal to $\mathbf{u}_2$. Note that both $\mathbf{u}_2$ and $\hat{\mathbf{u}}_2$ are automatically orthogonal to $\mathbf{u}_1$; hence $\mathbf{u}_1$, $\mathbf{u}_2$, $\hat{\mathbf{u}}_2$ form an orthonormal basis for E_3 (illustrate geometrially).

It is easy to find a different set of orthonormal eigenvectors for $\mathbf{A}$ which span E^n. If we set $x_1 = \frac{1}{2}$ in (7-25), then
$$x_3 = \frac{\sqrt{3}}{4}, \qquad x_2 = -\frac{3}{4};$$

and
$$\mathbf{u}_{2*} = \left[\frac{1}{2}, -\frac{3}{4}, \frac{\sqrt{3}}{4}\right]$$

is an eigenvector with eigenvalue λ_2. Now we annex

$$\mathbf{u}'_{2*}\mathbf{x} = \frac{1}{2}x_1 - \frac{3}{4}x_2 + \frac{\sqrt{3}}{4}x_3 = 0 \tag{7-27}$$

to (7–25) in order to obtain another eigenvector orthogonal to $\mathbf{u}_{2*}$. Substituting $x_2 = -\sqrt{3}\, x_3$ into (7–27), we obtain

$$\tfrac{1}{2}x_1 + \sqrt{3}\, x_3 = 0 \quad \text{or} \quad x_1 = -2\sqrt{3}\, x_3.$$

Since $\sum x_i^2 = 1$, $16x_3^2 = 1$; choosing the positive square root, we obtain

$$x_3 = \frac{1}{4}, \quad x_2 = -\frac{\sqrt{3}}{4}, \quad x_1 = -\frac{\sqrt{3}}{2},$$

whence

$$\hat{\mathbf{u}}_{2*} = \left[-\frac{\sqrt{3}}{2}, -\frac{\sqrt{3}}{4}, \frac{1}{4}\right]$$

is an eigenvector with eigenvalue λ_2 which is orthogonal to $\mathbf{u}_{2*}$; thus the set $\mathbf{u}_1, \mathbf{u}_{2*}, \hat{\mathbf{u}}_{2*}$ is an orthonormal basis (different from $\mathbf{u}_1, \mathbf{u}_2, \hat{\mathbf{u}}_2$) for E^n.

7–5 Diagonalization of symmetric matrices. Let the n eigenvalues of the nth-order symmetric matrix $\mathbf{A}$ be $\lambda_1, \ldots, \lambda_n$. In this listing, an eigenvalue is repeated a number of times equal to its multiplicity. Thus if one eigenvalue has multiplicity k, there will be k eigenvalues λ_j with the same numerical value. In the last section, we saw that each λ_j has a corresponding eigenvector $\mathbf{u}_j$ such that the set $\mathbf{u}_1, \ldots, \mathbf{u}_n$ is an orthonormal basis for E^n. There is at least one such set of $\mathbf{u}_j$, and there may be an infinite number of different sets. Since these eigenvectors $\mathbf{u}_j$ are orthonormal,

$$\mathbf{u}'_i \mathbf{u}_j = \delta_{ij} \qquad (\text{all } i, j). \tag{7-28}$$

Next, let us consider the matrix $\mathbf{Q} = (\mathbf{u}_1, \ldots, \mathbf{u}_n)$ whose columns are an orthonormal set of eigenvectors for $\mathbf{A}$. Matrix $\mathbf{Q}$ has the following property:

$$\mathbf{Q}'\mathbf{Q} = \|\mathbf{u}'_i \mathbf{u}_j\| = \|\delta_{ij}\| = \mathbf{I}, \tag{7-29}$$

whence

$$\mathbf{Q}^{-1} = \mathbf{Q}', \tag{7-30}$$

i.e., the inverse of $\mathbf{Q}$ is the transpose of $\mathbf{Q}$.

ORTHOGONAL MATRIX: *A matrix $\mathbf{Q}$ is called orthogonal if its inverse is its transpose, that is, $\mathbf{Q}^{-1} = \mathbf{Q}'$.*

Now
$$Q'AQ = \|u'_i A u_j\| = \|\lambda_j u'_i u_j\| = \|\lambda_j \delta_{ij}\|. \tag{7-31}$$

Thus $Q'AQ$ is a diagonal matrix whose diagonal elements are the eigenvalues of A. If we write
$$D = \|\lambda_j \delta_{ij}\|, \tag{7-32}$$
then
$$Q^{-1}AQ = Q'AQ = D, \tag{7-33}$$

and A is similar to D. We say that D is obtained by performing an orthogonal similarity transformation on A, and that the similarity transformation diagonalizes A.

We have just proved the important result that *any symmetric matrix A can be diagonalized by an orthogonal similarity transformation. Furthermore, the matrix Q which is used to diagonalize A has as its columns an orthonormal set of eigenvectors for A. The resulting diagonal matrix has as its diagonal elements the eigenvalues of A.*

EXAMPLES: (1) For the matrix
$$A = \begin{bmatrix} 2 & \sqrt{2} \\ \sqrt{2} & 1 \end{bmatrix} \quad \text{(see example p. 241),}$$
$$Q = (u_1, u_2) = \frac{1}{\sqrt{3}} \begin{bmatrix} -1 & \sqrt{2} \\ \sqrt{2} & 1 \end{bmatrix},$$
and
$$Q'AQ = \frac{1}{3} \begin{bmatrix} -1 & \sqrt{2} \\ \sqrt{2} & 1 \end{bmatrix} \begin{bmatrix} 2 & \sqrt{2} \\ \sqrt{2} & 1 \end{bmatrix} \begin{bmatrix} -1 & \sqrt{2} \\ \sqrt{2} & 1 \end{bmatrix}$$
$$= \frac{1}{3} \begin{bmatrix} -1 & \sqrt{2} \\ \sqrt{2} & 1 \end{bmatrix} \begin{bmatrix} 0 & 3\sqrt{2} \\ 0 & 3 \end{bmatrix} = \begin{bmatrix} 0 & 0 \\ 0 & 3 \end{bmatrix} = \begin{bmatrix} \lambda_1 & 0 \\ 0 & \lambda_2 \end{bmatrix}.$$

(2) For the symmetric matrix
$$A = \begin{bmatrix} 3 & 0 & 0 \\ 0 & 4 & \sqrt{3} \\ 0 & \sqrt{3} & 6 \end{bmatrix} \quad \text{(see example p. 245),}$$
$$Q = (u_1, u_2, \hat{u}_2) = \begin{bmatrix} 0 & 0 & 1 \\ \dfrac{1}{2} & -\dfrac{\sqrt{3}}{2} & 0 \\ \dfrac{\sqrt{3}}{2} & \dfrac{1}{2} & 0 \end{bmatrix},$$

$$\mathbf{Q'AQ} = \begin{bmatrix} 0 & \frac{1}{2} & \frac{\sqrt{3}}{2} \\ 0 & -\frac{\sqrt{3}}{2} & \frac{1}{2} \\ 1 & 0 & 0 \end{bmatrix} \begin{bmatrix} 3 & 0 & 0 \\ 0 & 4 & \sqrt{3} \\ 0 & \sqrt{3} & 6 \end{bmatrix} \begin{bmatrix} 0 & 0 & 1 \\ \frac{1}{2} & -\frac{\sqrt{3}}{2} & 0 \\ \frac{\sqrt{3}}{2} & \frac{1}{2} & 0 \end{bmatrix}$$

$$= \begin{bmatrix} 7 & 0 & 0 \\ 0 & 3 & 0 \\ 0 & 0 & 3 \end{bmatrix}.$$

We leave it for the reader to show that a similarity transformation involving the matrix $\mathbf{Q}_1 = (\mathbf{u}_1, \mathbf{u}_{2*}, \hat{\mathbf{u}}_{2*})$ also diagonalizes $\mathbf{A}$.

7–6 Characteristic value problems for nonsymmetric matrices. We shall not discuss in any great detail the theory of characteristic value problems for nonsymmetric matrices. The theory for nonsymmetric matrices is not nearly so simple as that for symmetric matrices. Let us first list the main points of difference between eigenvalue problems of symmetric and nonsymmetric matrices. If the nth-order matrix is not symmetric, then:

(1) It is not necessarily true that all the eigenvalues of $\mathbf{A}$ are real.

(2) It is not necessarily true that eigenvectors corresponding to different eigenvalues are orthogonal.

(3) Even if all eigenvalues are real, the eigenvectors of $\mathbf{A}$ may not span E^n.

(4) If eigenvalue λ_j has multiplicity k, the nullity of $\mathbf{A} - \lambda_j \mathbf{I}$ is not necessarily k.

(5) There may not exist any similarity transformation which diagonalizes $\mathbf{A}$.

Some of the problems will deal with these differences between symmetric and nonsymmetric matrices.

Although eigenvectors corresponding to different eigenvalues of a nonsymmetric matrix need not be orthogonal, they are linearly independent. In fact, any set of eigenvectors for the square matrix $\mathbf{A}$, no two of which correspond to the same eigenvalue, is linearly independent.* The proof is made by assuming that such a set is linearly dependent and by obtaining a contradiction. Let $\mathbf{x}_1, \ldots, \mathbf{x}_s$ be a set of eigenvectors for $\mathbf{A}$ such that $\mathbf{x}_j$ has eigenvalue λ_j, and no two λ_j are equal. Suppose that the set of $\mathbf{x}_j$ is linearly dependent, and that the matrix $\mathbf{X} = (\mathbf{x}_1, \ldots, \mathbf{x}_s)$ has rank $r < s$.

---

* It should be noted that this proof and the next are valid even if the eigenvalues are complex.

Number the eigenvectors so that $x_1, \ldots, x_r$ are linearly independent. Any x_j ($j = r + 1, \ldots, s$) can be written

$$x_j = \sum_{i=1}^{r} \alpha_i x_i, \qquad (7\text{--}34)$$

and at least one $\alpha_i \neq 0$ since $x_j \neq 0$. Then

$$A x_j = \sum_{i=1}^{r} \alpha_i A x_i,$$

or

$$\lambda_j x_j = \sum_{i=1}^{r} \alpha_i \lambda_i x_i. \qquad (7\text{--}35)$$

Multiply (7–34) by λ_j and subtract the result from (7–35) to obtain

$$0 = \sum_{i=1}^{r} \alpha_i (\lambda_i - \lambda_j) x_i. \qquad (7\text{--}36)$$

However, $\lambda_i - \lambda_j \neq 0$ for $j = r + 1, \ldots, s$, and at least one $\alpha_i \neq 0$. Thus Eq. (7–36) indicates that $x_1, \ldots, x_r$ are linearly dependent, which contradicts our original assumption. Therefore the set $x_1, \ldots, x_s$ cannot be linearly dependent.

Next we shall show that if an nth-order matrix A has n linearly independent eigenvectors, then there exists a similarity transformation which diagonalizes A. In fact, if $X = (x_1, \ldots, x_n)$ is a matrix whose columns are a set of n linearly independent eigenvectors, $X^{-1}AX$ is a diagonal matrix whose diagonal elements are the eigenvalues of A.

To prove this, let x_j be an eigenvector with eigenvalue λ_j (not all λ_j are necessarily different). Also write $D = \|\lambda_j \delta_{ij}\|$. Then

$$XD = (\lambda_1 X e_1, \lambda_2 X e_2, \ldots, \lambda_n X e_n) = (\lambda_1 x_1, \lambda_2 x_2, \ldots, \lambda_n x_n), \qquad (7\text{--}37)$$

and

$$AX = (A x_1, \ldots, A x_n) = (\lambda_1 x_1, \ldots, \lambda_n x_n).$$

Hence

$$AX = XD, \qquad (7\text{--}38)$$

or

$$D = X^{-1} A X.$$

Thus A is similar to the diagonal matrix D whose diagonal elements are the eigenvalues of A. If A is not symmetric, the matrix X is not in general an orthogonal matrix.

The above result shows that if the eigenvalues of **A** are all different, **A** can always be diagonalized by a similarity transformation. If the eigenvalues of **A** are not all different, **A** can be diagonalized if it has n linearly independent eigenvectors. If **A** does not have n linearly independent eigenvectors, **A** cannot be diagonalized by a similarity transformation (proof?). However, by a similarity transformation, any square matrix **A** can always be converted into a matrix with the following properties:

(1) All elements below the main diagonal vanish.

(2) The elements on the main diagonal are the eigenvalues of **A**, and equal eigenvalues appear in adjacent positions on the diagonal.

(3) The only elements above the main diagonal which do not vanish are those whose column index j is equal to $i+1$, where i is the row index. Any such nonvanishing element has the value unity. However, it can have the value unity only if the diagonal elements in positions i and $i+1$ are equal. Thus a 5th-order nonsymmetric matrix with $\lambda_1 = \lambda_2 = \lambda_3$, $\lambda_4 = \lambda_5$, $\lambda_4 \neq \lambda_3$ could be reduced to the unique form

$$\begin{bmatrix} \lambda_1 & \beta_1 & 0 & 0 & 0 \\ 0 & \lambda_2 & \beta_2 & 0 & 0 \\ 0 & 0 & \lambda_3 & 0 & 0 \\ 0 & 0 & 0 & \lambda_4 & \beta_3 \\ 0 & 0 & 0 & 0 & \lambda_5 \end{bmatrix},$$

where the value of the β_i is either 0 or 1. This is called the Jordan canonical form for the matrix **A**. We shall not attempt to prove the existence of a similarity transformation which will reduce a matrix to its Jordan canonical form.

7–7 Quadratic forms. The techniques of linear algebra are often useful in dealing with nonlinear expressions, such as, e.g., quadratic forms. A quadratic form in n variables $x_1, \ldots, x_n$ is an expression

$$F = \sum_{i=1}^{n} \sum_{j=1}^{n} a_{ij} x_i x_j = a_{11} x_1 x_1 + a_{12} x_1 x_2 + \cdots + a_{1n} x_1 x_n$$
$$+ a_{21} x_2 x_1 + \cdots + a_{2n} x_2 x_n + \cdots + a_{n1} x_n x_1 + \cdots + a_{nn} x_n x_n$$
(7–39)

which is a numerical function of the n variables. Equation (7–39) determines a unique value of F for any set of x_j. It is called a "quadratic" form because each term $a_{ij} x_i x_j$ contains either the square of a variable or the product of two different variables. Quadratic forms are important in (1) deriving sufficient conditions for maxima and minima in analysis; (2) quadratic programming; (3) approximating functions of n variables in the neighborhood of some point (this technique is used in statistics, for example, in response-surface analysis).

If we write $\mathbf{x} = [x_1, \ldots, x_n]$, $\mathbf{A} = \|a_{ij}\|$, Eq. (7–39) can be expressed in the form of

$$F = \sum_{i=1}^{n} x_i \sum_{j=1}^{n} a_{ij} x_j = \sum_{i=1}^{n} x_i (\mathbf{Ax})_i = \mathbf{x}'\mathbf{Ax}; \qquad (7\text{--}40)$$

i.e., if we use matrix notation, a quadratic form can be written $\mathbf{x}'\mathbf{Ax}$, and $\mathbf{A}$ is said to be the matrix associated with the form.

It should be noted that a_{ij}, a_{ji} [Eq. (7–39)] are both coefficients of $x_i x_j$ when $i \neq j$ (since $x_i x_j = x_j x_i$), that is, the coefficient of $x_i x_j$ is $a_{ij} + a_{ji}$ ($i \neq j$). If $a_{ij} \neq a_{ji}$, we can uniquely define new coefficients

$$b_{ij} = b_{ji} = \frac{a_{ij} + a_{ji}}{2} \qquad \text{(all } i, j\text{)}, \qquad (7\text{--}41)$$

so that $b_{ij} + b_{ji} = a_{ij} + a_{ji}$, and $\mathbf{B} = \|b_{ij}\| = \mathbf{B}'$; hence $\mathbf{B}$ is a symmetric matrix. This redefinition of the coefficients does not change the value F for any $\mathbf{x}$. Thus we can always assume that the matrix $\mathbf{A}$ associated with the quadratic form $\mathbf{x}'\mathbf{Ax}$ [Eq. (7–40)] is symmetric; if it is not, (7–41) can be used to convert it into a symmetric matrix. Note that the element a_{ij} of $\mathbf{A}$ is the coefficient of $x_i x_j$ in (7–39), so that if any $a_{ij} = 0$, the corresponding product of the variables $x_i x_j$ does not appear in the quadratic form.

EXAMPLES: (1) A quadratic form in one variable is the expression ax_1^2. (2) The most general quadratic form in two variables is

$$a_{11} x_1^2 + 2a_{12} x_1 x_2 + a_{22} x_2^2;$$

in matrix form, this can be written

$$(x_1, x_2) \begin{bmatrix} a_{11} & a_{12} \\ a_{12} & a_{22} \end{bmatrix} \begin{bmatrix} x_1 \\ x_2 \end{bmatrix}.$$

(3) The matrix

$$\begin{bmatrix} 4 & 2 \\ 4 & 6 \end{bmatrix}$$

associated with the quadratic form

$$4x_1^2 + 2x_1 x_2 + 4x_2 x_1 + 6x_2^2 = 4x_1^2 + 6x_1 x_2 + 6x_2^2$$

is not symmetric. However, without changing the value of the form, we can write $a_{12} = a_{21} = 3$ and obtain the symmetric matrix

$$\begin{bmatrix} 4 & 3 \\ 3 & 6 \end{bmatrix}.$$

7–8 Change of variables. It is often possible to simplify a quadratic form $\mathbf{x'Ax}$ by a change of variables $\mathbf{x} = \mathbf{Ry}$ or $\mathbf{y} = \mathbf{R^{-1}x}$, where $\mathbf{R}$ is, of course, a nonsingular matrix. We shall restrict ourselves to nonsingular transformations because these alone are one-to-one transformations; that is, a given $\mathbf{x}$ determines a unique $\mathbf{y}$, and a given $\mathbf{y}$ determines a unique $\mathbf{x}$. Such transformations are invertible, and we can go from $\mathbf{x}$ to $\mathbf{y}$ or from $\mathbf{y}$ to $\mathbf{x}$. Substitution of $\mathbf{x} = \mathbf{Ry}$ into $F = \mathbf{x'Ax}$ gives

$$F = (\mathbf{Ry})'\mathbf{ARy} = \mathbf{y'R'ARy} = \mathbf{y'By}, \qquad (7\text{–}42)$$

where $\mathbf{B} = \mathbf{R'AR}$. In terms of the new variables $\mathbf{y}$, the form $\mathbf{x'Ax}$ becomes $\mathbf{y'By}$, and $\mathbf{B} = \mathbf{R'AR}$. Note that if $\mathbf{A}$ is a symmetric matrix, $\mathbf{B}$ is also symmetric.

CONGRUENCE: *A square matrix* $\mathbf{B}$ *is said to be congruent to the square matrix* $\mathbf{A}$ *if there exists a nonsingular matrix* $\mathbf{R}$ *such that* $\mathbf{B} = \mathbf{R'AR}$.

If $\mathbf{B}$ is congruent to $\mathbf{A}$, then we say that $\mathbf{B}$ can be obtained by a congruence transformation* on $\mathbf{A}$. A congruence transformation is a special case of an equivalence transformation; i.e., if $\mathbf{B}$ is congruent to $\mathbf{A}$, $\mathbf{B}$ is equivalent to $\mathbf{A}$. The matrix $\mathbf{B}$ of the quadratic form $\mathbf{y'By}$ obtained by the nonsingular transformation of the variables $\mathbf{x} = \mathbf{Ry}$ in the form $\mathbf{x'Ax}$ is congruent to $\mathbf{A}$.

The determinant $|\mathbf{A}|$ is called the discriminant of the quadratic form $\mathbf{x'Ax}$. If $\mathbf{B} = \mathbf{R'AR}$ is congruent to $\mathbf{A}$, then the discriminant of the form $\mathbf{y'By}$ is

$$|\mathbf{B}| = |\mathbf{R'}|\,|\mathbf{A}|\,|\mathbf{R}| = |\mathbf{R}|^2|\mathbf{A}|;$$

that is, under a nonsingular change of the variables $\mathbf{x} = \mathbf{Ry}$, the discriminant of the new quadratic form assumes a magnitude of $|\mathbf{R}|^2$ times that of the original form. The determinant $|\mathbf{R}|$ is sometimes called the modulus of the transformation $\mathbf{x} = \mathbf{Ry}$. Note that the modulus of the transformation $\mathbf{y} = \mathbf{R^{-1}x}$ is the reciprocal of that of $\mathbf{x} = \mathbf{Ry}$ since $\mathbf{RR^{-1}} = \mathbf{I}$, and hence $|\mathbf{R^{-1}}| = 1/|\mathbf{R}|$.

---

* Note that if $\mathbf{B} = \mathbf{R'AR}$, then $\mathbf{B} = \mathbf{SAS'}$, where $\mathbf{S} = \mathbf{R'}$. Thus either $\mathbf{R'AR}$ or $\mathbf{RAR'}$ can be used to define congruence and a congruence transformation.

If we allow $\mathbf{x}$ to vary over all of E^n, then the set of values taken on by $F = \mathbf{x}'\mathbf{A}\mathbf{x}$ is called the range of the quadratic form. *Under a nonsingular transformation of variables, the range of a quadratic form remains unchanged.* To prove this statement, let us suppose that we have the form $\mathbf{x}'\mathbf{A}\mathbf{x}$ and make the change of variables $\mathbf{x} = \mathbf{R}\mathbf{y}$ or $\mathbf{y} = \mathbf{R}^{-1}\mathbf{x}$ to obtain the new form $\mathbf{y}'\mathbf{R}'\mathbf{A}\mathbf{R}\mathbf{y} = \mathbf{y}'\mathbf{B}\mathbf{y}$. Now it is only necessary to note that for any $\mathbf{x}$ there is a unique $\mathbf{y}$ (and, similarly, for any $\mathbf{y}$ there is a unique $\mathbf{x}$) such that

$$F = \mathbf{x}'\mathbf{A}\mathbf{x} = \mathbf{y}'\mathbf{B}\mathbf{y}. \qquad (7\text{--}43)$$

Hence $\mathbf{x}'\mathbf{A}\mathbf{x}$ and $\mathbf{y}'\mathbf{B}\mathbf{y}$ must have the same range. In general, this property will not hold if the matrix $\mathbf{R}$ is singular. For example, if $\mathbf{R} = \mathbf{0}$, the range of $\mathbf{y}'\mathbf{B}\mathbf{y}$ contains only the single number 0.

7–9 Definite quadratic forms. Some quadratic forms have the property $\mathbf{x}'\mathbf{A}\mathbf{x} > 0$ for all $\mathbf{x}$ except $\mathbf{x} = \mathbf{0}$; some are negative for all $\mathbf{x}$ except $\mathbf{x} = \mathbf{0}$; and some can assume both positive and negative values. We introduce the following definitions:

POSITIVE DEFINITE QUADRATIC FORM: *The quadratic form $\mathbf{x}'\mathbf{A}\mathbf{x}$ is said to be positive definite if it is positive (>0) for every $\mathbf{x}$ except $\mathbf{x} = \mathbf{0}$.*

POSITIVE SEMIDEFINITE QUADRATIC FORM: *The quadratic form $\mathbf{x}'\mathbf{A}\mathbf{x}$ is said to be positive semidefinite if it is non-negative (≥ 0) for every $\mathbf{x}$, and there exist points $\mathbf{x} \neq \mathbf{0}$ for which $\mathbf{x}'\mathbf{A}\mathbf{x} = 0$.*

Negative definite and semidefinite forms are defined by interchanging the words "negative" and "positive" in the above definitions. If $\mathbf{x}'\mathbf{A}\mathbf{x}$ is positive definite (semidefinite), then $\mathbf{x}'(-\mathbf{A})\mathbf{x}$ is negative definite (semidefinite).

INDEFINITE FORMS: *A quadratic form $\mathbf{x}'\mathbf{A}\mathbf{x}$ is said to be indefinite if the form is positive for some points $\mathbf{x}$ and negative for others.*

A symmetric matrix $\mathbf{A}$ is often said to be positive definite, positive semidefinite, negative definite, etc., if the respective quadratic form $\mathbf{x}'\mathbf{A}\mathbf{x}$ is positive definite, positive semidefinite, negative definite, etc.

EXAMPLES: (1) $F = 3x_1^2 + 5x_2^2$, $F = 2x_1^2 + 3x_2^2 + x_3^2$, $F = x_1^2$ are positive definite forms in two, three, and one variable, respectively.

(2) $F = 4x_1^2 + x_2^2 - 4x_1x_2 + 3x_3^2 = (2x_1 - x_2)^2 + 3x_3^2$ is positive semidefinite since it is never negative and vanishes if $x_2 = 2x_1$, $x_3 = 0$.

(3) $F = -2x_1^2 - x_2^2$, $F = -x_1^2 - x_2^2$, $F = -x_1^2$ are negative definite forms in two, two, and one variable, respectively.

(4) $F = 4x_1^2 - 3x_2^2$ is indefinite since it is positive when $x_1 = 1$, $x_2 = 1$ and negative when $x_1 = 0$, $x_2 = 1$.

A *positive (negative) definite form remains positive (negative) definite when expressed in terms of a new set of variables provided the transformation of the variables is nonsingular.* Thus if $x'Ax$ is positive (negative) definite and R is nonsingular, then $y'R'ARy = y'By$ ($x = Ry$) is positive (negative) definite. The proof is simple: Since we know that the range of $y'By$ is the same as that of $x'Ax$, it is only necessary to show that $y = 0$ is the only y for which $y'By = 0$. Now the form $x'Ax = 0$ only if $x = 0$. However, $y = R^{-1}x$ and hence $y = 0$ is the only value of y for which $x = 0$. Of course, semidefinite and indefinite forms remain semidefinite and indefinite, respectively, under a nonsingular transformation of variables.

7–10 Diagonalization of quadratic forms. Given a quadratic form $x'Ax$, let us consider the nonsingular transformation of variables $x = Qy$, where the columns of matrix Q are an orthonormal set of eigenvectors for A. The matrix Q is therefore an orthogonal matrix, and the transformation of variables is called an orthogonal transformation. In terms of the variables y, the quadratic form becomes

$$y'Q'AQy = y'Dy,$$

and $D = \|\lambda_j \delta_{ij}\|$ is a diagonal matrix whose diagonal elements are the eigenvalues of A, in agreement with Section 7–5, where we showed* that $D = Q'AQ$. Thus

$$y'Dy = \sum_{j=1}^{n} \lambda_j y_j^2. \qquad (7\text{–}44)$$

Only the squares of the variables appear; there are no cross products $y_i y_j$ ($i \neq j$).

A quadratic form containing only the squares of the variables is said to be *in diagonal form*. Furthermore, we say that the transformation of variables $x = Qy$ has diagonalized the quadratic form $x'Ax$. A quadratic form will be in diagonal form if the matrix associated with the form is a diagonal matrix. We have proved that *by an orthogonal transformation of variables every quadratic form $x'Ax$ may be reduced to a diagonal form* (7–44). *Furthermore, in the transformation of variables $x = Qy$, the matrix Q has as its columns a set of orthonormal eigenvectors of A which span E^n*. In the diagonal form (7–44), the coefficient of y_j^2 is the eigenvalue λ_j of A. It should be noted at this point that it does not automatically follow that if a quadratic form $x'Ax$ has been reduced to diagonal form by a change of

---

* If Q is an orthogonal matrix, a similarity transformation $Q^{-1}AQ$ is also a congruence transformation $Q'AQ$.

variables $\mathbf{x} = \mathbf{Ry}$, the coefficients of the y_i^2 are the eigenvalues of $\mathbf{A}$. Later a transformation of variables in which $\mathbf{R}$ is not orthogonal will be introduced which will diagonalize the form; however, the coefficients of the y_i^2 will not, in general, be the eigenvalues of $\mathbf{A}$.

If we know the eigenvalues of $\mathbf{A}$, we can immediately determine whether the form $\mathbf{x}'\mathbf{Ax}$ is positive definite, indefinite, etc. We can do this because: (a) the range of a form is unchanged under a nonsingular transformation of variables; (b) a positive or negative definite form remains positive or negative definite under a nonsingular transformation of variables; (c) the transformation $\mathbf{x} = \mathbf{Qy}$ discussed above reduces $\mathbf{x}'\mathbf{Ax}$ to the diagonal form (7–44).

If each eigenvalue of $\mathbf{A}$ is positive (negative), then the only value of $\mathbf{y}$ for which (7–44) vanishes is $\mathbf{y} = \mathbf{0}$. Hence (7–44) is positive (negative) definite and by (b) $\mathbf{x}'\mathbf{Ax}$ is positive or negative definite. Suppose that all the eigenvalues of $\mathbf{A}$ are non-negative (nonpositive), but one or more of the eigenvalues are zero, say $\lambda_n = 0$. Then (7–44) will always be non-negative (nonpositive). However, if we set $y_1 = y_2 = \cdots = y_{n-1} = 0$, Eq. (7–44) vanishes for any value of y_n; hence there exist $\mathbf{y} \neq \mathbf{0}$ for which (7–44) is zero. For any $\mathbf{y} \neq \mathbf{0}$, there is an $\mathbf{x} = \mathbf{Qy} \neq \mathbf{0}$ such that $\mathbf{x}'\mathbf{Ax} = 0$, and by (a) the form $\mathbf{x}'\mathbf{Ax}$ is positive (negative) semidefinite. If $\mathbf{A}$ has both positive and negative eigenvalues, (7–44) is indefinite and by (a) $\mathbf{x}'\mathbf{Ax}$ is indefinite. These results show that:

(1) $\mathbf{x}'\mathbf{Ax}$ *is positive (negative) definite if and only if every eigenvalue of* $\mathbf{A}$ *is positive (negative)*.

(2) $\mathbf{x}'\mathbf{Ax}$ *is positive (negative) semidefinite if and only if all eigenvalues of* $\mathbf{A}$ *are non-negative (nonpositive), and at least one of the eigenvalues vanishes*.

(3) $\mathbf{x}'\mathbf{Ax}$ *is indefinite if and only if* $\mathbf{A}$ *has both positive and negative eigenvalues*.

EXAMPLE: Consider the quadratic form

$$F = 2x_1^2 + 2\sqrt{2}\, x_1 x_2 + x_2^2 = \mathbf{x}'\mathbf{Ax}; \qquad (7\text{–}45)$$

the symmetric matrix $\mathbf{A}$ is then

$$\mathbf{A} = \begin{bmatrix} 2 & \sqrt{2} \\ \sqrt{2} & 1 \end{bmatrix}.$$

To diagonalize the form, we use the transformation of variables $\mathbf{x} = \mathbf{Qy}$, where the columns of $\mathbf{Q}$ are the orthonormal eigenvectors of $\mathbf{A}$. The eigenvectors of $\mathbf{A}$ (Section 7–3, p. 242) are

$$\mathbf{u}_1 = \left[-\frac{1}{\sqrt{3}}, \sqrt{\frac{2}{3}}\right] \quad \text{and} \quad \mathbf{u}_2 = \left[\sqrt{\frac{2}{3}}, \frac{1}{\sqrt{3}}\right].$$

Thus
$$\mathbf{Q} = \frac{1}{\sqrt{3}} \begin{bmatrix} -1 & \sqrt{2} \\ \sqrt{2} & 1 \end{bmatrix},$$
and the transformation of variables is

$$x_1 = \frac{1}{\sqrt{3}}(-y_1 + \sqrt{2}\, y_2), \qquad y_1 = \frac{1}{\sqrt{3}}(-x_1 + \sqrt{2}\, x_2),$$
$$x_2 = \frac{1}{\sqrt{3}}(\sqrt{2}\, y_1 + y_2), \qquad y_2 = \frac{1}{\sqrt{3}}(\sqrt{2}\, x_1 + x_2). \tag{7-46}$$

Note that $\mathbf{Q}^{-1} = \mathbf{Q}'$; hence it is very easy to find the inverse transformation for an orthogonal transformation of variables. Since the eigenvalues of $\mathbf{A}$ are $\lambda_1 = 0$, $\lambda_2 = 3$, the form F becomes $F = 3y_2^2$ under this transformation of variables. The form is therefore positive semidefinite. The point $\mathbf{y} = [2, 0]$ causes the form to vanish. The $\mathbf{x}$ corresponding to this $\mathbf{y}$ is $\mathbf{x} = [-2/\sqrt{3}, 2\sqrt{2/3}]$, and of course, (7-45) vanishes for this value of $\mathbf{x}$. It is suggested that the reader introduce the new variables into (7-45) by direct substitution, and show that F reduces to $F = 3y_2^2$.

7-11 Diagonalization by completion of the square. Another procedure for diagonalizing quadratic forms is a generalization of the familiar technique of completing the square, learned in elementary algebra. Consider the quadratic form in two variables

$$F = a_{11}x_1^2 + 2a_{12}x_1x_2 + a_{22}x_2^2. \tag{7-47}$$

If either a_{11} or a_{22} is not zero, we can assume without loss of generality that a_{11} is not zero. Then (7-47) can be written

$$F = a_{11}\left[x_1^2 + \frac{2a_{12}}{a_{11}}x_1x_2 + \left(\frac{a_{12}}{a_{11}}\right)^2 x_2^2 - \left(\frac{a_{12}}{a_{11}}\right)^2 x_2^2 + \frac{a_{22}}{a_{11}}x_2^2\right]$$
$$= a_{11}\left\{\left(x_1 + \frac{a_{12}}{a_{11}}x_2\right)^2 + \left[\frac{a_{22}}{a_{11}} - \left(\frac{a_{12}}{a_{11}}\right)^2\right]x_2^2\right\}. \tag{7-48}$$

If we introduce the transformation of variables,

$$\begin{aligned} y_1 &= x_1 + \frac{a_{12}}{a_{11}}x_2, \\ y_2 &= x_2, \end{aligned} \quad \text{or} \quad \mathbf{y} = \mathbf{S}\mathbf{x} = \begin{bmatrix} 1 & \frac{a_{12}}{a_{11}} \\ 0 & 1 \end{bmatrix}\mathbf{x}, \tag{7-49}$$

(7-48) becomes

$$F = a_{11}y_1^2 + \left[a_{22} - \frac{a_{12}^2}{a_{11}}\right]y_2^2, \tag{7-50}$$

and the form (7–47) has been diagonalized. The transformation of variables is nonsingular ($|\mathbf{S}| = 1$), but it is not orthogonal. The coefficients of y_1^2, y_2^2 in (7–50) are not, in general, the eigenvalues of $\mathbf{A}$.

In the event that a_{11}, a_{22} both vanish, the above procedure will not work. When $a_{11} = a_{22} = 0$, (7–47) becomes

$$F = 2a_{12}x_1x_2. \tag{7-51}$$

Now make the transformation

$$\begin{aligned}x_1 &= y_1 + y_2, \\ x_2 &= y_1 - y_2,\end{aligned} \quad \text{or} \quad \mathbf{x} = \begin{bmatrix} 1 & 1 \\ 1 & -1 \end{bmatrix}\mathbf{y}.$$

This is a nonsingular transformation which reduces (7–51) to

$$F = 2a_{12}(y_1^2 - y_2^2).$$

Hence in this case also the form has been diagonalized. The procedure just outlined can be generalized to diagonalize any quadratic form.

In the text, we shall discuss only reductions for positive definite and negative definite forms, while the generalization to arbitrary forms will be the subject of some of the problems. Let

$$F = \mathbf{x}'\mathbf{A}\mathbf{x} = \sum_{i,j} a_{ij}x_ix_j \tag{7-52}$$

be a positive definite quadratic form. The terms in (7–52) involving x_1 are

$$a_{11}x_1^2 + 2a_{12}x_1x_2 + \cdots + 2a_{1n}x_1x_n. \tag{7-53}$$

Since the form is positive definite, it must be positive when $x_2 = x_3 = \cdots = x_n = 0$ and $x_1 \neq 0$. Then $F = a_{11}x_1^2$, and a_{11} must be positive. Hence (7–53) can be written

$$a_{11}\left(x_1^2 + 2\sum_{k=2}^{n}\frac{a_{1k}}{a_{11}}x_1x_k\right)$$

$$= a_{11}\left[x_1^2 + 2\sum_{k=2}^{n}\frac{a_{1k}}{a_{11}}x_1x_k + \left(\sum_{k=2}^{n}\frac{a_{1k}}{a_{11}}x_k\right)^2 - \left(\sum_{k=2}^{n}\frac{a_{1k}}{a_{11}}x_k\right)^2\right]$$

$$= a_{11}\left[\left(x_1 + \sum_{k=2}^{n}\frac{a_{1k}}{a_{11}}x_k\right)^2 - \left(\sum_{k=2}^{n}\frac{a_{1k}}{a_{11}}x_k\right)^2\right], \tag{7-54}$$

and the following transformation of variables suggests itself:

$$v_1 = x_1 + \sum_{k=2}^{n}\frac{a_{1k}}{a_{11}}x_k, \quad v_2 = x_2, \ldots, v_n = x_n, \tag{7-55}$$

or $\mathbf{v} = \mathbf{S}_1\mathbf{x}$, where

$$\mathbf{S}_1 = \begin{bmatrix} 1 & \frac{a_{12}}{a_{11}} & \cdots & \frac{a_{1n}}{a_{11}} \\ 0 & 1 & \cdots & 0 \\ \vdots & & & \vdots \\ 0 & 0 & \cdots & 1 \end{bmatrix} \quad \text{and} \quad |\mathbf{S}_1| = 1.$$

Thus a nonsingular transformation of variables has reduced F to

$$F = a_{11}v_1^2 + \sum_{i,j=2}^{n} b_{ij}v_iv_j, \tag{7-56}$$

and $a_{11} > 0$. Furthermore, (7-56) is positive definite.

This procedure is now repeated. Since (7-56) is positive definite, $b_{22} > 0$. We complete the square for the variable v_2, define another transformation of variables,

$$w_1 = v_1, \quad w_2 = v_2 + \sum_{k=3}^{n} \frac{b_{2k}}{b_{22}} v_k, \quad w_3 = v_3, \ldots, w_n = v_n, \tag{7-57}$$

or $\mathbf{w} = \mathbf{S}_2\mathbf{v}$, $|\mathbf{S}_2| = 1$, and obtain the form

$$F = a_{11}w_1^2 + b_{22}w_2^2 + \sum_{i,j=3}^{n} c_{ij}w_iw_j, \tag{7-58}$$

with $a_{11}, b_{22} > 0$. Repeating this process $n-1$ times, we arrive at

$$F = a_{11}y_1^2 + b_{22}y_2^2 + c_{33}y_3^2 + \cdots + z_{nn}y_n^2; \tag{7-59}$$

the coefficient of each y_j^2 is positive. The last step yields two square terms, and all cross products disappear. The nonsingular transformation $\mathbf{y} = \mathbf{S}\mathbf{x}$ which reduces (7-52) to (7-59) is

$$\mathbf{S} = \mathbf{S}_{n-1}\mathbf{S}_{n-2}\cdots\mathbf{S}_2\mathbf{S}_1, \tag{7-60}$$

and $|\mathbf{S}| = 1$ since each $|\mathbf{S}_i| = 1$.

The same procedure can be used to diagonalize a negative definite form. It should be noted that $\mathbf{S}$ is a triangular matrix, i.e., all elements below the main diagonal vanish; hence $\mathbf{S}$ is not orthogonal. The elements a_{11}, b_{22}, etc., in (7-59) are not, in general, the eigenvalues of $\mathbf{A}$.

7-12 Another set of necessary and sufficient conditions for positive and negative definite forms. Since we often wish to establish whether a quadratic form $\mathbf{x}'\mathbf{A}\mathbf{x}$ is positive definite without determining the eigenvalues of $\mathbf{A}$, we shall now develop another set of conditions that will enable us to make such a decision. If the order n of **A** is large, these conditions are not easily applicable to practical problems, but they are useful in theoretical work.

A set of necessary and sufficient conditions for the form $\mathbf{x'Ax}$ *to be positive definite is*

$$a_{11} > 0; \quad \begin{vmatrix} a_{11} & a_{12} \\ a_{21} & a_{22} \end{vmatrix} > 0; \quad \begin{vmatrix} a_{11} & a_{12} & a_{13} \\ a_{21} & a_{22} & a_{23} \\ a_{31} & a_{32} & a_{33} \end{vmatrix} > 0; \quad \ldots; \quad |\mathbf{A}| > 0. \tag{7-61}$$

If these n minors of **A** are positive, $\mathbf{x'Ax}$ is positive definite; and $\mathbf{x'Ax}$ is positive definite only if these minors are positive.

To prove the necessity, assume that $\mathbf{x'Ax}$ is positive definite. Then there exists a nonsingular transformation (see Section 7-11) $\mathbf{y} = \mathbf{Sx}$ or $\mathbf{x} = \mathbf{Ry}$, $\mathbf{S} = \mathbf{R}^{-1}$ with modulus unity which reduces the form to

$$\mathbf{y'Dy} = \sum_{i=1}^{n} d_i y_i^2 \quad (d_i > 0, \quad i = 1, \ldots, n). \tag{7-62}$$

However, $\mathbf{D} = \mathbf{R'AR}$, and since $|\mathbf{R}| = 1/|\mathbf{S}| = 1$,

$$|\mathbf{D}| = d_1 d_2 \cdots d_n = |\mathbf{R}|^2 |\mathbf{A}| = |\mathbf{A}|; \tag{7-63}$$

thus $|\mathbf{A}| > 0$.

Next we set $x_n = 0$. The resulting form in $n - 1$ variables is also positive definite. The matrix of its coefficients is obtained by crossing off the last row and column of **A**. If this form is diagonalized by the method described in Section 7-11, we obtain $\sum_{i=1}^{n-1} d_i y_i^2$, where the d_i are, in fact, the same as in (7-62), the only difference being that the term $d_n y_n^2$ does not appear. Thus

$$\begin{vmatrix} a_{11} & \cdots & a_{1,n-1} \\ \vdots & & \vdots \\ a_{n-1,1} & \cdots & a_{n-1,n-1} \end{vmatrix} = d_1 d_2 \cdots d_{n-1} > 0. \tag{7-64}$$

If we now set $x_n = x_{n-1} = 0$ in the original form, then $y_n = y_{n-1} = 0$ in (7-62); hence

$$\begin{vmatrix} a_{11} & \cdots & a_{1,n-2} \\ \vdots & & \vdots \\ a_{n-2,1} & \cdots & a_{n-2,n-2} \end{vmatrix} = d_1 d_2 \cdots d_{n-2} > 0. \tag{7-65}$$

Continuing in this way, we see that if $\mathbf{x'Ax}$ is positive definite, Eq. (7-61) holds.

To prove the sufficiency, let us suppose that (7–61) holds for the form $\mathbf{x'Ax}$. We wish to show that $\mathbf{x'Ax}$ is positive definite. Since $a_{11} > 0$, we can perform a nonsingular transformation $\mathbf{x} = \mathbf{R}_1\mathbf{v}$ with unit modulus, of the type discussed in Section 7–11, and obtain

$$a_{11}v_1^2 + \sum_{i,j=2}^{n} b_{ij}v_iv_j. \tag{7-66}$$

To demonstrate that $b_{22} > 0$, we begin by noting that the coefficients b_{ij} are independent of the values of the variables v_i (or x_i). Thus, if we set $x_i = v_i = 0$ $(i = 3, \ldots, n)$, (7–66) becomes a form in two variables whose discriminant is $a_{11}b_{22}$. When the above variables are set to zero, the original form $\mathbf{x'Ax}$ reduces to a form in two variables whose discriminant is

$$\begin{vmatrix} a_{11} & a_{12} \\ a_{21} & a_{22} \end{vmatrix} > 0.$$

The discriminants are equal since the modulus of the transformation is unity. Thus $a_{11}b_{22} > 0$, and $b_{22} > 0$ since $a_{11} > 0$.

Another nonsingular transformation with unit modulus reduces the form (7–66) to

$$a_{11}w_1^2 + b_{22}w_2^2 + \sum_{i,j=3}^{n} c_{ij}w_iw_j. \tag{7-67}$$

Setting $w_i = v_i = x_i = 0$ $(i = 4, \ldots, n)$, we can see that c_{33} is positive. Thus

$$a_{11}b_{22}c_{33} = \begin{vmatrix} a_{11} & a_{12} & a_{13} \\ a_{21} & a_{22} & a_{23} \\ a_{31} & a_{32} & a_{33} \end{vmatrix} > 0,$$

and $c_{33} > 0$ since $a_{11}, b_{22} > 0$. This process can be continued until we obtain

$$\sum_{i=1}^{n} d_iy_i^2 \quad (d_i > 0, \quad i = 1, \ldots, n). \tag{7-68}$$

This form is clearly positive definite. Hence the original form is also positive definite because it may be obtained from (7–68) by a nonsingular transformation of variables.

Equation (7–61) represents only one set of necessary and sufficient conditions for a positive definite form. The variables do not have any specific property that made us call x_1 the first variable (it could have been x_7 or any other variable); that is, we can renumber the variables in any way we choose. Thus the determinants formed by permutations of the subscripts will also serve as a set of necessary and sufficient conditions.

In general,

$$a_{kk} > 0; \quad \begin{vmatrix} a_{kk} & a_{kj} \\ a_{jk} & a_{jj} \end{vmatrix} > 0; \quad \begin{vmatrix} a_{kk} & a_{kj} & a_{ki} \\ a_{jk} & a_{jj} & a_{ji} \\ a_{ik} & a_{ij} & a_{ii} \end{vmatrix} > 0; \quad \ldots, |\mathbf{A}| > 0, \tag{7-69}$$

where $(k, j, i \ldots)$ is any permutation of the set of integers $(1, 2, \ldots, n)$, represents a set of necessary and sufficient conditions ensuring that $\mathbf{x}'\mathbf{A}\mathbf{x}$ will be positive definite. Note that the permutation of subscripts never affects the sign of $|\mathbf{A}|$ since both rows and columns of $\mathbf{A}$ are interchanged in the process.

The determinants in (7-61) are found from $\mathbf{A}$ as follows:

$$\begin{vmatrix} a_{11} & a_{12} & a_{13} & \cdots & a_{1n} \\ a_{21} & a_{22} & a_{23} & \cdots & a_{2n} \\ a_{31} & a_{32} & a_{33} & \cdots & a_{3n} \\ \vdots & & & & \vdots \\ a_{n1} & a_{n2} & a_{n3} & \cdots & a_{nn} \end{vmatrix}.$$

These determinants are called the *naturally ordered principal minors* of $\mathbf{A}$. *A necessary and sufficient condition for the form $\mathbf{x}'\mathbf{A}\mathbf{x}$ or the symmetric matrix $\mathbf{A}$ to be positive definite is that the naturally ordered principal minors of $\mathbf{A}$ are all positive.* It should be observed that if $\mathbf{A}$ is positive definite, $|\mathbf{A}| > 0$, and hence $\mathbf{A}$ is nonsingular.

Using the criteria for positive definite forms, we can easily derive a set of necessary and sufficient conditions ensuring that a form $\mathbf{x}'\mathbf{A}\mathbf{x}$ or the symmetric matrix $\mathbf{A}$ will be negative definite. If $\mathbf{x}'\mathbf{A}\mathbf{x}$ is negative definite, then $\mathbf{x}'(-\mathbf{A})\mathbf{x}$ is positive definite; furthermore, we recall that $|-\mathbf{A}| = (-1)^n|\mathbf{A}|$. A set of necessary and sufficient conditions for $\mathbf{x}'\mathbf{A}\mathbf{x}$ to be negative definite, or equivalently, for $\mathbf{x}'(-\mathbf{A})\mathbf{x}$ to be positive definite then follows immediately from (7-61), i.e.,

$$a_{11} < 0; \quad \begin{vmatrix} a_{11} & a_{12} \\ a_{21} & a_{22} \end{vmatrix} > 0; \quad \begin{vmatrix} a_{11} & a_{12} & a_{13} \\ a_{21} & a_{22} & a_{23} \\ a_{31} & a_{32} & a_{33} \end{vmatrix} < 0; \quad \ldots; \quad (-1)^n|\mathbf{A}| > 0, \tag{7-70}$$

where the a_{ij} are the elements of $\mathbf{A}$ (*not* $-\mathbf{A}$).

EXAMPLE: The form $3x_1^2 + 4x_1x_2 + 2x_2^2$, which can be written

$$(x_1, x_2) \begin{bmatrix} 3 & 2 \\ 2 & 2 \end{bmatrix} \begin{bmatrix} x_1 \\ x_2 \end{bmatrix},$$

is positive definite since

$$a_{11} = 3 > 0; \quad \begin{vmatrix} a_{11} & a_{12} \\ a_{21} & a_{22} \end{vmatrix} = \begin{vmatrix} 3 & 2 \\ 2 & 2 \end{vmatrix} = 2 > 0.$$

If we introduce the transformation of variables (see Section 7–11),

$$y_1 = x_1 + \tfrac{2}{3}x_2, \quad y_2 = x_2 \quad \text{or} \quad x_1 = y_1 - \tfrac{2}{3}y_2, \quad x_2 = y_2,$$

the form becomes

$$3(y_1 - \tfrac{2}{3}y_2)^2 + 4(y_1 - \tfrac{2}{3}y_2)y_2 + 2y_2^2$$
$$= 3y_1^2 - 4y_1y_2 + \tfrac{4}{3}y_2^2 + 4y_1y_2 - \tfrac{8}{3}y_2^2 + 2y_2^2$$
$$= 3y_1^2 + \tfrac{2}{3}y_2^2.$$

The form has been reduced to a sum of squares whose coefficients are all positive.

7–13 Simultaneous diagonalization of two quadratic forms. The analysis of certain problems in mechanics and economics can be considerably simplified by introducing a nonsingular transformation of variables which will simultaneously diagonalize two quadratic forms. In general, one of the two forms will be positive definite (or negative definite), and for these two cases it is indeed possible to find such a nonsingular transformation of variables. We shall now show how this can be done.

Let $F_1 = \mathbf{x}'\mathbf{A}\mathbf{x}$ be a positive definite quadratic form in n variables, and $F_2 = \mathbf{x}'\mathbf{B}\mathbf{x}$ any other quadratic form in n variables. We wish to find a nonsingular transformation $\mathbf{x} = \mathbf{R}\mathbf{y}$ which simultaneously diagonalizes both forms. Let $\mathbf{Q}_1$ be an orthogonal matrix whose columns are an orthonormal set of eigenvectors for $\mathbf{A}$. Introducing the transformation of variables $\mathbf{x} = \mathbf{Q}_1\mathbf{w}$, we find

$$F_1 = \mathbf{w}'\mathbf{Q}_1'\mathbf{A}\mathbf{Q}_1\mathbf{w} = \mathbf{w}'\mathbf{D}\mathbf{w} = \sum_{j=1}^{n} \lambda_j w_j^2, \tag{7-71}$$

$$F_2 = \mathbf{w}'\mathbf{Q}_1'\mathbf{B}\mathbf{Q}_1\mathbf{w} = \mathbf{w}'\hat{\mathbf{B}}\mathbf{w}, \tag{7-72}$$

where $\mathbf{D} = \|\lambda_j \delta_{ij}\|$ (the λ_j being the eigenvalues of $\mathbf{A}$) and $\hat{\mathbf{B}} = \mathbf{Q}_1'\mathbf{B}\mathbf{Q}_1$. Since F_1 is positive definite, each $\lambda_j > 0$.

Next we introduce the nonsingular (but not orthogonal) transformation of variables

$$w_j = |\lambda_j|^{-1/2} z_j \quad \text{or} \quad \mathbf{w} = \mathbf{H}\mathbf{z}, \quad \text{where} \quad \mathbf{H} = \| |\lambda_j|^{-1/2} \delta_{ij}\|. \tag{7-73}$$

Then (7–71) and (7–72) become

$$F_1 = \mathbf{z}'\mathbf{H}'\mathbf{D}\mathbf{H}\mathbf{z} = \mathbf{z}'\mathbf{I}\mathbf{z} = \sum_{j=1}^{n} z_j^2, \tag{7-74}$$

$$F_2 = \mathbf{z}'\mathbf{H}'\hat{\mathbf{B}}\mathbf{H}\mathbf{z} = \mathbf{z}'\mathbf{C}\mathbf{z}, \quad \text{where} \quad \mathbf{C} = \mathbf{H}'\hat{\mathbf{B}}\mathbf{H}. \tag{7-75}$$

Finally, we construct an orthogonal matrix $\mathbf{Q}_2$ whose columns are an orthonormal set of eigenvectors for $\mathbf{C}$ and introduce the transformation of variables $\mathbf{z} = \mathbf{Q}_2\mathbf{y}$. In terms of the variables $\mathbf{y}$, (7–74) and (7–75) become

$$F_1 = \mathbf{y}'\mathbf{Q}_2'\mathbf{I}\mathbf{Q}_2\mathbf{y} = \mathbf{y}'\mathbf{Q}_2'\mathbf{Q}_2\mathbf{y} = \mathbf{y}'\mathbf{I}\mathbf{y} = \sum_{j=1}^{n} y_i^2 \qquad (\mathbf{Q}_2'\mathbf{Q}_2 = \mathbf{I}) \qquad (7\text{–}76)$$

and

$$F_2 = \mathbf{y}'\mathbf{Q}_2'\mathbf{C}\mathbf{Q}_2\mathbf{y} = \mathbf{y}'\hat{\mathbf{D}}\mathbf{y}, \qquad \text{where } \hat{\mathbf{D}} = \|\hat{\lambda}_j\,\delta_{ij}\|; \qquad (7\text{–}77)$$

the $\hat{\lambda}_j$ are the eigenvalues of $\mathbf{C}$. Thus both forms have been diagonalized by the nonsingular transformation of variables $\mathbf{x} = \mathbf{R}\mathbf{y}$, where

$$\mathbf{R} = \mathbf{Q}_1\mathbf{H}\mathbf{Q}_2. \qquad (7\text{–}78)$$

This transformation is not orthogonal, and hence the congruence transformation which diagonalizes $\mathbf{A}$, $\mathbf{B}$ is not a similarity transformation. This transformation which diagonalizes both matrices reduces $\mathbf{A}$ to the identity matrix, not to a matrix with the eigenvalues of $\mathbf{A}$ as the diagonal elements.

If $\mathbf{A}$ were negative definite instead of positive definite, the same transformation of variables would reduce $\mathbf{A}$ to $-\mathbf{I}$. It is important to note why a transformation which simultaneously diagonalizes both forms can always be found only if one of the forms is positive (negative) definite. The key to a simultaneous diagonalization is to find a congruence transformation which will reduce $\mathbf{A}$ to an identity matrix. If, after introducing the transformation $\mathbf{x} = \mathbf{Q}_1\mathbf{w}$, we were to perform the transformation $\mathbf{w} = \mathbf{Q}_3\mathbf{y}$, where $\mathbf{Q}_3$ contained an orthonormal set of eigenvectors for $\hat{\mathbf{B}}$, F_1 may not be in diagonal form because $\mathbf{Q}_3'\mathbf{D}\mathbf{Q}_3$ may not be a diagonal matrix. If $\mathbf{A}$ is not positive (negative) definite, we cannot make a transformation of the type (7–73), which reduces $\mathbf{A}$ to the identity matrix. For example, if $\mathbf{A}$ is indefinite and each $\lambda_j \neq 0$, the transformation (7–73) reduces $\mathbf{A}$ to a diagonal matrix $\mathbf{G}$ with diagonal elements ± 1, and with at least one element being -1. However, $\mathbf{Q}_2'\mathbf{G}\mathbf{Q}_2$ may not be diagonal. If a $\lambda_j = 0$, we cannot make the transformation (7–73) for w_j. Again, it will be impossible to convert $\mathbf{A}$ to an identity matrix; hence if the transformation diagonalizing F_2 is introduced, F_1 will not remain diagonal.

7–14 Geometric interpretation; coordinates and bases. Sections 7–14 through 7–16 will be devoted to the geometric interpretation of a number of concepts discussed in the present chapter. We shall begin by extending the notion of generalized coordinate systems, introduced in Chapter 2. We noted there that any set of basis vectors for E^n can be thought of as defining a coordinate system for E^n. If $\mathbf{a}_1, \ldots, \mathbf{a}_n$ is a basis for E^n, then any vector $\mathbf{x}$ in E^n can be written as a linear combination of the basis vectors

$$\mathbf{x} = \sum_{j=1}^{n} \alpha_j \mathbf{a}_j, \qquad (7\text{--}79)$$

and the α_j are called the coordinates of $\mathbf{x}$ with respect to the coordinate system defined by the basis vectors $\mathbf{a}_1, \ldots, \mathbf{a}_n$. If the $\mathbf{a}_j$ are mutually orthogonal, they define an orthogonal coordinate system.

Let us consider another basis, $\mathbf{b}_1, \ldots, \mathbf{b}_n$, for E^n. This basis also defines a coordinate system for E^n, and any $\mathbf{x}$ in E^n can be written

$$\mathbf{x} = \sum_{j=1}^{n} \beta_j \mathbf{b}_j, \qquad (7\text{--}80)$$

where the β_j are the coordinates of $\mathbf{x}$ with respect to the coordinate system defined by the basis vectors $\mathbf{b}_j$. Now it is possible to write any $\mathbf{b}_j$ of the second basis as a linear combination of $\mathbf{a}_1, \ldots, \mathbf{a}_n$, that is,

$$\mathbf{b}_j = \sum_{i=1}^{n} s_{ij} \mathbf{a}_i \qquad (j = 1, \ldots, n); \qquad (7\text{--}81)$$

and if

$$\mathbf{A} = (\mathbf{a}_1, \ldots, \mathbf{a}_n), \quad \mathbf{B} = (\mathbf{b}_1, \ldots, \mathbf{b}_n), \quad \mathbf{S} = \|s_{ij}\| = (\mathbf{s}_1, \ldots, \mathbf{s}_n),$$
$$(7\text{--}82)$$

then

$$\mathbf{B} = \mathbf{AS} = (\mathbf{As}_1, \ldots, \mathbf{As}_n). \qquad (7\text{--}83)$$

Equation (7–83) tells us how the two sets of basis vectors are related.

Next we wish to relate the coordinates β_j of $\mathbf{x}$ relative to the coordinate system defined by the $\mathbf{b}_j$ to the coordinates α_j of the coordinate system defined by the $\mathbf{a}_j$. We write

$$\boldsymbol{\alpha} = [\alpha_1, \ldots, \alpha_n], \qquad \boldsymbol{\beta} = [\beta_1, \ldots, \beta_n].$$

Then

$$\mathbf{x} = \mathbf{A}\boldsymbol{\alpha} = \mathbf{B}\boldsymbol{\beta}; \qquad (7\text{--}84)$$

using (7–83), we obtain

$$\mathbf{x} = \mathbf{A}\boldsymbol{\alpha} = \mathbf{AS}\boldsymbol{\beta},$$

or, since $\mathbf{A}$ and $\mathbf{S}$ are nonsingular (why?),

$$\boldsymbol{\beta} = \mathbf{S}^{-1}\boldsymbol{\alpha}. \qquad (7\text{--}85)$$

Equation (7–85) gives the relation between the coordinates in the two coordinate systems, and (7–83) expresses the relation between the basis vectors for the two systems. If we write (7–83) as

$$\mathbf{B}' = \mathbf{S}'\mathbf{A}', \tag{7–86}$$

it follows that (7–86) can be obtained from (7–85) by replacing $\boldsymbol{\beta}$ by $\mathbf{B}'$, $\mathbf{S}^{-1}$ by $\mathbf{S}'$, and $\boldsymbol{\alpha}$ by $\mathbf{A}'$. Because of the relation between (7–85) and (7–86) it is often stated that the matrix giving the transformation of the coordinates is the reciprocal transpose of that giving the transformation of the basis vectors. Thus we have developed the general equations describing the change from one coordinate system to another in E^n. In terms of the notation of Chapter 2, we would write $\mathbf{x_a} = \boldsymbol{\alpha}$, $\mathbf{x_b} = \boldsymbol{\beta}$ since $\boldsymbol{\alpha}$ can be thought of as the representation of $\mathbf{x}$ in the coordinate system defined by the $\mathbf{a}_j$, etc., for $\boldsymbol{\beta}$.

Equation (7–85) suggests that for any nth-order nonsingular matrix $\mathbf{R}$, the transformation $\mathbf{y} = \mathbf{R}\mathbf{x}$ can be imagined to relate the coordinates of a given vector in two different coordinate systems for E^n. In one coordinate system, the vector can be represented by $\mathbf{x}$ (for example, in the orthogonal coordinate system defined by the unit vectors $\mathbf{e}_j$), and in the other coordinate system, the vector is represented by $\mathbf{y}$. If $\mathbf{A}$ contains as columns the basis vectors for the coordinate system where the vector is represented by $\mathbf{x}$, then the matrix $\mathbf{B}$ whose columns are the basis vectors for the coordinate system where the vector is represented by $\mathbf{y}$ is given by $\mathbf{B} = \mathbf{A}\mathbf{R}^{-1}$.

This interpretation of $\mathbf{y} = \mathbf{R}\mathbf{x}$ differs from that in Section 4–1. There we suggested that $\mathbf{y}$ and $\mathbf{x}$ could be considered to be different points in E^n referred to the same coordinate system; that is, the coordinate system remains fixed but the vectors change. Thus $\mathbf{y} = \mathbf{R}\mathbf{x}$ has a dual interpretation: It can be thought of as relating the coordinates of the same point

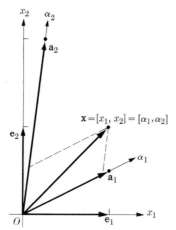

FIGURE 7–1

(vector) in two different coordinate systems or as moving the point **x** into another point **y**, both points being referred to the same coordinate system. The change of coordinate interpretation is called the *alias* interpretation, while the interpretation of having the point move elsewhere is called the *alibi*.

EXAMPLE: The unit vectors $\mathbf{e}_1 = [1, 0]$, $\mathbf{e}_2 = [0, 1]$ define an orthogonal coordinate system in E^2 with coordinates x_1, x_2. Any vector **x** can be written $\mathbf{x} = x_1\mathbf{e}_1 + x_2\mathbf{e}_2$. The set of basis vectors $\mathbf{a}_1 = [1, \frac{1}{2}]$, $\mathbf{a}_2 = [\frac{1}{4}, 2]$ also defines a coordinate system for E^2 (see Fig. 7–1) which is not orthogonal since $\mathbf{a}_1$ and $\mathbf{a}_2$ are not orthogonal. Such a coordinate system is often called oblique.

Let α_1, α_2 be the coordinates for **x** in the coordinate system defined by $\mathbf{a}_1, \mathbf{a}_2$, that is, $\mathbf{x} = \alpha_1\mathbf{a}_1 + \alpha_2\mathbf{a}_2$. The matrix whose columns are $\mathbf{e}_1$, $\mathbf{e}_2$ is the identity matrix **I**. If $\mathbf{A} = (\mathbf{a}_1, \mathbf{a}_2)$, then the matrix **S** which relates the two sets of basis vectors is

$$\mathbf{S} = \mathbf{A} = \begin{bmatrix} 1 & \frac{1}{4} \\ \frac{1}{2} & 2 \end{bmatrix},$$

since $\mathbf{A} = \mathbf{IS} = \mathbf{S}$. The matrix relating $\boldsymbol{\alpha} = [\alpha_1, \alpha_2]$ and $\mathbf{x} = [x_1, x_2]$ is $\mathbf{S}^{-1}$, that is, $\boldsymbol{\alpha} = \mathbf{S}^{-1}\mathbf{x}$. Thus

$$\begin{bmatrix} \alpha_1 \\ \alpha_2 \end{bmatrix} = \tfrac{8}{15}\begin{bmatrix} 2 & -\frac{1}{4} \\ -\frac{1}{2} & 1 \end{bmatrix}\begin{bmatrix} x_1 \\ x_2 \end{bmatrix} \quad \text{or} \quad \begin{aligned} \alpha_1 &= \tfrac{16}{15}x_1 - \tfrac{2}{15}x_2, \\ \alpha_2 &= -\tfrac{4}{15}x_1 + \tfrac{8}{15}x_2. \end{aligned} \qquad (7\text{–}87)$$

Consider the vector **x** whose representation in the system defined by $\mathbf{e}_1$, $\mathbf{e}_2$ is $\mathbf{x} = [1, 1]$. The coordinates of **x** in the system defined by $\mathbf{a}_1$, $\mathbf{a}_2$ are [from (7–87)] $\alpha_1 = 14/15$, $\alpha_2 = 4/15$. Note that since $\mathbf{a}_1$ does not have unit length, α_1 is not the distance measured along the α_1-axis from the origin to the point where a line drawn through **x** parallel to $\mathbf{a}_2$ intersects the α_1-axis. This distance is $|\alpha_1\mathbf{a}_1|$ which is $\alpha_1|\mathbf{a}_1|$, because in our example, α_1 is positive. Since $|\mathbf{a}_1| = (1/2)\sqrt{5}$, the distance is $(14/15)[(1/2)\sqrt{5}] = (7/15)\sqrt{5}$.

7–15 Equivalence and similarity. Let us consider a linear transformation T which maps points **x** in E^n into points **y** in E^m so that $\mathbf{y} = T(\mathbf{x})$. The problems of Chapter 4 show that for fixed bases (i.e., given coordinate systems) in E^n and E^m, there exists a unique $m \times n$ matrix **A** such that $\mathbf{y} = \mathbf{Ax}$. The matrix **A** is the representation of the linear transformation T with respect to the given coordinate systems.*

---

* Here **x**, **y** refer not only to points in E^n and E^m, respectively, but also to the representation of these points in the coordinate systems for which the representation of T is **A**.

Now, suppose that we introduce new coordinate systems into E^n and E^m. Let the new coordinates, $\hat{x}$, in E^n be related to the original ones, x, by $x = F\hat{x}$, and the new coordinates, $\hat{y}$, in E^m to the original ones, y, by $y = G\hat{y}$. Then in terms of the new coordinates, the transformation $y = Ax$ becomes

$$G\hat{y} = AF\hat{x} \quad \text{or} \quad \hat{y} = G^{-1}AF\hat{x} = B\hat{x}, \qquad (7\text{–}88)$$

where $B = G^{-1}AF$. The matrix B is the representation of the linear transformation T in the new coordinate systems and is equivalent to the matrix A.

The preceding discussion shows that equivalent matrices can be thought of as representing the same linear transformation in various coordinate systems. In fact, if A represents the linear transformation T for given coordinate systems in E^n and E^m, and if B is any other matrix equivalent to A, then there exist a coordinate system in E^n and a coordinate system in E^m such that B is the representation of T for these coordinate systems.

Next let us concentrate our attention on the special case where $m = n$, so that T maps points x in E^n into points y in E^n. Assume that the points x, y are referred to the same coordinate system in E^n. Then for this coordinate system, there exists a unique matrix A such that $y = Ax$. If a new coordinate system is introduced into E^n and the new coordinates, $\hat{x}$, $\hat{y}$, are related to the old ones by $x = S\hat{x}$, $y = S\hat{y}$ (note that the same S appears in both equations, since both sets of vectors are referred to the same coordinate system), then $y = Ax$ becomes

$$S\hat{y} = AS\hat{x} \quad \text{or} \quad \hat{y} = S^{-1}AS\hat{x} = Bx. \qquad (7\text{–}89)$$

Thus the matrix $B = S^{-1}AS$ which represents T in the new coordinate system is similar to the matrix A which represented T in the old coordinate system. If T "looked like" A in the original coordinate system, it "looks like" $S^{-1}AS$ in the new coordinate system.

Note: Similarity transformations are of frequent occurrence in physics. Consider an anisotropic dielectric (nonconducting) material. If an electric field is imposed on this material, there is a separation of charge, and the material becomes polarized. The electric field f and the polarization p are both vector quantities. Because the material is not isotropic, the direction of the polarization vector will not, in general, lie along the same line as the electric field vector. The vector p will be related to f by an equation of the form $p = Ef$, where E is a third-order symmetric matrix, called the dielectric tensor. The matrix E depends on the coordinate system to which the vectors p, f are referred. Suppose that $p = Ef$ when p, f are referred to a coordinate system with coordinates x_1, x_2, x_3. Now a transformation is made to a new coordinate system with coordinates y_1, y_2, y_3, and $x =$

Sy. Then in terms of this new coordinate system, the dielectric tensor is $\hat{\mathbf{E}} = \mathbf{S}^{-1}\mathbf{E}\mathbf{S}$, and $\mathbf{p} = \hat{\mathbf{E}}\mathbf{f}$ when $\mathbf{p}$, $\mathbf{f}$ are referred to the **y**-coordinate system.

7–16 Rotation of coordinates; orthogonal transformations. Orthogonal coordinate systems are probably used more frequently in practice than any other type. By "rotating" such an orthogonal coordinate system it is often possible to simplify the equations of interest. Let us study the rotation of orthogonal coordinates in E^2. Consider Fig. 7–2. Imagine that we begin with the x_1x_2-coordinate system. Another y_1y_2-coordinate system is obtained by rotating the x_1x_2-system through the angle θ, as shown. The matrix relating the y_1y_2-coordinates to the x_1x_2-coordinates will now be found. The vectors which define the x_1x_2-coordinate system are $\mathbf{e}_1 = [1, 0]$, $\mathbf{e}_2 = [0, 1]$. The orthonormal vectors defining the y_1y_2-coordinate system will be denoted by $\boldsymbol{\epsilon}_1$, $\boldsymbol{\epsilon}_2$. Any vector $\mathbf{v}$ can be written

$$\mathbf{v} = x_1\mathbf{e}_1 + x_2\mathbf{e}_2 = y_1\boldsymbol{\epsilon}_1 + y_2\boldsymbol{\epsilon}_2.$$

The component of $x_1\mathbf{e}_1$ along the y_1-axis is $x_1 \cos \theta$, and the component of $x_2\mathbf{e}_2$ along the y_1-axis is $x_2 \sin \theta$. The sum of these components must be y_1. Hence

$$y_1 = x_1 \cos \theta + x_2 \sin \theta.$$

Similarly,
$$y_2 = -x_1 \sin \theta + x_2 \cos \theta.$$

The transformation of coordinates in matrix form can be written $\mathbf{y} = \mathbf{Q}\mathbf{x}$, where

$$\mathbf{Q} = \begin{bmatrix} \cos \theta & \sin \theta \\ -\sin \theta & \cos \theta \end{bmatrix}. \tag{7–90}$$

The matrix $\mathbf{Q}$ is orthogonal since $\mathbf{Q}'\mathbf{Q} = 1$ (recall that $\sin^2 \theta + \cos^2 \theta = 1$).

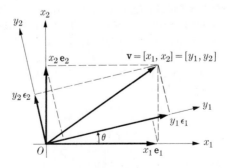

Figure 7–2

The basis vectors $\boldsymbol{\epsilon}_1, \boldsymbol{\epsilon}_2$ for the rotated coordinate system are related to $\mathbf{e}_1, \mathbf{e}_2$ by $(\boldsymbol{\epsilon}_1, \boldsymbol{\epsilon}_2) = (\mathbf{e}_1, \mathbf{e}_2)\mathbf{Q}'$ because $\mathbf{Q}^{-1} = \mathbf{Q}'$. Thus $(\boldsymbol{\epsilon}_1, \boldsymbol{\epsilon}_2) = \mathbf{Q}'$ or $\boldsymbol{\epsilon}_1 = [\cos\theta, \sin\theta]$, $\boldsymbol{\epsilon}_2 = [-\sin\theta, \cos\theta]$.

The notion of rotating orthogonal coordinates can be generalized to E^n. First, however, we shall develop a few more properties of orthogonal matrices: An orthogonal matrix has been defined as one whose inverse is its transpose; that is, $\mathbf{Q}$ is orthogonal if $\mathbf{Q}^{-1} = \mathbf{Q}'$. If we denote the columns of $\mathbf{Q}$ by $\mathbf{q}_j$, then since $\mathbf{Q}'\mathbf{Q} = \mathbf{I}$, it follows that

$$\mathbf{q}'_i \mathbf{q}_j = \delta_{ij}, \tag{7-91}$$

and the columns of $\mathbf{Q}$ form an orthonormal basis for E^n. If the rows of $\mathbf{Q}$ are denoted by $\mathbf{q}^i$, then since $\mathbf{Q}\mathbf{Q}' = \mathbf{I}$,

$$\mathbf{q}^j (\mathbf{q}^i)' = \delta_{ij}, \tag{7-92}$$

and the rows of $\mathbf{Q}$ also form an orthonormal basis for E^n. Any matrix whose columns are an orthonormal set of vectors is an orthogonal matrix, and we have just shown that the rows therefore are also an orthonormal set of vectors. Note that if $\mathbf{Q}$ is orthogonal, $\mathbf{Q}'$ and $\mathbf{Q}^{-1}$ are also orthogonal.

From $\mathbf{Q}'\mathbf{Q} = \mathbf{I}$, it follows that

$$|\mathbf{Q}'\mathbf{Q}| = |\mathbf{Q}'|\,|\mathbf{Q}| = |\mathbf{Q}|^2 = 1 \quad \text{or} \quad |\mathbf{Q}| = \pm 1. \tag{7-93}$$

The determinant of an orthogonal matrix can have only the values ± 1.

If $\mathbf{Q}_1$ and $\mathbf{Q}_2$ are nth-order orthogonal matrices, then $\mathbf{Q}_1\mathbf{Q}_2$ is also an orthogonal matrix since

$$(\mathbf{Q}_1\mathbf{Q}_2)'\mathbf{Q}_1\mathbf{Q}_2 = \mathbf{Q}'_2\mathbf{Q}'_1\mathbf{Q}_1\mathbf{Q}_2 = \mathbf{Q}'_2\mathbf{I}\mathbf{Q}_2 = \mathbf{Q}'_2\mathbf{Q}_2 = \mathbf{I}, \tag{7-94}$$

and hence

$$(\mathbf{Q}_1\mathbf{Q}_2)^{-1} = (\mathbf{Q}_1\mathbf{Q}_2)'.$$

It will be noted that for the $\mathbf{Q}$ of (7-90), $|\mathbf{Q}| = 1$. This $\mathbf{Q}$ provided the transformation of coordinates on rotation of axes through an angle θ in E^2. An interesting geometrical interpretation can also be given to orthogonal matrices with $|\mathbf{Q}| = -1$. Consider the orthogonal coordinate system defined by $\mathbf{e}_1, \mathbf{e}_2$, with coordinates x_1, x_2. We introduce a transformation of coordinates $\mathbf{y} = \mathbf{Q}\mathbf{x}$, where

$$\mathbf{Q} = \begin{bmatrix} 1 & 0 \\ 0 & -1 \end{bmatrix}.$$

$\mathbf{Q}$ is orthogonal and $|\mathbf{Q}| = -1$. Thus $y_1 = x_1$, and $y_2 = -x_2$. The new basis vectors $\boldsymbol{\epsilon}_1, \boldsymbol{\epsilon}_2$ are $\boldsymbol{\epsilon}_1 = \mathbf{e}_1$, $\boldsymbol{\epsilon}_2 = -\mathbf{e}_2$. The coordinate systems

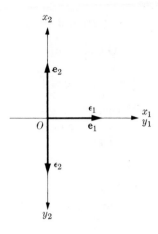

FIGURE 7-3

are shown in Fig. 7-3. We cannot obtain the new coordinate system by rotating the coordinate system defined by $\mathbf{e}_1, \mathbf{e}_2$; instead we reflect the x_2-axis in the origin, that is, we replace x_2 by $y_2 = -x_2$. Problem 7-57 will ask the reader to show that any second-order orthogonal matrix can be written as (7-90) or as

$$\mathbf{Q} = \begin{bmatrix} 1 & 0 \\ 0 & -1 \end{bmatrix} \begin{bmatrix} \cos\theta & \sin\theta \\ -\sin\theta & \cos\theta \end{bmatrix},$$

so that if $|\mathbf{Q}| = 1$, $\mathbf{y} = \mathbf{Q}\mathbf{x}$ can be interpreted as a rotation of axes, and if $|\mathbf{Q}| = -1$, $\mathbf{y} = \mathbf{Q}\mathbf{x}$ can be interpreted as a rotation plus a reflection. The latter is sometimes referred to as an improper rotation.

Consider the transformation $\boldsymbol{\beta} = \mathbf{Q}\boldsymbol{\alpha}$ in E^n, where $\mathbf{Q}$ is an orthogonal matrix. This can be interpreted as a change of coordinates. If the $\boldsymbol{\alpha}$ are coordinates relative to an *orthonormal basis* $\mathbf{a}_1, \ldots, \mathbf{a}_n$, then $\mathbf{A} = (\mathbf{a}_1, \ldots, \mathbf{a}_n)$ is an orthogonal matrix. The basis vectors for the system whose coordinates are $\boldsymbol{\beta}$ will be denoted by $\mathbf{b}_1, \ldots, \mathbf{b}_n$. Then if $\mathbf{B} = (\mathbf{b}_1, \ldots, \mathbf{b}_n)$, $\mathbf{B} = \mathbf{A}\mathbf{Q}'$, and $\mathbf{B}$ is an orthogonal matrix since the product of orthogonal matrices is an orthogonal matrix. Thus $\mathbf{b}_1, \ldots, \mathbf{b}_n$ form an orthonormal set and define an orthogonal coordinate system. In the coordinate system defined by the $\mathbf{a}_j$, the length of any vector $\mathbf{x} = \sum \alpha_i \mathbf{a}_i$ is the square root of $\mathbf{x}'\mathbf{x} = \sum \alpha_i \alpha_j \mathbf{a}_i' \mathbf{a}_j = \sum_{j=1}^n \alpha_j^2 = \boldsymbol{\alpha}'\boldsymbol{\alpha}$. In the coordinate system defined by the $\mathbf{b}_j$, the length $\mathbf{x} = \sum_{j=1}^n \beta_j \mathbf{b}_j$ is then the square root of $\sum_{j=1}^n \beta_j^2 = \boldsymbol{\beta}'\boldsymbol{\beta}$. Thus

$$\boldsymbol{\beta}'\boldsymbol{\beta} = \sum_{j=1}^n \beta_j^2 = \sum_{j=1}^n \alpha_j^2 = \boldsymbol{\alpha}'\boldsymbol{\alpha}.$$

This relation also follows directly from $\boldsymbol{\beta} = \mathbf{Q}\boldsymbol{\alpha}$ since

$$\boldsymbol{\beta}'\boldsymbol{\beta} = \boldsymbol{\alpha}'\mathbf{Q}'\mathbf{Q}\boldsymbol{\alpha} = \boldsymbol{\alpha}'\mathbf{I}\boldsymbol{\alpha} = \boldsymbol{\alpha}'\boldsymbol{\alpha}. \tag{7-95}$$

The above discussion suggests that any transformation $\boldsymbol{\beta} = \mathbf{Q}\boldsymbol{\alpha}$ ($\mathbf{Q}$ orthogonal) can be considered to give the transformation of coordinates on rotation of an orthogonal coordinate system in E^n. The rotation will be either proper or improper, depending on whether $|\mathbf{Q}| = 1$ or $|\mathbf{Q}| = -1$.

We have seen that any quadratic form $\mathbf{x}'\mathbf{A}\mathbf{x}$ can be diagonalized by a transformation of variables $\mathbf{x} = \mathbf{Q}\mathbf{y}$ or $\mathbf{y} = \mathbf{Q}'\mathbf{x}$, where $\mathbf{Q}$ is an orthogonal matrix whose columns are an orthonormal set of eigenvectors for $\mathbf{A}$. This transformation of variables can be interpreted geometrically as a rotation of axes. The points $\mathbf{x}$ may be thought of as being referred to the orthogonal coordinate system defined by the unit vectors $\mathbf{e}_1, \ldots, \mathbf{e}_n$. The orthonormal basis vectors $\mathbf{u}_1, \ldots, \mathbf{u}_n$ which define the coordinate system for the coordinates $\mathbf{y}$ are obtained from $(\mathbf{u}_1, \ldots, \mathbf{u}_n) = \mathbf{IQ} = \mathbf{Q}$. Hence the $\mathbf{u}_j$ are the set of orthonormal eigenvectors for $\mathbf{A}$. A set of orthonormal eigenvectors for $\mathbf{A}$ defines a coordinate system in which $\mathbf{x}'\mathbf{A}\mathbf{x}$ is diagonal. The $\mathbf{u}_j$ are said to define the principal axes for the quadratic form.

References

1. G. Birkhoff and S. MacLane, *A Survey of Modern Algebra*. New York: Macmillan, 1941.
2. F. B. Hildebrand, *Methods of Applied Mathematics*. Englewood Cliffs: Prentice-Hall, 1952.
3. H. Margenau and G. Murphy, *The Mathematics of Physics and Chemistry*. New York: Van Nostrand, 1943.
4. S. Perlis, *Theory of Matrices*. Reading, Mass.: Addison-Wesley, 1952.
5. R. Thrall and L. Thornheim, *Vector Spaces and Matrices*. New York: Wiley, 1957.

Problems

7-1. Consider the polynomial

$$f(\lambda) \equiv (-\lambda)^n + b_{n-1}(-\lambda)^{n-1} + \cdots + b_1(-\lambda) + b_0.$$

By induction or by some different method, prove that

$$b_{n-1} = \sum_{i=1}^{n} \lambda_i = \lambda_1 + \lambda_2 + \cdots + \lambda_n,$$

$$b_{n-2} = \sum_{j>i} \lambda_i \lambda_j = \lambda_1 \lambda_2 + \cdots + \lambda_1 \lambda_n + \lambda_2 \lambda_3 + \cdots + \lambda_2 \lambda_n + \cdots + \lambda_{n-1} \lambda_n,$$

$$\vdots$$

$$b_{n-r} = \sum_{k>\cdots>j>i} \lambda_i \lambda_j \cdots \lambda_k, \quad \text{(each term a product of } r \text{ of the } \lambda_i)$$

$$\vdots$$

$$b_0 = \lambda_1 \lambda_2 \cdots \lambda_{n-1} \lambda_n,$$

where the λ_i are the n roots of $f(\lambda) = 0$.

7-2. Consider the equation

$$|\mathbf{A} - \lambda \mathbf{I}| = (-\lambda)^n + b_{n-1}(-\lambda)^{n-1} + \cdots + b_1(-\lambda) + b_0.$$

By induction or some different method prove that

$$b_{n-1} = \sum_{i=1}^{n} a_{ii} = a_{11} + a_{22} + \cdots + a_{nn},$$

$$b_{n-2} = \sum_{j>i} \begin{vmatrix} a_{ii} & a_{ij} \\ a_{ji} & a_{jj} \end{vmatrix},$$

$$\vdots$$

$$b_{n-r} = \text{sum of } n!/r!(n-r)! = \begin{array}{l} \text{principal minors of } r\text{th order which} \\ \text{preserve the natural column order,} \end{array}$$

$$\vdots$$

$$b_0 = |\mathbf{A}|.$$

7-3. Find the eigenvalues and a set of orthonormal eigenvectors for the matrix

$$\mathbf{A} = \begin{bmatrix} 3 & 5 \\ 5 & 7 \end{bmatrix}.$$

7–4. Find the eigenvalues and two different orthonormal sets of eigenvectors for the matrix
$$\mathbf{A} = \begin{bmatrix} 7 & 0 & 0 \\ 0 & 3 & 0 \\ 0 & 0 & 3 \end{bmatrix}.$$

7–5. Find the eigenvalues and a set of orthonormal eigenvectors for
$$\mathbf{A} = \begin{bmatrix} 3 & 0 & 0 \\ 0 & 2 & 5 \\ 0 & 5 & 4 \end{bmatrix}.$$

7–6. For the matrix of Problem 7–5, form the matrix $\mathbf{Q}$ whose columns are a set of orthonormal eigenvectors for $\mathbf{A}$. Show that $\mathbf{Q'AQ}$ is a diagonal matrix whose diagonal elements are the eigenvalues of $\mathbf{A}$.

7–7. Prove that if all eigenvalues of an nth-order symmetric matrix $\mathbf{A}$ are different from zero, the rank of $\mathbf{A}$ is n. Prove that if 0 is an eigenvalue of $\mathbf{A}$ of multiplicity k, then $r(\mathbf{A}) = n - k$.

7–8. Show that it is impossible for any 2×2 symmetric matrix of the form
$$\mathbf{A} = \begin{bmatrix} a_1 & b \\ b & a_2 \end{bmatrix} \quad (b \neq 0)$$
to have two identical eigenvalues.

7–9. In Problem 2–18 it was shown that if we have two subspaces S'_n, S''_n of E^n having dimensions d', d'', respectively, and if $S'_n \cap S''_n = \{\mathbf{0}\}$, then d, the dimension of $S'_n + S''_n$, is given by $d = d' + d''$. Show that under these assumptions, any vector $\mathbf{a}$ in $S'_n + S''_n$ can be written uniquely as $\mathbf{a} = \mathbf{u}_1 + \mathbf{u}_2$, $\mathbf{u}_1 \in S'_n$, $\mathbf{u}_2 \in S''_n$. Then $\mathbf{u}_1$, $\mathbf{u}_2$ are called the projections of $\mathbf{a}$ on the subspaces S'_n, S''_n, respectively, and $S'_n + S''_n$ is called the *direct sum* of S'_n and S''_n. Illustrate this geometrically when S'_n is the subspace generated by [1, 2], and S''_n the subspace generated by [−1, 3]. What is the projection of [3, 3] on S'_n and on S''_n?

More generally, if we have k subspaces of E^n, $S^{(1)}_n, \ldots, S^{(k)}_n$, and $S'_n \cap S''_n = \{\mathbf{0}\}$, where S'_n is any partial sum of the $S^{(j)}_n$, and S''_n is any partial sum of the $S^{(j)}_n$ containing a set of $S^{(j)}_n$ different from those in S'_n, then $S_n = \sum_{j=1}^{k} S^{(j)}_n$ is called the direct sum of the $S^{(j)}_n$. This condition requires that $\mathbf{0}$ is the only vector common to any different partial sums of the $S^{(j)}_n$. In particular, this condition requires that $S^{(i)}_n \cap S^{(j)}_n = \{\mathbf{0}\}$, $i \neq j$, $S^{(i)}_n + S^{(j)}_n \cap S^{(k)}_n = \{\mathbf{0}\}$, $i \neq j \neq k$, etc.

Show that if $\mathbf{a}$ is any vector in S_n, it can be written uniquely as $\mathbf{a} = \sum_{j=1}^{k} \mathbf{u}_j$, $\mathbf{u}_j \in S^{(j)}_n$; $\mathbf{u}_j$ is called the projection of $\mathbf{a}$ on the subspace $S^{(j)}_n$. The notion of a direct sum is a generalization of the basis concept. Give a geometrical illustration of a direct sum in E^3.

Let $S^{(j)}_n$ be the subspace generated by the eigenvectors for the nth-order symmetric matrix $\mathbf{A}$ corresponding to the eigenvalue λ_j. Show that E^n is the direct sum of these subspaces.

7-10. Prove that if $Q_1, \ldots, Q_n$ are orthogonal, then

$$\hat{Q} = \begin{bmatrix} Q_1 & 0 & \cdots & 0 \\ 0 & Q_2 & \cdots & 0 \\ \vdots & & & \vdots \\ 0 & 0 & \cdots & Q_n \end{bmatrix}$$

is also orthogonal.

7-11. Compute the number of independent elements in an nth-order orthogonal matrix.

7-12. Consider the nth-order symmetric matrix A. Let λ_1 be an eigenvalue of A, and u_1 an eigenvector with eigenvalue λ_1. Show that there exist vectors $v_1, \ldots, v_{n-1}$ such that $Q_1 = (u_1, v_1, \ldots, v_{n-1})$ is an orthogonal matrix. Then show that

$$Q_1' A Q_1 = \begin{bmatrix} \lambda_1 & 0 \\ 0 & A_1 \end{bmatrix},$$

where A_1 is a symmetric matrix of order $n - 1$. Show that the remaining $n - 1$ eigenvalues of A are the $n - 1$ eigenvalues of A_1. Let λ_2 be any eigenvalue of A_1 (and hence of A). Assume that $\hat{u}_2$ is an eigenvector of A_1 with eigenvalue λ_2. Consider $P = (\hat{u}_2, \hat{v}_1, \ldots, \hat{v}_{n-2})$, where $\hat{v}_1, \ldots, \hat{v}_{n-2}$ are chosen such that P is orthogonal. Prove that

$$P' A_1 P = \begin{bmatrix} \lambda_2 & 0 \\ 0 & A_2 \end{bmatrix},$$

where A_2 is a symmetric matrix of order $n - 2$. Now form the matrix

$$Q_2 = \begin{bmatrix} 1 & 0 \\ 0 & P \end{bmatrix}.$$

Show that Q_2 is orthogonal and that

$$Q_2' \begin{bmatrix} \lambda_1 & 0 \\ 0 & A_1 \end{bmatrix} Q_2 = \begin{bmatrix} \lambda_1 & 0 & 0 \\ 0 & \lambda_2 & 0 \\ 0 & 0 & A_2 \end{bmatrix}.$$

Next prove by induction or some different method that there exists an orthogonal matrix Q such that $Q'AQ$ is a diagonal matrix with the eigenvalues of A as its diagonal elements. Many texts use this method of proving that A is similar to a diagonal matrix. Does this method of proof demonstrate that (1) Q contains an orthonormal set of eigenvectors for A; (2) if an eigenvalue has multiplicity k, then the eigenvectors of A span a k-dimensional subspace of A?

7-13. Prove that if **A** is a symmetric matrix and **Q** an orthogonal matrix such that $\mathbf{D} = \mathbf{Q'AQ}$ is diagonal, then **Q** must have as its columns a set of orthonormal eigenvectors of **A**. Prove also that the diagonal elements of **D** must be the eigenvalues of **A**.

7-14. Prove that two commuting nth-order symmetric matrices **A**, **B** are simultaneously diagonalizable by an orthogonal similarity transformation, that is, there exists an orthogonal matrix **Q** such that $\mathbf{Q'AQ}$ and $\mathbf{Q'BQ}$ are diagonal. Hint: Let $\lambda_1, \ldots, \lambda_r$ denote the *different* eigenvalues of **A** and assume that λ_j has multiplicity m_j. Then there exists an orthogonal matrix $\mathbf{Q}_1$ such that

$$\mathbf{D}_1 = \mathbf{Q}_1'\mathbf{AQ}_1 = \begin{bmatrix} \lambda_1 \mathbf{I}_{m_1} & 0 & \cdots & 0 \\ 0 & \lambda_2 \mathbf{I}_{m_2} & \cdots & 0 \\ \vdots & & & \vdots \\ 0 & 0 & \cdots & \lambda_r \mathbf{I}_{m_r} \end{bmatrix}.$$

Show that $\mathbf{D}_1$ commutes with $\mathbf{Q}_1'\mathbf{BQ}_1$ and prove that $\mathbf{Q}_1'\mathbf{BQ}_1$ must have the form (see Problem 3-11)

$$\mathbf{Q}_1'\mathbf{BQ}_1 = \begin{bmatrix} \mathbf{B}_1 & 0 & \cdots & 0 \\ 0 & \mathbf{B}_2 & \cdots & 0 \\ \vdots & & & \vdots \\ 0 & 0 & \cdots & \mathbf{B}_r \end{bmatrix}.$$

Show that each $\mathbf{B}_i$ is symmetric. Let $\mathbf{P}_i$ be an orthogonal matrix such that $\mathbf{P}_i'\mathbf{B}_i\mathbf{P}_i$ is diagonal. Then consider the matrix

$$\mathbf{Q}_2 = \begin{bmatrix} \mathbf{P}_1 & 0 & \cdots & 0 \\ 0 & \mathbf{P}_2 & \cdots & 0 \\ \vdots & & & \vdots \\ 0 & 0 & \cdots & \mathbf{P}_r \end{bmatrix}.$$

Show that $\mathbf{Q} = \mathbf{Q}_1\mathbf{Q}_2$ is orthogonal and simultaneously diagonalizes **A** and **B**.

7-15. Consider the matrices

$$\mathbf{A} = \begin{bmatrix} 3 & 4 & 0 & 0 \\ 4 & 2 & 0 & 0 \\ 0 & 0 & 1 & 0 \\ 0 & 0 & 0 & 1 \end{bmatrix}, \quad \mathbf{B} = \begin{bmatrix} 1 & 0 & 0 & 0 \\ 0 & 1 & 0 & 0 \\ 0 & 0 & -3 & 2 \\ 0 & 0 & 2 & 5 \end{bmatrix}.$$

Show that **A**, **B** commute. Find the matrix **Q** such that $\mathbf{Q'AQ}$, $\mathbf{Q'BQ}$ are diagonal.

7-16. Let **A** be an nth-order nonsingular symmetric matrix and **u** an eigenvector of **A** with eigenvalue λ. Show that **u** is an eigenvector of $\mathbf{A}^{-1}$ with eigenvalue $1/\lambda$, i.e., if λ is an eigenvalue of **A**, then $1/\lambda$ is an eigenvalue of $\mathbf{A}^{-1}$. Furthermore **A**, $\mathbf{A}^{-1}$ have the same set of eigenvectors. Can **A** have an eigenvalue 0?

7-17. Show that the eigenvalues of the transpose of the square (not necessarily symmetric) matrix **A** are the same as those of **A**. Are the eigenvectors of **A**′ the same as those of **A**?

7-18. Demonstrate that if **x** satisfies $\mathbf{Ax} = \lambda \mathbf{x}$, then $\mathbf{A}^n \mathbf{x} = \lambda^n \mathbf{x}$, so that if λ is an eigenvalue of **A**, then λ^n is an eigenvalue of $\mathbf{A}^n$; furthermore, **A**, $\mathbf{A}^n$ have the same set of eigenvectors. Show directly that if $\mathbf{Q'AQ} = \mathbf{D}$ is diagonal, then $\mathbf{Q'A}^n\mathbf{Q} = \mathbf{D}^n$.

7-19. Prove that if **x** satisfies $\mathbf{Ax} = \lambda \mathbf{x}$ and $P(\mathbf{A})$ is a matrix polynomial in **A**, then $P(\mathbf{A})\mathbf{x} = P(\lambda)\mathbf{x}$.

7-20. Show that if **x** is an eigenvector of **A** with eigenvalue λ_i and **y** is an eigenvector of **A**′ with eigenvalue $\lambda_j (\lambda_i \neq \lambda_j)$, then $\mathbf{y'x} = 0$. Note that **A** does not need to be symmetric.

7-21. Consider the matrix
$$\mathbf{A} = \begin{bmatrix} 3 & 2 \\ 4 & 7 \end{bmatrix}.$$

Find the eigenvalues of **A** and a set of eigenvectors for **A** and **A**′. Show that in this special case the results of Problem 7-20 hold.

7-22. If λ_1 is the largest eigenvalue of the symmetric matrix **A**, prove that

$$\lambda_1 = \max_{\mathbf{x}} \frac{\mathbf{x'Ax}}{\mathbf{x'x}}$$

when **x** is allowed to range over all of E^n. Hint: Express **x** as a linear combination of the eigenvectors of **A**.

7-23. Consider the symmetric matrix **A**. Let $\mathbf{x}_0$ be any vector in E^n. Compute $\mathbf{x}_1 = \mathbf{Ax}_0$, $\mathbf{x}_2 = \mathbf{Ax}_1$, etc., $\mathbf{x}_n \doteq \mathbf{Ax}_{n-1}$. If $|\lambda_1|$ is the "largest" eigenvalue of **A** and has multiplicity 1 (is not repeated), then show that as $n \to \infty$, $\mathbf{x}_n$ becomes proportional to $\lambda_1^n \mathbf{u}_1$, where $\mathbf{u}_1$ is the eigenvector of **A** with eigenvalue λ_1, provided that $\mathbf{x}_0$ is not orthogonal to $\mathbf{u}_1$. How can this result be used to compute numerically the largest eigenvalue of **A** and its corresponding eigenvector? Hint: Write $\mathbf{x}_0$ as a linear combination of the eigenvectors of **A**.

7-24. In Problem 7-23, show that if the multiplicity of λ_1 is greater than unity, the process can lead to one eigenvector with eigenvalue λ.

7-25. Compute the exact value for the largest eigenvalue of
$$\mathbf{A} = \begin{bmatrix} 7 & 2 \\ 2 & 6 \end{bmatrix}$$

and its eigenvector. Using the technique described in Problem 7-23, try to determine approximately the eigenvalue and eigenvector; start with $\mathbf{x}_0 = [1, 0]$.

7-26. Consider the inhomogeneous eigenvalue problem

$$\mathbf{Ax} - \lambda \mathbf{x} = \mathbf{b} \quad (\mathbf{A} \text{ symmetric}, \ \mathbf{b} \neq \mathbf{0}).$$

Show that there is a solution $\mathbf{x}$ provided that λ is not one of the eigenvalues λ_j of $\mathbf{A}$. Show that any solution $\mathbf{x}$ can be written

$$\mathbf{x} = \sum_{j=1}^{n} \left(\frac{\mathbf{u}_j' \mathbf{b}}{\lambda_j - \lambda} \right) \mathbf{u}_j,$$

where the $\mathbf{u}_j$ form an orthonormal set of eigenvectors of $\mathbf{A}$. Hint: Write $\mathbf{x} = \sum \alpha_j \mathbf{u}_j$ and evaluate the α_j. The problem is called inhomogeneous because $\alpha \mathbf{x}$ is not a solution if $\mathbf{x}$ is for all scalars α.

7-27. In Problem 7-26, let

$$\mathbf{A} = \begin{bmatrix} 4 & 1 \\ 1 & 3 \end{bmatrix}, \quad \mathbf{b} = \begin{bmatrix} 2 \\ 1 \end{bmatrix}.$$

Write $\mathbf{x}$ as a function of λ.

7-28. Find the eigenvalues and eigenvectors of unit length for

$$\mathbf{A} = \begin{bmatrix} 2 & 4 \\ 3 & 1 \end{bmatrix}.$$

Show that the eigenvectors are linearly independent, but not orthogonal.

7-29. For the matrix $\mathbf{A}$ of Problem 7-28, find a matrix $\mathbf{P}$ such that $\mathbf{P}^{-1}\mathbf{AP}$ is a diagonal matrix whose diagonal elements are the eigenvalues of $\mathbf{A}$. Carry out the multiplication to show that $\mathbf{P}^{-1}\mathbf{AP}$ is diagonal.

7-30. Find the eigenvalues and eigenvectors of

$$\mathbf{A} = \begin{bmatrix} 2 & -1 \\ 1 & 0 \end{bmatrix}.$$

In this case, the two eigenvalues are equal, but there is only a single linearly independent eigenvector. The nullity of $\mathbf{A} - \lambda \mathbf{I}$ is not the multiplicity of the eigenvalue λ.

7-31. By working directly with $|\mathbf{A} - \lambda \mathbf{I}|$, prove that similar matrices $\mathbf{A}$, $\mathbf{B}$ have the same characteristic polynomial.

7-32. Show that the symmetric matrix $\mathbf{A}$ is positive definite if and only if there exists a nonsingular matrix $\mathbf{P}$ such that $\mathbf{A} = \mathbf{P}'\mathbf{P}$. What is $\mathbf{P}$? Hint: See Section 7-13.

7-33. Write the following matrix in the form $\mathbf{A} = \mathbf{P}'\mathbf{P}$:

$$\mathbf{A} = \begin{bmatrix} 3 & 1 \\ 1 & 2 \end{bmatrix}.$$

7-34. Show that if the symmetric matrix $\mathbf{A}$ is positive (negative) semidefinite, then $|\mathbf{A}| = 0$, and $\mathbf{A}$ is singular.

7-35. Write the following quadratic forms in simplified matrix notation $\mathbf{x}'\mathbf{A}\mathbf{x}$, with $\mathbf{A}$ being a symmetric matrix:

(a) $3x_1^2 + 2x_1x_2 + 4x_2^2$;

(b) $3x_1^2 + x_2^2 + 5x_3^2 + 4x_1x_2 + 2x_1x_3 + 6x_2x_3$;

(c) $4x_1^2 + 3x_2^2$.

Check to determine whether each of the forms is positive definite.

7-36. Prove that a nonsingular transformation of variables can reduce any quadratic form $\mathbf{x}'\mathbf{A}\mathbf{x}$ not identically zero to a form with the leading coefficient $a_{11} \neq 0$. Hint: Examine the two-variable case in Section 7-11.

7-37. Using the results of Problem 7-36, prove that a nonsingular transformation of variables can reduce any quadratic form $\mathbf{x}'\mathbf{A}\mathbf{x}$ to $\sum \gamma_j y_j^2$, where each γ_j can be positive, negative, or zero.

7-38. Prove in detail that the nonsingular transformation introduced in Section 7-11 to diagonalize positive definite forms is such that $\mathbf{S}$ is a triangular matrix.

7-39. Find a triangular matrix $\mathbf{R}$ which diagonalizes

$$F = 9x_1^2 + 2x_1x_2 + 2x_2^2.$$

Determine the resulting diagonal form of F.

7-40. Find an orthogonal transformation of variables which diagonalizes F of Problem 7-39. Determine the diagonal form of F under this transformation.

7-41. Find the transformation of variables which diagonalizes

$$F = 4x_1^2 - 2x_2^2 + x_3^2 - 2x_1x_2 + 4x_1x_3 - 5x_2x_3,$$

using the technique of completing the square developed in Problem 7-37. What is the diagonal form of F?

7-42. Prove that if $\mathbf{x}'\mathbf{A}\mathbf{x}$ is positive definite and $\mathbf{A}^{-1}$ exists, then $\mathbf{x}'\mathbf{A}^{-1}\mathbf{x}$ is also positive definite. Hint: Consider the transformation $\mathbf{x} = \mathbf{A}^{-1}\mathbf{y}$.

7-43. Given the form $F = a_{11}x_1^2 + 2a_{12}x_1x_2 + a_{22}x_2^2$. For what values of a_{11}, a_{12}, a_{22}, and F does this form describe a circle, ellipse, hyperbola, a pair of lines, a point, or all of E^2?

7-44. According to Section 7-11, there exists a nonsingular, nonorthogonal transformation of variables which reduces a positive definite form $\mathbf{x}'\mathbf{A}\mathbf{x}$ to a sum of squares $\sum \alpha_j y_j^2$. By induction or by any other method prove that

$$\alpha_1 = a_{11}; \quad \alpha_2 = \frac{\begin{vmatrix} a_{11} & a_{12} \\ a_{21} & a_{22} \end{vmatrix}}{a_{11}}; \quad \alpha_3 = \frac{\begin{vmatrix} a_{11} & a_{12} & a_{13} \\ a_{21} & a_{22} & a_{23} \\ a_{31} & a_{32} & a_{33} \end{vmatrix}}{\begin{vmatrix} a_{11} & a_{12} \\ a_{21} & a_{22} \end{vmatrix}}; \quad \ldots;$$

$$\alpha_n = \frac{|\mathbf{A}|}{\begin{vmatrix} a_{11} & \cdots & a_{1,n-1} \\ \vdots & & \vdots \\ a_{n-1,1} & \cdots & a_{n-1,n-1} \end{vmatrix}}.$$

7-45. Show that if any quadratic form $\mathbf{x}'\mathbf{A}\mathbf{x}$ is diagonalized to $\sum \gamma_j y_j^2$ by a nonsingular transformation of variables, the number of positive and negative γ_j is uniquely determined and is independent of the particular transformation which diagonalized the form. This result is called Sylvester's law of inertia. Hint: Assume the contrary! Suppose that there exist two nonsingular transformations $\mathbf{x} = \mathbf{R}_1 \mathbf{y}$, $\mathbf{x} = \mathbf{R}_2 \mathbf{z}$ such that we obtain

$$\alpha_1 y_1^2 + \cdots + \alpha_p y_p^2 - \alpha_{p+1} y_{p+1}^2 - \cdots - \alpha_r y_r^2 \quad (\alpha_j > 0 \text{ for all } j)$$

and

$$\beta_1 z_1^2 + \cdots + \beta_q z_q^2 - \beta_{q+1} z_{q+1}^2 - \cdots - \beta_r z_r^2 \quad (\beta_i > 0 \text{ for all } i).$$

(Why can we assume that r is the same in both cases?). Assume that $q < p$. Set $z_i = 0$ $(i = 1, \ldots, q)$ and $y_j = 0$ $(j = p+1, \ldots, n)$. Then $q + n - p < n$ homogeneous linear equations restrict the values of the x_i. There exists a solution $\mathbf{x} \neq \mathbf{0}$. However, the first form must be ≥ 0, and the second ≤ 0. Hence the form $\mathbf{x}'\mathbf{A}\mathbf{x}$ must vanish from any such $\mathbf{x}$, and $\alpha_1 y_1^2 + \cdots + \alpha_p y_p^2 = 0$. This implies $\mathbf{y} = \mathbf{0}$, which in turn implies $\mathbf{x} = \mathbf{R}_1 \mathbf{y} = \mathbf{0}$, and this contradicts the fact that there was a solution $\mathbf{x} \neq \mathbf{0}$. Fill in the details. Note that according to this theorem all congruence transformations $\mathbf{P}'\mathbf{A}\mathbf{P}$ ($\mathbf{P}$ nonsingular) which diagonalize the symmetric matrix $\mathbf{A}$ give the same number of positive and negative diagonal elements.

7-46. Find a nonsingular transformation of variables which simultaneously diagonalizes

$$\mathbf{A} = \begin{bmatrix} 2 & 1 \\ 1 & 3 \end{bmatrix}, \quad \mathbf{B} = \begin{bmatrix} -5 & -6 \\ -6 & 4 \end{bmatrix}.$$

Give the diagonal form of $\mathbf{A}$ and $\mathbf{B}$.

7-47. Show that the eigenvalues of the matrix $\mathbf{C}$ in Eq. 7-75 are the roots of $|\mathbf{B} - \lambda \mathbf{A}| = 0$.

7-48. A bilinear form in the variables $y_1, \ldots, y_n$ and $x_1, \ldots, x_n$ is defined to be the expression

$$\sum_{i=1}^{n} \sum_{j=1}^{n} a_{ij} y_i x_j = \mathbf{y}'\mathbf{A}\mathbf{x}.$$

Show that it no longer follows that $\mathbf{A}$ can be assumed to be symmetric. Furthermore, show that by the nonsingular transformation of variables $\mathbf{y} = \mathbf{R}_1 \mathbf{v}$, $\mathbf{x} = \mathbf{R}_2 \mathbf{u}$, the bilinear form can be reduced to $\mathbf{v}'\mathbf{u}$ if $\mathbf{A}$ is nonsingular. What happens when $\mathbf{A}$ is singular?

7-49. A nonhomogeneous quadratic function of n variables $x_1, \ldots, x_n$ is defined to be the expression

$$F = \sum_{i=1}^{n} \sum_{j=1}^{n} a_{ij} x_i x_j + \sum_{j=1}^{n} b_j x_j + c = \mathbf{x}'\mathbf{A}\mathbf{x} + \mathbf{b}\mathbf{x} + c,$$

where $\mathbf{A} = \|a_{ij}\|$, $\mathbf{b} = (b_1, \ldots, b_n)$. Consider a change of variables $\mathbf{x} = \mathbf{y} + \mathbf{r}$, where $\mathbf{r} = [r_1, \ldots, r_n]$, that is, a translation. Express the form F in terms of the variables **y** and show that the translation leaves the matrix **A** unchanged. Why is the function called nonhomogeneous?

7-50. For the nonhomogeneous function defined in Problem 7-49, show that by an affine transformation of variables, $\mathbf{x} = \mathbf{Q}\mathbf{y} + \mathbf{r}$, with **Q** orthogonal, F can be reduced to one of the following forms:

$$F = \lambda_1 y_1^2 + \cdots + \lambda_r y_r^2 + h y_{r+1} \quad (h > 0)$$

or

$$F = \lambda_1 y_1^2 + \cdots + \lambda_r y_r^2 + g,$$

where the λ_i are the eigenvalues of **A**, and each $\lambda_i \neq 0$ (the eigenvalues are numbered in such a way that the nonzero values appear first). Hint: Apply $\mathbf{x} = \mathbf{Q}_1 \mathbf{z}$ to diagonalize $\mathbf{x}'\mathbf{A}\mathbf{x}$, and complete the square to eliminate the linear terms for nonzero λ_i. Next take care of the remaining linear terms. Note that h is needed to make the final transformation orthogonal. If we wish to obtain an orthogonal transformation of variables $\mathbf{x} = \mathbf{Q}\mathbf{y}$ which will take $\mathbf{f}\mathbf{x}$ into $d y_1$ ($d > 0$), then $\mathbf{f}\mathbf{Q}\mathbf{y} = d y_1$. What is the vector $\mathbf{f}\mathbf{Q}$? What is d? Does this yield the first column of **Q**? How can we obtain the remaining columns of **Q**?

7-51. Reduce the following inhomogeneous quadratic function to one of the forms discussed in Problem 7-50:

$$F = 2x_1^2 - 4x_1 x_2 - 5x_2^2 + 3x_1 + 4x_2 + 7.$$

7-52. If F_1 is a positive definite quadratic form in the variables $x_1, \ldots, x_k$ and F_2 is a positive definite quadratic form in the variables $x_{k+1}, \ldots, x_n$, consider the quadratic form $F = F_1 + F_2$. If $\mathbf{A}_1, \mathbf{A}_2$ are the matrices associated with F_1, F_2, respectively, what is the matrix **A** associated with F? Prove that F is positive definite.

7-53. Given any square matrix **A**, show that if $|a_{ii}| > \sum_{j \neq i} |a_{ij}|$ for each i, then **A** is nonsingular. Thus show that if λ is any eigenvalue of **A**, then $|a_{ii} - \lambda| \leq \sum_{j \neq i} |a_{ij}|$ for one or more i. Thus, bounds on the eigenvalues of **A** can be found. Use this procedure to determine bounds on the eigenvalues of

$$\mathbf{A} = \begin{bmatrix} 3 & 1 & 5 \\ 1 & -2 & 4 \\ 5 & 4 & 9 \end{bmatrix}.$$

Hint: To prove the first part of the problem, assume that $\mathbf{A}\mathbf{x} = \mathbf{0}$ has a solution $\mathbf{x} \neq \mathbf{0}$. Let x_i be the largest component of **x** in absolute value. Consider the ith equation. Is a contradiction obtained?

7-54. A linear transformation T which takes vectors **x** in E^n into vectors **y** in E^n, $\mathbf{y} = T(\mathbf{x})$, is called orthogonal if the scalar product is preserved, that is, $\hat{\mathbf{y}}'\mathbf{y} = \hat{\mathbf{x}}'\mathbf{x}$. Show that the matrix **Q** representing T is orthogonal if **y**, **x** are referred to orthonormal bases (orthogonal coordinates) in E^n.

7-55. It was shown in the text that $\mathbf{y} = \mathbf{Qx}$, where

$$\mathbf{Q} = \begin{bmatrix} \cos\theta & \sin\theta \\ -\sin\theta & \cos\theta \end{bmatrix}$$

can be interpreted as a rotation of an orthogonal coordinate system through an angle θ. Give the alibi interpretation of $\mathbf{y} = \mathbf{Qx}$ and show that it rotates vectors through the angle $-\theta$.

7-56. Show that for some θ, any orthogonal matrix $\mathbf{Q}$ of order 2 with $|\mathbf{Q}| = 1$ can be written in the form given in Problem 7-55.

7-57. Show that any orthogonal matrix $\mathbf{Q}$ of order 2 with $|\mathbf{Q}| = -1$ can be written

$$\mathbf{Q}_1 = \begin{bmatrix} 1 & 0 \\ 0 & -1 \end{bmatrix} \begin{bmatrix} \cos\theta & \sin\theta \\ -\sin\theta & \cos\theta \end{bmatrix} \quad \text{or} \quad \mathbf{Q}_2 = \begin{bmatrix} -1 & 0 \\ 0 & 1 \end{bmatrix} \begin{bmatrix} \cos\theta & \sin\theta \\ -\sin\theta & \cos\theta \end{bmatrix}.$$

What is the geometrical interpretation of $\mathbf{Q}_1$ and $\mathbf{Q}_2$? Note that if in $\mathbf{Q}_2$, θ is replaced by $\theta + \pi$, $\mathbf{Q}_1$ is obtained. Thus $\mathbf{Q}_2$ does not really differ from $\mathbf{Q}_1$. Illustrate this graphically.

7-58. Consider the oblique coordinate system for E^2 determined by the basis vectors $\mathbf{a}_1, \mathbf{a}_2$, where γ is the angle between $\mathbf{a}_1, \mathbf{a}_2$, and assume that this coordinate system is rotated through the angle θ. Denote the new coordinates by β_1, β_2. Show that

$$\begin{bmatrix} \alpha_1 \\ \alpha_2 \end{bmatrix} = \frac{1}{\sin\gamma} \begin{bmatrix} \sin(\gamma-\theta) & -\sin\theta \\ \sin\theta & \sin(\gamma+\theta) \end{bmatrix} \begin{bmatrix} \beta_1 \\ \beta_2 \end{bmatrix}.$$

Is the matrix representing this transformation an orthogonal matrix?

7-59. Suppose that we begin with an orthogonal coordinate system whose coordinates are x_1, x_2, x_3. Now we shall perform three counterclockwise rotations: (1) through an angle θ about the x_3-axis to yield a new set of coordinates $y_1, y_2, y_3 = x_3$; (2) through an angle ϕ about the y_1-axis to yield the set of coordinates $z_1 = y_1, z_2, z_3$; and (3) through an angle ζ about the z_3-axis to yield the set of coordinates $v_1, v_2, v_3 = z_3$. Find the matrix which relates the coordinates v_1, v_2, v_3 to x_1, x_2, x_3, and show that it is orthogonal. The angles θ, ϕ, ζ are called eulerian angles; they are of considerable use in rigid body mechanics.

7-60. Consider a rectangular coordinate system with coordinates x_1, x_2, x_3. A counterclockwise rotation through an angle θ about the x_3-axis yields a new set of coordinates $y_1, y_2, y_3 = x_3$. The new coordinates are related to the original set by the matrix $\mathbf{Q}_1$. Find $\mathbf{Q}_1$. Next, a counterclockwise rotation through an angle ϕ about the y_2-axis gives the coordinates $z_1, z_2 = y_2, z_3$. The z-coordinates are related to the y-coordinates by the matrix $\mathbf{Q}_2$. Find $\mathbf{Q}_2$. Show that $\mathbf{Q}_1\mathbf{Q}_2 \neq \mathbf{Q}_2\mathbf{Q}_1$. What does this mean geometrically?

7–61. A dyadic $\mathfrak{B}$ of nth order is defined as the generalized quadratic form

$$\mathfrak{B} = \sum_{i,j=1}^{n} b_{ij}\mathbf{e}_i\mathbf{e}_j,$$

where the $\mathbf{e}_i$ are the unit vectors for E^n. The combination $\mathbf{e}_i\mathbf{e}_j$ does not denote a scalar product. The two unit vectors are merely written side-by-side, and the notation $\mathbf{e}_i\mathbf{e}_j$ does not imply any additional significance. However, it is not true that $\mathbf{e}_i\mathbf{e}_j = \mathbf{e}_j\mathbf{e}_i$, that is, the order of the subscripts is important. For this reason, the matrix $\mathbf{B} = \|b_{ij}\|$ can no longer be assumed to be a symmetric matrix. The left scalar product of a column vector $\mathbf{v}$ and the dyadic $\mathfrak{B}$ is defined as

$$\mathbf{v} \cdot \mathfrak{B} = \sum_{i,j=1}^{n} b_{ij}(\mathbf{v}'\mathbf{e}_i)\mathbf{e}_j = \sum_{i,j=1}^{n} b_{ij}v_i\mathbf{e}_j,$$

where $(\mathbf{v}'\mathbf{e}_i)$ is the scalar product of $\mathbf{v}$ and $\mathbf{e}_i$. Thus the scalar product $\mathbf{v} \cdot \mathfrak{B}$ is a vector. The right scalar product is defined to be

$$\mathfrak{B} \cdot \mathbf{v} = \sum_{i,j=1}^{n} b_{ij}\mathbf{e}_i(\mathbf{e}_j'\mathbf{v}) = \sum_{i,j=1}^{n} b_{ij}v_j\mathbf{e}_i.$$

Under what conditions is $\mathbf{v} \cdot \mathfrak{B} = \mathfrak{B} \cdot \mathbf{v}$? Show that $\mathbf{v}_1 \cdot \mathfrak{B} \cdot \mathbf{v}_2$ is uniquely defined. What is $\mathbf{x} \cdot \mathfrak{B} \cdot \mathbf{x}$? Evaluate $\mathbf{e}_i \cdot \mathfrak{B}$, $\mathfrak{B} \cdot \mathbf{e}_i$, $\mathbf{e}_i \cdot \mathfrak{B} \cdot \mathbf{v}$, $\mathbf{e}_i \cdot \mathfrak{B} \cdot \mathbf{e}_j$, $\mathbf{v} \cdot \mathfrak{B} \cdot \mathbf{e}_j$.

7–62. Show that there is a complete equivalence between dyadics and matrices for the operations defined in Problem 7–61. That is, show that $\mathfrak{B}$ is completely described by $\mathbf{B}$ and that

$$\mathbf{v} \cdot \mathfrak{B} \to \mathbf{v}'\mathbf{B}, \quad \text{or} \quad \mathfrak{B} \cdot \mathbf{v} \to \mathbf{B}\mathbf{v}, \quad \text{or} \quad \mathbf{v} \cdot \mathfrak{B} \cdot \mathbf{v} \to \mathbf{v}'\mathbf{B}\mathbf{v}.$$

7–63. Plot $F = 3x_1^2 + 2x_1x_2 + 4x_2^2$ for several different values of F. Find the principal axes for this quadratic form and illustrate geometrically.

Problems Involving Complex Numbers

7–64. Show that if $\mathbf{A}$ is an nth-order matrix with real elements, there does not exist a matrix $\mathbf{P}$ (with real or complex elements) such that $\mathbf{P}^{-1}\mathbf{A}\mathbf{P}$ is diagonal unless $\mathbf{A}$ has n linearly independent eigenvectors.

7–65. Find the eigenvectors and eigenvalues of the matrix

$$\mathbf{A} = \begin{bmatrix} 2 & 4 \\ -2 & 3 \end{bmatrix}.$$

7–66. Show that if $\mathbf{A}$ is an nth-order matrix with real elements and there exists a matrix $\mathbf{P}$ such that $\mathbf{D} = \mathbf{P}^{-1}\mathbf{A}\mathbf{P}$ is diagonal, then $\mathbf{P}$ contains as columns a set of n linearly independent eigenvectors of $\mathbf{A}$, and the diagonal elements of $\mathbf{D}$ are the eigenvalues of $\mathbf{A}$.

7-67. Let **Q** be an orthogonal matrix with real elements. Show that if λ is an eigenvalue of **Q**, so is $1/\lambda$.

7-68. The eigenvalues of an orthogonal matrix with real elements can be complex. Show that the absolute value of any eigenvalue is unity, that is, $\lambda^*\lambda = 1$, and demonstrate that there exists a real θ such that $\lambda = e^{i\theta} = \cos\theta + i\sin\theta$. Hint: $\mathbf{Qx} = \lambda\mathbf{x}$ and $(\mathbf{x}^*)'\mathbf{Q}' = \lambda^*(\mathbf{x}^*)'$.

7-69. If λ is an eigenvalue of the square matrix **A** with real elements, show that λ^* is also an eigenvalue of **A**. What is the relation between the corresponding eigenvectors?

7-70. If **Q** is a real 3×3 orthogonal matrix ($|\mathbf{Q}| = 1$), show that one of its eigenvalues is $\lambda = 1$. This result has an important implication for the theory of mechanics. It means that in any arbitrary combination of rotations of a rigid body with one point fixed in space, one vector remains unaltered. We can then accomplish all rotations by a single rotation, using the unchanged vector as the axis of rotation. This is known as Euler's theorem.

7-71. Prove that the eigenvalues of a Hermitian matrix are real (see Problem 3-83 for the definition of a Hermitian matrix).

7-72. Prove that the eigenvectors corresponding to different eigenvalues of a Hermitian matrix are orthogonal in the sense that the Hermitian scalar product vanishes.

7-73. Prove that if **H** is a Hermitian matrix, then there exists at least one set of n eigenvectors $\mathbf{u}_j$ which satisfies

$$\bar{\mathbf{u}}_i \mathbf{u}_j = \delta_{ij} \qquad (\text{all } i, j).$$

Show that if $\mathbf{U} = (\mathbf{u}_1, \ldots, \mathbf{u}_n)$, then $\mathbf{D} = \mathbf{U}^{-1}\mathbf{H}\mathbf{U}$ is a diagonal matrix, its diagonal elements being the eigenvalues of **H**. Show that $\bar{\mathbf{U}} = \mathbf{U}^{-1}$. Such a matrix is called a unitary matrix. Thus any Hermitian matrix can be diagonalized by a unitary similarity transformation. For matrices with complex elements, unitary matrices play a role similar to that played by orthogonal matrices for matrices whose elements are real.

7-74. Find the eigenvalues and an orthogonal set of eigenvectors of unit length for

$$\mathbf{H} = \begin{bmatrix} 2 & 4+i \\ 4-i & 3 \end{bmatrix}.$$

Let **U** be a matrix whose columns are the eigenvectors of **H**. Show that $\mathbf{D} = \mathbf{U}^{-1}\mathbf{H}\mathbf{U}$ is diagonal, and that the diagonal elements of **D** are the eigenvalues of **H**.

INDEX

INDEX

additivity, 2, 133
adjoint, 95
affine subspace, 178
Aitken, A. C., 120, 183
alias, 267
alibi, 267
Allen, R. G. D., 120, 183
angle, 32
associate matrix, 131
augmented matrix, 168

basic solutions, 178
basis, 39, 52
   change of, 41
   orthogonal, 45
Berger, W. J., 120
Birkhoff, G., 52, 120, 183, 272
block multiplication, 81
Bôcher, M., 183
bounded, 194
boundary point, 192
Brillouin, L., 16

Cauchy expansion, 128
characteristic equation, 237
   value, 237
   vector, 237
closed set, 192
cofactor, 91
Collar, A. R., 152
complement, 193
complementary cofactor, 97
   minor, 97
complex numbers, 57
component, 26
cone, 219
congruence, 253
congruent matrices, 253
connected set, 196
convex combination, 202, 206
   cone, 221
   hull, 207

polyhedral cone, 224
polyhedron, 209
   set, 202
coordinates, 30, 48, 264
Corben, H., 16
Courant, R., 120
Cramer's rule, 166

d-c circuit, 12
decomposable matrix, 158
degeneracy, 180
determinant, 87
   of product of rectangular matrices, 100
   of product of square matrices, 99
diagonal matrix, 75
diagonalization, of symmetric matrix, 247
   of quadratic form, 255, 257
dimension, 33
   of cone, 224
   of subspace, 50
direct sum, 274
distance, 31
domain, 135
Dorfman, R., 15
Duncan, W. J., 152
Dwyer, P. S., 183
dyadic, 283

echelon matrix, 148
edge, 219
eigenvalue, 237
eigenvector, 237
elementary column operations, 147
elementary matrix, 145
elementary row operation, 144
element, of a matrix, 60
   of a set, 188
equality, of matrices, 62
   of sets, 188
   of vectors, 28

equivalent matrices, 152
equivalence transformation, 152
euclidean space, 33
expansion by cofactors, 90
extreme point, 202
extreme supporting half-space, 226
extreme supporting hyperplane, 226

face, 227
facet, 227
Ferrar, W. L., 120, 183
Fraser, D., 15
Frazer, R. A., 152
Frenkel, J., 16

Gale, D., 230
Gass, S., 15
gaussian elimination, 162
Gauss-Jordan reduction, 164
generalized coordinates, 48
Gerstenhaber, M., 230
Goldman, A. J., 230
Goldstein, H., 16
Gram determinant, 184
Guillemin, E., 15, 120

Hadley, G., 15
half-line, 222
half-space, 201
  extreme supporting, 226
Halmos, P. R., 52, 120
Hermitian matrix, 131
  scalar product, 58
Hilbert, D., 120
Hildebrand, F., 16, 120, 183, 272
Hoffman, A. J., 230
homogeneity, 2, 134
homogeneous linear equations, 39, 173
hyperplane, 197
  extreme supporting, 226
  supporting, 212
hypersphere, 191

identity matrix, 73
image, 132
indefinite matrix, 254

indefinite quadratic form, 254
independent set of vectors, 35
inside, of a hypersphere, 191
interior point, 192
intersection, of sets, 189
  of subspaces, 55
inverse matrix, 103
  computation by partitioning, 107
  numerical technique for computing, 157
  of a product, 105
  product form, 111
  right or left, 104
  of the transpose, 106
inversion, 86
isomorphism, 2

Jordan canonical form, 251

Kaplan, W., 16
Kemeny, J., 16, 230
Klein, L., 15
Koopmans, T. C., 230
Kronecker delta, 73
Kruskal, J. G., 230
Kuhn, H. W., 230

Laplace expansion, 95
latent root, 237
least squares, 12
Le Corbeiller, P., 15, 120
length, 32
Leontief interindustry model, 3
line, 195
linear algebra, 2
linear combination, 30
linear dependence, 35
linear equations, 2, 162
linear programming, 6
linear transformations, 133
  product of, 137

MacLane, S. A., 52, 120, 183, 272
mapping, 132
Margenau, H., 272

matrix, 60
　addition, 62
　augmented, 168
　multiplication, 64
　multiplication by a scalar, 62
　series, 116
minor, 92
Mirkil, H., 230
modulus of a matrix, 129
Morgenstern, O., 15, 120
Murphy, G., 272

negative cone, 220
negative definite matrix, 254
negative definite quadratic form, 254
negative semidefinite matrix, 254
negative semidefinite quadratic form, 254
neighborhood, 192
non-negative orthant, 205
nonsingular, 103
normal, 197
normalized eigenvector, 241
n-tuple, 25
null matrix, 75
null space, 176
null vector, 28
nullity, 176

open set, 192
operator, 134
order, of a determinant, 87
　of a square matrix, 60
orthant, non-negative, 205
orthogonal complement, 56
　coordinates, 33
　cone, 223
　matrix, 247
　transformation, 269, 281
orthonormal basis, 46
orthonormal vectors, 241

parallel hyperplanes, 199
parallelogram law, 20
partitioning of matrices, 79
Perlis, S., 52, 120, 183, 272

permutation, 86
point, 26
point set, 189
polar cone, 220
polynomial, characteristic, 237
positive definite matrix, 254
　quadratic form, 254
positive semidefinite matrix, 254
　quadratic form, 254
principal minor, 262
product, of linear transformations, 137
　of matrices, 66
　of a matrix by a scalar, 62
　of vectors, 31
　of a vector by a scalar, 29

quadratic form, 251

range, 135
rank of a matrix, 138
　determinantal criterion, 140
　of product, 138
rectangular coordinates, 17
region, 196
regression analysis, 10
row rank, 142

Saibel, E., 120
Samuelson, P., 15, 16
scalar, 20
scalar matrix, 75
scalar product, 22, 31
Schmidt orthogonalization process, 46
Schwarz inequality, 33, 56
secular equation, 237
set, 188
similar matrices, 239
similarity transformation, 239
simplex, 209
simultaneous linear equations, 162
singular matrix, 103
skew-symmetric matrix, 78
Solow, R., 15
Snell, L., 16, 230
span, 39, 52

Stehle, P., 16
submatrix, 79
subset, 188
subspace, 50
sum check, 160
sum, of cones, 220
 of matrices, 62
 of subspaces, 55
 of vectors, 29
sum vector, 28, 34
supporting hyperplane, 212
Sylvester's law, 280
symmetric matrix, 78

Thrall, R. M., 53, 120, 183, 272
Thompson, G. L., 230
Tornheim, L., 53, 120, 183, 272
transpose, 76
 of product, 77
transposition, 86

triangle inequality, 31
triangular matrix, 125
trivial solution, 173
Tucker, A. W., 230

unit vectors, 22, 27
union, 189
unitary matrix, 284

vector(s), 17, 26
 addition of, 20, 29
 component of, 21, 22, 26
 equality, 19, 28
 multiplication by a scalar, 20, 29
 orthonormal, 241
vector space, 50
vertex, 220
von Neumann, J., 230

Weyl, H., 230

THE COLLEGE OF STATEN ISLAND LIBRARY
SUNNYSIDE CAMPUS